Istkhar Rao
Editor

Insects as Humans' Frenemies

From Friends to Foes

nova
science publishers
www.novapublishers.com

DOI: https://doi.org/10.52305/LAXP0217

NOTICE TO THE READER

The Publisher has taken reasonable care in the preparation of this book but makes no expressed or implied warranty of any kind and assumes no responsibility for any errors or omissions. No liability is assumed for incidental or consequential damages in connection with or arising out of information contained in this book. The Publisher shall not be liable for any special, consequential, or exemplary damages resulting, in whole or in part, from the readers' use of, or reliance upon, this material. Any parts of this book based on government reports are so indicated and copyright is claimed for those parts to the extent applicable to compilations of such works.

Independent verification should be sought for any data, advice or recommendations contained in this book. In addition, no responsibility is assumed by the Publisher for any injury and/or damage to persons or property arising from any methods, products, instructions, ideas or otherwise contained in this publication.

This publication is designed to provide accurate and authoritative information with regards to the subject matter covered herein. It is sold with the clear understanding that the Publisher is not engaged in rendering legal or any other professional services. If legal or any other expert assistance is required, the services of a competent person should be sought. FROM A DECLARATION OF PARTICIPANTS JOINTLY ADOPTED BY A COMMITTEE OF THE AMERICAN BAR ASSOCIATION AND A COMMITTEE OF PUBLISHERS.

Library of Congress Cataloging-in-Publication Data

ISBN: 979-8-89530-320-7 (Hardcover)
ISBN: 979-8-89530-410-5 (eBook)

Published by Nova Science Publishers, Inc. † New York

To My Beloved Father

Even though you are not here with us anymore, you are still there in every memory, every experience, and everything I do. Even though you did not say much, your deeds spoke louder than words. I will always be grateful for how your uncomplicated lifestyle, silent strength, and unfailing love have moulded me.
I can feel your love and your care, and I cannot forget your unconditional sacrifices even when you are not here. Even if you are not physically present here, you will always be a part of my heart. I am incredibly thankful for every second I spent with you.

I miss you more than words can express.

With All My Love

Contents

Preface

Insects are everywhere around us if we take a moment to look. They are important because of their diversity, ecological role, and influence on agriculture, human health, and natural resources. Insects possess a remarkable ability to thrive in a multitude of environments, which contributes to their status as the most dominant group within the animal kingdom. Their ecological importance is profound, as they provide a wide array of vital services that bolster ecosystems and agriculture alike. However, their duality also reveals the complex relationship they maintain with humans; they can serve both as invaluable allies and as formidable adversaries. This dichotomy highlights the beneficial roles insects play—such as pollination, decomposition, and serving as food for other species—alongside their potential to cause harm, including spreading diseases and damaging crops. The interplay of these two roles underscores the significance of insects in our world, making them both fascinating and multifaceted creatures. Insects play a vital role in our ecosystem. They pollinate crops, fruits, and vegetables, helping to ensure food production. In addition, they disperse seeds and recycle nutrients, which maintains soil fertility and structure. Many organisms rely on insects as a food source. They also break down debris, aiding decomposers in their work. Furthermore, insects produce valuable products such as silk, honey, lac, and cosmetics. Some insects act as predators, feeding on harmful pests, while others live as parasitoids within their hosts. Insects can serve as bioindicators, reflecting the health of soils and water. They are also the foundation of complex ecological food webs and are beautiful creatures that enrich our natural world. In their relationship with humans, insects can be quite harmful. They act as pests that damage crops, serve as vectors for diseases that affect plants and animals, and can compromise stored grains and households. Due to their dual role as both beneficial and detrimental, insects are the subject of extensive study around the world.

The present book explores the multifaceted roles of insects in the life of humans in light of current findings. The chapters in this book discuss the major roles of insects concerning humans. The introductory chapter establishes that insects serve as competitors to humans, effectively summarising their beneficial and harmful impacts. The following chapters delve into the significant roles that insects play across various agricultural settings, highlighting their impact on cereal crops, vegetable crops, and orchard ecosystems. Additionally, the discussions extend to the challenges posed by these pests in stored grains and forests, where they can cause considerable damage. Each segment examines the specific ways these insects interact with their environments, emphasising the complexities of pest management in both cultivated and natural landscapes. The chapters also include detailed discussions on pesticide management and integrated pest management (IPM) strategies. The concept of integrated pest management is examined, highlighting its holistic approach to managing pest populations sustainably and reducing reliance on chemical inputs. In the context of human health, the significance of two key mosquito genera, *Anopheles* and *Culex*, is emphasised. *Anopheles* mosquitoes are particularly noteworthy as they are the primary vectors responsible for transmitting malaria, a disease that poses a substantial global health threat. On the other hand, *Culex* mosquitoes are known for their role in spreading various viruses, including West Nile virus and encephalitis. Understanding the impact of these mosquitoes is crucial for developing effective strategies for disease prevention and management.

Insects play a vital role as predators in their ecosystems, and understanding their impact is essential. Chapters that explore their predatory functions provide crucial insights into the balance of nature and the intricate relationships within food webs. These discussions illuminate the significance of insects in controlling pest populations, supporting biodiversity, and maintaining ecological harmony. In environmental monitoring, insects play a crucial role as bioindicators, providing valuable information about the health of ecosystems and the quality of habitats. This chapter focuses on the different ways insects serve as bioindicators, emphasising their importance in evaluating ecological conditions and offering insights into environmental changes. Food is a fundamental aspect of human survival, and within this realm, the exploration of insects as viable food sources, known as Entomophagy, highlights their crucial role in our diets. Additionally, the other insect-derived resources, such as honey, silk, and lac, showcase the diverse and significant contributions that these remarkable creatures make to human life and culture. Their nutritional value and economic importance illustrate why we should acknowledge and incorporate these alternative sources into our understanding of sustainable food practices. Insects play a significant role in the breakdown of plastic, which is one of the most pressing forms of pollution in our environment. A dedicated chapter explores how these remarkable creatures contribute to the degradation of plastic materials, highlighting their importance in addressing this critical environmental issue. In addition to these chapters, the book also delves into the fascinating role of insects as invaluable forensic assistants, detailing how they can aid in criminal investigations. One chapter also explores their use as experimental models, shedding light on how these tiny creatures contribute to scientific research across various fields. Furthermore, the text addresses the pressing issue of climate change and its effects on insect populations, highlighting the vulnerabilities these essential organisms face in a rapidly changing environment. Together, these sections aim to provide a comprehensive understanding of the multifaceted roles of insects and the significant threats they encounter.

With a comprehensive understanding of current research, I am confident that this book, which covers a wide array of subjects, will serve as an invaluable resource for students, researchers, and practitioners across various fields. It will particularly benefit individuals engaged in agriculture and horticulture, those studying forensic science, experts in medical entomology, and enthusiastic amateur scientists. Additionally, it will appeal to anyone interested in exploring the profound effects insects have on human life, offering insights that are both practical and academic.

I would like to express my heartfelt gratitude to all the authors, reviewers, and supporters who have played a vital role in bringing this book to fruition. A special acknowledgement goes to the dedicated scientists and researchers in the field of entomology and its related disciplines. Their tireless efforts and unwavering commitment, often extending across the globe, contribute invaluable insights and knowledge to our understanding of this intricate field. It is through their generous sharing of expertise and their willingness to invest their precious time that I have been inspired and empowered to compile this work into its present form. Their passion for advancing our understanding truly elevates the collective knowledge within the scientific community.

Istkhar Rao

About the Editor

Dr. Istkhar Rao

Specialised in Nematology and Entomology, Dr. Istkhar is a post-graduate from Chaudhary Charan Singh University, Meerut, India. After completing his M.Sc. and M.Phil. in Zoology from the same university, Dr. Istkhar worked as a senior research fellow at the ICAR-Indian Agricultural Research Institute, Pusa, New Delhi, India, on an insect-entomopathogenic nematode-bacterial tripartite interaction project in the Nematology Laboratory. Later, he worked at the ICMR-National Institute of Malaria Research, Dwarka, New Delhi, India, as a research assistant. By returning to his parent laboratory at Chaudhary Charan Singh University, Meerut, he earned his highest degree by working on "bacterial endotoxin activities associated with entomopathogenic nematodes on haemolymph of *Helicoverpa armigera*." Simultaneously, he got a senior research fellowship in the CSIR-sponsored project "Development and production of bio-formulation of potential strains of entomopathogenic nematodes against white grub infesting sugarcane in Western Uttar Pradesh." After completing his PhD, he joined Jaipur National University Jaipur as an assistant professor and later joined Banasthali Vidyapith, Banasthali, Rajasthan, a world-renowned university for women's education. Currently, he is working as an assistant professor there in the Department of Bioscience and Biotechnology, Banasthali Vidyapith, Banasthali, Rajasthan, India, focusing on research on entomopathogenic nematodes' taxonomy and biological control of insect pests and vectors of diseases. He has an ongoing ICMR-funded project for the biological control of mosquitoes, vectors of malaria, and dengue using entomopathogenic nematodes. He has published one edited book, several national and international research articles and book chapters in reputed journals and books, and he has presented his research work at many national and international scientific events, where he has been awarded many best paper presentation awards. He has delivered invited lectures and provided hands-on training programmes. He has guided several M.Sc. dissertations, and currently four research scholars are working under his guidance on different topics of entomology and nematology.

Chapter 1

Insects as Human Competitors: From Benefits to Detriments

Ayushi Chaudhary
Parul Chauhan
Sakshi Singh
and Istkhar Rao*

Department of Bioscience and Biotechnology, Banasthali Vidyapith, Banasthali, Rajasthan, India

Abstract

Insects have played a crucial role in the lives of humans since ancient times. Early in the evolution of human society from hunter-gatherer to urban society, the association of mites and insects with humans started to develop. The very first alliance of humans was with beetles, carrion-feeding insects, and flies, which are attracted to the garbage dumps and preparation sites at camps. As human living space increased and improved later, the wood-infesting beetles and bed bugs became permanent. When pottery allowed people to store water nearby or inside the home, mosquitoes adapted to the living conditions. When major pest relationships become irreversible, it is unknown whether the bugs or mites cut ties to their natural environment and develop a long-term association with the human environment. However, it is possible to examine some of the mite and bug species that live near people, estimate the extent of their dependence, and make an educated guess as to when they might have had the chance to transfer. Numerous different insects and mites are connected to people and their living environments. Some of them, like honeybees and silkworms, were purposefully introduced there. Cat fleas, carpet beetles, and dust mites are a few that were unintentionally introduced to pets and possessions. Some just followed as people made their way from the edge of the woodland to the cave and eventually to the city's tallest housing complex. These include the yellow jacket, bed bugs, and house flies. All of these mites and insects have evolved to withstand fluctuations in temperature, humidity, and the availability of food and shelter that are unique to the human habitat. One indicator of their pest status in the human environment is how far they have adapted. In addition to the fact that their very existence in a human dwelling is unpleasant, most of the mites and insects pose a health threat to humans; stings and painful bites are also inflicted by some of them, as well as damage to the material.

Keywords: pests, insect-borne diseases, parasites, bioindicator, food, predators

---

* Corresponding Author's Email: istkharrao@gmail.com; istkhar@banasthali.in.

In: Insects as Humans' Frenemies: From Friends to Foes
Editor: Istkhar Rao
ISBN: 979-8-89530-320-7

1. Introduction

The human-insect relationship is described as complex and lengthy. Humans are afflicted with a variety of diseases throughout history, the majority of which are brought on by insects. Insects attack our crops, domestic animals are plagued by them, and stored foods are infested by them. Among an estimated 5–10 million insect species, probably less than 1% of them interact, formally or informally, with humans (Grimaldi et al., 2005). Possibly 10,000 pests make up the problem, both alone and in combination with microbes, and inflict serious harm or death to humans, forests, agricultural products, and manufactured goods (Pimentel et al., 1997). All over the world, the damage done to fibres and foods is mainly caused by insects, including plant pathogens, pests, birds, and weeds. It is estimated that approximately 12% of the damage is done by mites and insects, out of a total of 40% (Pimentel et al., 2000), as shown in Figure 1. Post-harvest losses are not included in these calculations; 20% is assumed to be the case and persist despite the use of approximately 3 million metric tonnes of pesticides worth more than US$31 billion in value, including pesticides worth approximately $9 billion (Pimentel, 2002; Gillott, 2005). In the United States alone, in the years 1945–1989, crop losses due to pest damage increased from 7% to 13%, even if the amount of pesticide used increased by a factor of ten, which is about 120,000 metric tonnes per year (Pimentel et al., 1992).

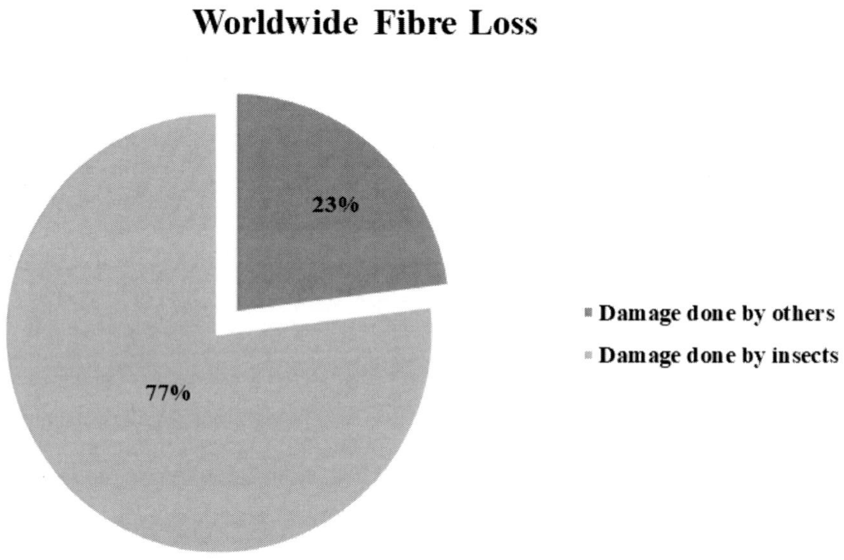

Figure 1. Worldwide fibre loss due to insects.

On the other hand, the advantages that insects offer outweigh the disadvantages brought about by their pollination activities, including their significance as makers of silk, wax, and honey, and their function in biological control. The fact that insects do more good than harm is surprising for the common person to believe, who has only insights into insects like cockroaches, houseflies, mosquitoes, and some of the garden pests and knowledge of the variety of pests that infect their crops and livestock. A lot of humans would probably not go past the honeybee on a list of beneficial insects, and it is possible that even silk moths would be overlooked as one of the many species that act as pollinators or consume harmful insects that could become pests (Pedigo et al., 2021).

Since ancient times, people have understood the value of insects in their lifestyle. Humans have consumed insects and/or their byproducts for thousands of years. Since over 5000 years ago, silk has been made from silk moth pupae (Zhang et al., 2018). Historically, locust swarms may have served as a significant seasonal food source for people, but they took on new meaning as humans began to live a farming life rather than a hunting one. But with a few uncommon cases, for instance, the comparatively easy-to-manage and labour-intensive silk moth and honey bee, humans had neither the desire nor the ability until lately, due to a lack of both technology and fundamental knowledge, to try and significantly alter the surroundings of insects either to raise the population of helpful insects or to lower the population of nuisance insects (Skendžić et al., 2021). Such efforts are necessary because of several characteristics of modern human evolution. Among these are a sharp rise in population, increased geographic human mobility, agriculture-related goods, a pattern of urbanisation, and, connected to the requirement to feed more people, a move towards monoculture as a method of agriculture. Urban environments' relative crowdedness makes it easier for insects that parasitise people to find a host (often a requirement for reproduction) and move from one host to another. Therefore, urbanisation encourages the spread of human diseases brought on by insects, for example, malaria, plague, and typhoid, whose effects on human populations are already documented (Gillott, 2005). For instance, the plague killed nearly 50% of the population of the Roman Empire in the sixth century A.D. Then, in the mid-1300s, the "Black Death" (bubonic plague) destroyed a comparable percentage of the population in England (Southwood, 1977).

2. Mythology, Literature and Insects

Insects have been employed by many cultures as totems or symbols to explain origin tales throughout the world. According to a South American tribe's legend, a beetle created the universe when God made men and women from the earth's leftover granules. *Lasiorhyncus barbicornis* is one of the most horrifyingly shaped weevils in New Zealand, with a weird giraffe-like morphology. Given how much it resembles the design of their canoes, the weevil was given the name Tuwhaipapa by the Maori people of New Zealand, the canoe's newly constructed deity. The use of insect iconography increased throughout antiquity, particularly in the Middle East. Hornets, for instance, were an emblem of the first Egyptian dynasty (from around 3100 BC) (Hogue, 1987), because they were menacing and ferocious. Early Egyptians' religious practices placed a high value on the sacred scarab. Not only was the word "psyche" the original Greek word for "butterfly," but to the soul, it is also a symbolic allusion. In their stories and plays, fleas, flies, and beetles are often included, like in the well-known story of the ant and the grasshopper by Aesop. The Greeks employed several insects as symbols for traits ranging from tenacity to insignificance. In early Christian symbolism, insects were associated with wickedness and foulness; in contrast, eastern faiths that place a strong emphasis on spiritual harmony with nature view insects as a good luck charm. Mediaeval knights frequently had bees on their shields. Many European families used insects as their coats of arms, including moths, grasshoppers, and beetles. Insects are used as mascots or as names by contemporary "clans" like collegiate and professional sports teams.

Around 6,000 years ago, hunting honey scenes first appeared in a Spanish cave (Chand et al., 2021). Subsequently, skewered locust banquet images decorated the walls of an Egyptian tomb. In ancient Greece, insects were depicted on coinage and jewellery. From 1600 to 1100 BC, sacred scarab ornaments as well as other insect jewellery first found in Egypt were frequently worn as pendants, headdresses, necklaces, and amulets (Allen, 2005). Japanese pendants, known as netsuke, have been carved for centuries in many insect forms. Insects were decorated on writing sets, cups, and decorative boxes during the Middle Ages. Insects were used by mediaeval monks to create their illuminations, the precious holy manuscripts' brilliantly decorated borders. As scientific knowledge and geographic exploration grew, insect subjects moved from the margins to the centre of the page in descriptions of newly discovered species. Franz Kafka's Metamorphosis is one of literature's best-known allusions to insects, whereby the main character changes into a bug that resembles a beetle. Cicadas, beetles, dragonflies, and butterflies are some of the insects that are carved from stone or made from glass and metal in today's fashionable insect jewellery. Beetle corpses are strong enough to be used as earrings and pendants, but in tropical nations, butterfly-wing art is frequently offered to tourists. On postal stamps, many nations honour insects (Wright, 2005). Even rock band names and car nicknames have been influenced by insects. They are also the source of words with negative connotations like dirty, lousy, beetle-browed, and nitpicker. Other insects have appeared on the big screen, causing horror and disgust in B-movies or serving as the inspiration for amusing, sometimes sympathetic, characters in blockbuster films and animated movies.

3. Diversity of Insects

It is a considerable topic of debate as to how many insect species there are all over the world. The estimated number can range from two million to eighty million (Erwin 2004). The estimation of these high numbers has been questioned because of the lack of evidence and assumptions made about the various ranges of undescribed species. According to the various sources of the estimation, it had concluded that worldwide, five million to fifteen million is more acceptable (Stork 1993). The total number of the described species of insects is about 1,060,704 in volume; to date, it is about 1.4 million–1.8 million of the total biota (Hammond 1992). Moreover, only 20% of the named species are known. On Earth, insects are more numerous. They are found everywhere in niches; only except the oceanic benthic zone. In the calculation of the total known biodiversity of arthropods, they are 65%. Henceforth, from the beginning of 400 mya, species of insects are dominant over all the known biodiversity on Earth, out of the total of 100 million species ever lived (Grimaldi & Engel, 2005). Considering their biomass and interactions with other terrestrial creatures, the most significant class of terrestrial animals is the insect family, which is so crucial that if everything vanished, humanity probably wouldn't last a few months. Moreover, these insects are the main competitors to humans on this planet.

4. Insects and Their Importance

Agricultural ecosystem services provided by beneficial insects include pollination and the organic control of plant pests. It tries to improve ecological services provided by insects from a conservation standpoint, improving the ecosystem services provided by beneficial insects in agricultural settings to crops. These helpful insects have helped to sustain human cultures and civilisations in various ways, which include controlling the number of numerous dangerous pest species, producing natural products, recycling organic nutrients, and disposing of waste. The extent to which we depend on insects for survival and the nature of a world devoid of insects should be taken into consideration. Certain important roles played by insects and their importance are illustrated in Figure 2 and examples are marked in Table 1.

Table 1. The role of insects with examples

S. No.	Role of Insects	Examples
1	Pollinator	Bumblebees, monarch butterfly
2	Predator	Thrips, mantis
3	Food	Caterpillars, beetle grubs
4	Pests	Cotton bollworms, click beetles
5	Carrier/ vector for diseases	Houseflies, mosquitoes
6	Bioindicators	Honeybees, butterflies
7	Biocontrol agents	Ladybird beetles, bugs

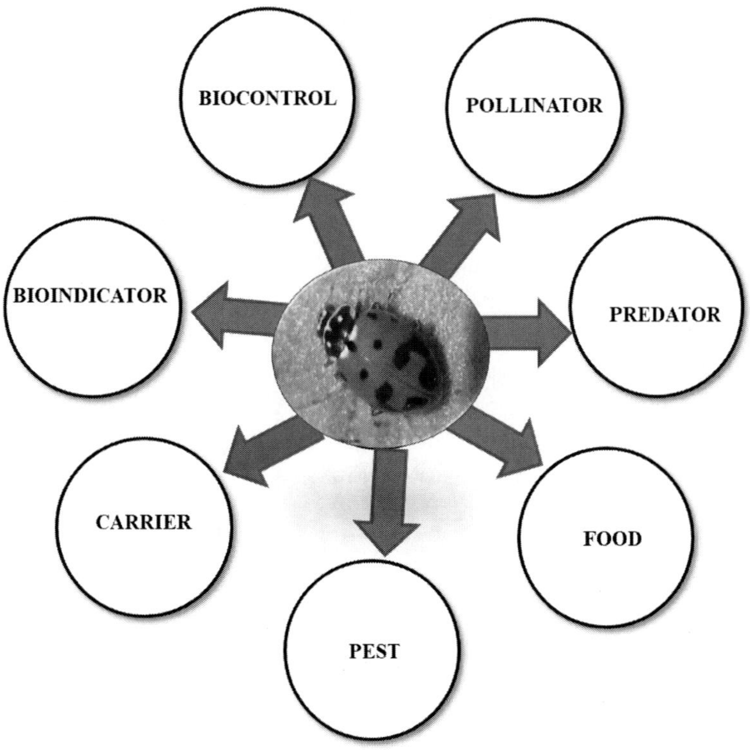

Figure 2. The importance of insects.

4.1. Insects in Ecology

All terrestrial ecosystems' biological base is provided by insects. They perform various crucial roles, such as pollination in plants, cycling nutrients, maintaining soil structure, dispersal of seeds, population control for other species, and providing major food sources for different organisms. For agricultural ecosystems to remain robust and productive, soil insects are crucial. In almost all representations of a terrestrial or freshwater food chain, insects are a major component, even though the food-web structures in these two environments differ greatly. As a source of food for a wide variety of predators, insects are very important. Fish eat the larvae of aquatic insects, and many stream fish appear to be constrained by the presence or abundance of such prey, at least during certain seasons. Trout consumes many adult mayflies during the time of year when they emerge, and the sport of fly fishing is founded on this phenomenon. Many lizards get the majority of their nutrition from insects. Insects make up the majority of an amphibian's diet because many of them are carnivorous, especially once they reach adulthood. Many different families of birds eat primarily insects, at least for a portion of the year. Swifts, swallows, and martins are almost entirely reliant on flying insects for survival. The appearance of damselflies is related to the yellow-headed blackbird's ability to successfully raise young in the Cariboo region of British Columbia. American anteaters, sloth bears, sun bears, and other animals are particularly connected to ant and termite colonies, and a lot of predatory mammals eat insects. The British badger frequently excavates wasp nests to eat the grubs. Black bears in north-central Minnesota in North America consume ants in the spring to receive vital amino acids, trace minerals, and fast sources of protein that are not accessible in other spring foods. In many parts of the world, insects provide a significant additional source of calories and protein for people, and in more than 70 families and 260 genera, there are more than 500 species of insects that are being consumed as food. Insects that belong to the main order are consumed, whereas mainly the species that are found habitually in the location are preferred, for example, termites, occasionally swarms (locusts), as well as huge species like Saturniid moth larvae. When other food sources may be insufficient, their seasonal abundance at specific periods of the year makes them extremely crucial. Insects could become increasingly important to food security in the future.

The overall number of insects that are other insects' natural enemies cannot be determined with any degree of certainty; however, there might be just as many, if not more, as prey or hosts as entomophagous insects. All major insect orders have the behaviour of feeding on other insects. Here, both parasitoids and predators are mentioned, both of which play a role in the efficient and natural control of insects. Approximately 400,000 species of known insects eat primarily plants. Thus, at least 25% of all species of living things on Earth are phytophagous insects. The majority of bug species fall under the phytophagous classification of the Orthoptera, Lepidoptera, and Hemiptera notable orders. As critters that consume plants, insects have a greater impact than all other animals combined. Insects are a prime factor under natural conditions, which are known to regulate the majority of all plants, mainly flowering plants because these are especially insect-attack-prone. More than any other animal, insects have extensively abused and profited from their food source. When the moth, *Cactoblastis cactorum* berg, was used to manage the prickly pear cactus in Australia between 1920 and 1925, this ability was harnessed. But occasionally, plants use their predators against them. Several truly insectivorous flowering plant species belong to a variety of different families.

4.2. Insects as Pollinators

Insect pollinators are insects that feed on flowering plants to receive nourishment like nectar and pollen from the plants. Insects that visit flowers can pollinate them by transferring the male gametes (found in pollen) to the female gametes while foraging. Insects, especially adult hymenopterans, play an important role in pollinating many plants. Specifically, bees are considered the most significant pollinators of blooming plants due to the protein-rich pollen and nectar they collect. Without the intricate webs of interactions formed by pollinators like bees, many blooming plants would be unable to reproduce sexually. While some solitary bees may show strong seasonal preferences for certain flowers, the vast majority of eusocial bees are floral generalists. Since certain solitary species have shorter adult activity periods than eusocial species, they can meet all of their nutritional needs during the few weeks when pollen and nectar are available, making them floral specialists. Spring and early summer are prime times for native plant pollination, when solitary bees like halictine and colletine, as well as the invasive *Apis mellifera*, prefer to work the blossoms of plants like eucalyptus and other Myrtaceae. Many additional families of hymenoptera graze at flowers for nectar and pollen and likely carry out passive pollination as well. Therefore, bees and other flower-visiting hymenopterans are very important for protecting and managing their habitats. For the majority of blooming plants around the world, insect-mediated pollination is a crucial stage in reproduction that involves various species of plants, including cucurbitaceous vegetables, coriander, sunflower, apple, cardamom, alfalfa, gingelly, and many more. For major fruit sets and seed production, most of the plants are dependent on pollination. Worldwide, it is reported that because of insects' pollination, 35% of crops are produced. Diptera, the flower-visiting flies, have been known as experienced pollinators of different crops, which involve *Dacus carota* (carrot), and *Brassica* spp. (mustard), *Allium ampeloprasum* (leek), and *Prunis dulcis* (almond).

4.3. Predators and Parasitoids

Insects, parasitoids, and predators that feed on and attack different insects, especially insect pests of plants, are termed natural enemies. Natural enemies participate in a sort of pest control known as natural biological control through this kind of feeding. Nearly 33% of natural pest control in cultivated systems is carried out by natural enemies (Ofuya et al., 2023). Predaceous natural enemies are classified as members of various insect orders and are typically described as being larger than their insect prey, mobile, free-living, and capable of consuming various prey during their whole life cycle. However, the parasitoids are primarily members of the orders Hymenoptera and Diptera, and they are thought to have more specialised host ranges than predators (Fei et al., 2023). Adult free-living parasitoids look for a host; they parasitise various stages of their host's life, like eggs, larvae, pupa, and adults, depending on the species of parasite. Parasitoids can deposit a single egg (solitary) or several eggs (gregarious) on or inside their host, and the young parasite(s) feed on their host to grow, mature, kill, and become an adult parasite that lives on their own. Natural enemies in agricultural settings may be able to stop crop pests before they cause significant economic damage. By increasing the mortality of pests that are most vulnerable to herbivores, predators, and parasitoids, we can slow or stop the expansion of pest populations (Sánchez-Bayo, 2021).

4.4. Insects as Food and Feed

All over the world, as reported, there are more than twenty thousand species of insects consumed as food, which are mostly in Latin America, Africa, and Asia. Several centuries ago, entomophagy (insect eating habits) was ditched by the Western world. Primarily, from nature, insects were harvested, and because of some environmental reasons, in comparison with temperate zones, the harvesting yield was less than in the tropics. Due to the smaller size of the insects, in winter times they are unavailable, and their occurrence is also less clumped. For human consumption, insects have good potential for sustainability and nutritive efficacy. Nowadays, it is very difficult to estimate the percentage of humans consuming insects; moreover, entomophagy has again become to trend in Western countries, especially among young people (Mancini et al., 2019; van Huis, 2020). Consumers in Western nations are more interested in farmed insects since they can be completely controlled in a close habitat (a farm) and for the goods' safety. Even if there is a limited possibility that insects would spread zoonotic illnesses, allergies, namely those connected to crustaceans (invertebrates), may still arise. Insect-based food, for example, protein snacks, whole burgers, and roasted burgers, as well as previously well-known insect-supplemented food items like pasta, cookies, bread, and pizza, drives the entire market crazy because of humans' readiness to eat these items and can reach out to our tables in the future (Murefu et al., 2019). For various animals, insects are part of their natural diets. The use of insects in livestock production as feed is still very limited because of their laws and regulations, as those insects are not included in the list of ingredients. Perhaps insects can be used as supplementation feed, dead or alive, and as feed ingredients such as protein meal, full meal (whole body), oil, and other derived products. Their utilisation has only been focused on three main sectors by researchers: pig farming, aquaculture, and aviculture (van Huis, 2020). Feed nutritional values are precise for farmed animals; the projects of researchers are mainly focused on substituting already-derived fat or protein ingredients with insect ones. Moreover, the utilisation of insects' fat oil or protein meal can be replaced by fish-wheat-maize-soy meal-oil to cause less impact on the environment, that is, the water, land, and sea, and also retain animal welfare as various omnivorous or carnivorous animals that are fed on exclusive plant-based raw products. Every day, reviews and research articles are published and growing, whereas increasing knowledge about the production section of insects is still needed.

5. Insects as Pests

Insects have troubled people ever since they first evolved and competed with them for resources like food and water. There is a long list where insects acted as pests causing damages to humans (Figure 3). They acted as carriers of microbes that spread diseases to them or the living things they valued. As the human population rose and became more urbanised, the impact of such insects significantly increased. Urbanisation created convenient conditions for the spread of human parasitic insects and the diseases they transmit. The long-term, extensive cultivation of the same crop throughout a region aided in the rapid population growth of some plant-eating species and the spread of plant diseases. Additionally, the spread of nuisance insects and diseases transmitted by insects is aided by modern transportation. Some attempts at pest removal have failed, causing even more financial harm.

5.1. In Agriculture

Pest insects in agriculture around the world severely reduce agricultural yields, and many of them act as disease vectors for both humans and animals. As a result, several attempts are made to manage insect infestations. One crucial component of contemporary pest management strategies is chemical pesticides (Cooper & Dobson, 2007). Insects that consume dead trees become pests when they infiltrate the structural timbers of buildings. This is true for insects that can degrade the keratin in feathers and hair, such as dermestid beetles and certain tineid moths, which are ecologically latecomers to carcasses. These insects can cause issues for people if they get into clothing made of carpets, skins, wool, or fur. Ripe grain in the fields is invaded by specific beetles and moths in various hot, dry locations, such as the plains of India or North Africa. At the time of harvest, these insects proliferate in grain bins. They are transportable in commercial contexts around the globe and have evolved into common pests of items like tobacco, dried fruit, and grains that have been preserved. To prevent the introduction of these insects from nations that export grains, disinfestation and quarantine techniques are applied (Bhargava & Kumawat, 2010). Many insects graze on plants; additionally, if the plants are crucial to agriculture, people have to frequently compete with these insects. Some factors restrict the number of insect populations, such as unfavourable weather conditions, parasites and predators, and bacterial, viral, and fungal illnesses, among many other variables that work to stabilise insect populations (Raven & Wagner, 2021). Agricultural practices that promote the growing of single crops in ever-larger regions, which offer practically endless food resources, have eliminated some of these controlling elements and allowed the rate of expansion in the number of insects that attack those crops. This raises the possibility of serious insect pest infestations. In several natural forests that resemble massive monocultures, there have always seemed to be sporadic outbreaks of damaging insects (Nair, 2007).

Certain monocultures in agriculture accidentally introduce non-native insect pests into a crop without also bringing along its entire complement of natural foes. Oyster shell scale (*Lepidosaphes ulmi*) on apples and other fruit trees has caused this in the United States (Crouch et al., 2021). Citrus's "cotton-cushion scale" (*Icerya purchasi*), *Pyrausta nubilalis*, most notably known as *Ostrinia nubilalis*, the European corn borer, and others are also included in this category. *Leptinotarsa decemlineata*, known as the Colorado potato beetle, which, starting in 1840, caused terrible damage to the domesticated potato in the United States, was a natural insect of a semi-arid region. The insect, which consumed the plant known as buffalo burr, adapted to a freshly introduced diet that was rich in potatoes and consequently got away from all prior controlling factors (Considine, 2012). By identifying the main predators or parasites of an alien insect pest in its place of origin and introducing them as control agents, similar situations have frequently been managed. The cottony-cushion scale is a well-known illustration that, in 1886, posed a danger to the California citrus sector (Hajek and Eilenberg 2018). The vedalia beetle (*Rodolia cardinalis*), a ladybird predator, was brought over from Australia, and within a year or two, the scale bug almost completely vanished. Every nation where the scale insect had established itself without predators saw a replay of its triumph. Early in the 1940s, the European spruce sawfly (*Gilpinia hercyniae*) was found in eastern Canada

(Payette et al., 200); enormous damage has been done by them; it was entirely regulated by the occurrence of a viral illness; and it was possibly unintentionally imported into Europe. This incident raised awareness of the possible benefits of employing insecticides to control pest populations.

5.2. In Forests

Moths and beetles, both adult and larval, inhabit temperate woodlands; a few flies, and to a lesser extent, horntail and sawfly wasps, attack groves of thriving, deteriorating, or dead trees (Wermelinger, 2021). Although these insects are a crucial component of the system that cycles nutrients, their presence in trees that are maintained for timber can turn them into pests very quickly. These insects eat through limbs and trunks, whereas others target roots, cones, buds, seeds, and foliage in their attacks. Living trees may experience diminished vigour, decreased development, or even death. Forest pests typically target trees that are already stressed from lack of water, inadequate nourishment, or damage from disease or fire. Due to their compromised defence mechanisms, plant diseases spread by these insects frequently kill trees (Ferrell, 1996). Forest pests affect the value of timber and other goods by causing stunted growth, reduced productivity, and other negative effects; they are also accountable for the destruction of watersheds, the loss of habitats, and the rise in fire danger. Erosion and flooding are also caused by habitat alteration. In plantations, old-growth forests, and pure stands, pest outbreaks are more likely to occur. Forest pest outbreaks can be brought on by hail storms, floods, ice storms, illness, high winds, fire, and drought (Vogt et al., 2020).

5.3. Structural Pests

Carpenter ants, termites, and wood-boring beetles can and do infest dry, treated wood used to construct houses and sheds. They eat away at wood, creating tunnels as they go, until the wood is hollowed out, leaving only a false exterior. Termites, which destroy wood all around the world, are the most devastating structural pests. While termites covertly tunnel, generations of anobiids ("death watch") have merely left the occasional clump of frass in door and window jams, and false powder post (bostrichid) beetles leave behind shot-like exit holes on wooden surfaces, indicating their presence (Govorushko, 2019).

5.4. Household Pests

Before the emergence of humanity, in addition to eating dispersed seeds and grains, insects were also scavenging for rotting fruit, decaying flesh, feathers, and hair. When we first started processing and storing these minerals for our purposes, our bond with these insects was irrevocably established. The scavengers adapted to life among humans very fast, becoming almost cosmopolitan (distributed globally) as a result of trade and other human endeavours (Nayak & Daglish, 2018). The carpet beetle and clothing moth larvae destroy woollen skins, garments, and rugs. Powderpost beetles harm furniture, cabinets, and flooring in addition to destroying them. The cigarette beetle (*Lasioderma serricorne*), a large pest, causes substantial harm to cereals, legumes, spices, and grain products (Hill, 2002). The drugstore beetle (*Stegobium paniceum*) feeds on crackers, herbs, and sweets in addition to spices and legumes. The sawtoothed grain beetle (*Oryzaephilus surinamensis*), an omnivore, infests nuts, preserved meats, sugar, dried fruits, bread, cereal, and pasta. One of the most significant pests of food stored in households and stores is *Tribolium confusum* (confused flour beetle). Stored grains, to be more specific, rice, are infested by the *Sitophilus oryzae* (rice weevil), whereas barley and wheat products are infected by one of the closely related grain weevils known as *Sitophilus granaries*. *Plodia interpunctella*, an Indian meal moth, feeds on coarsely crushed grains as larvae, like cornmeal, dried fruits, pet food, crackers, and nuts, leaving sheets of silk in their wake (Evans, 2008). Some species of cockroaches, like *Periplaneta americana* (American), *Blatta orientalis* (Oriental), *Supella longipalpa* (brown-banded), and *Blattella germanica* (German), infect not only stored food but also clothing and paper products. In the search for water and food, ants usually invade homes (Pratt, 1976).

5.5. Domestic Animals Pests

Fly infestations, flea infestations, lice harassment, livestock damage, pet and poultry weakening, and decreased production of eggs, meat, wool, and milk along with damaged skins result in economic loss. Some nuisance insects choose their hosts based on very particular criteria, whereas others have attacked numerous domestic and wild animal species. General blood-feeding pests include black flies, horn flies, blowflies, screwworms, mosquitoes, and others. Female heel flies' venomous bites can send huge herds of animals into a stampede (Hill & Hill, 1997). Lice and horn flies that feed on blood may be so numerous that their hosts have anaemia due to blood loss. Screwworm maggots are parasites that dwell in animal wounds, where they consume living tissues, occasionally killing the host. Stable flies, blowflies, and horseflies spread anthrax and other keratitis- and mastitis-causing infections. The majority, if not all, of the lives of parasitic fleas and lice, are spent on the bodies of their mammal and bird hosts. Cattle lice infestations can cause excruciating itching and, as a result of repeated scratching against posts and fences, blood and hair loss, as well as raw skin (Antonelli & Ramsay, 2014). Animals who are infected become lethargic and fail to gain weight. Swinepox and other contagious swine diseases are spread by hog lice. Parasitic sheep kings are flat, wingless flies that pierce sheep's skin and drain their blood, causing the sheep to scratch, bite, and rub. Some species, including cockroaches, fleas, and beetles, are the intermediary hosts for pathogenic organisms, such as heartworm and poultry tapeworm, and if consumed, can infect domesticated pets and animals (Rawson, 1934).

5.6. Pests at Campsites

In the beginning, a home base would have been something material, like a straightforward circle made of stones or wood. Later, a more permanent home base was created and inhabited by members of one or more families. It would then transform into a true home, cherished for what it was and what it included, and those on two legs, four legs, six legs, or eight legs should be defended from intruders. Campsites were only occupied for brief periods, but excavations of some of these prehistoric sites reveal that there were clear attempts to move plant and animal detritus to a nearby location (Robinson, 2020). The edges of primitive campsites frequently include large quantities of bone pieces. Most likely, the location where the gang habitually dropped its waste was where the debris was gathered. The amount of organic waste was presumably small and was mainly made up of fruit and nut fragments, nut shells, and possibly small animal bones. There might have been different areas to deposit human faeces. If the campsite or the makeshift shelter were permitted to accumulate this debris, it would undoubtedly start to deteriorate. The unpleasant odour would eventually draw flies and possibly scavenging rodents into the living area. Primitive tribes probably quickly realised that leaving the camp area was a wise housekeeping move that would help everyone (Biehler, 2013).

5.7. Food Transport and Storage

It is believed that early humans began to rely more and more on meat to meet their protein needs, as finding certain types of plant food became more challenging. To regularly include meat in one's diet. Bands of early humans gradually developed new techniques for food collection, both to harvest energy-rich plant food. They implemented a labour division, where some individuals (mostly males) specialised in gathering plants while others (likely females and the young) specialised in acquiring meat through hunting or scavenging. To ensure that everyone in the social unit had a balanced diet, some of the foods obtained were not immediately consumed but rather preserved and sent back to the base camp. A significant stage in the evolution of human society and the interaction with some insect (and possibly rodent) pests was the transportation of plant and animal food from the gathering location to the home base (Heeps, 2016). The different insects that already ate downed seeds and dry meat fruit would have been. Scourge material kept hanging around the base camp. The capacity to gather and transport objects, particularly food, distinguished species of *Homo* from some of their more ape-like relatives. This essential act gave *Homo* a new level of independence from its surroundings and possibly promoted some degree of group collaboration. The value of group cooperation can significantly improve when food and water can be stored and transported to a new site. Early humans would not have been able to carry enough plant food from the field to their home base without the use of a carrying mechanism (Pontzer, 2017). Even for short-term storage, the storage containers take on equal importance to any potential contents if surplus food can be obtained. It might be done with anything as simple as a piece of tree bark, but such a thing would have been required. Some of these ancient jars would have been crafted from leaves and shrub branches. Reeds from the edge of a river, sometimes bound with vines, or perhaps organic containers like bird nests Their service to carry food and tools might later be converted into something permanent and elaborate. The interaction between the pests and humans would be useful for the future purpose of storing food in vessels (Yen, 2015).

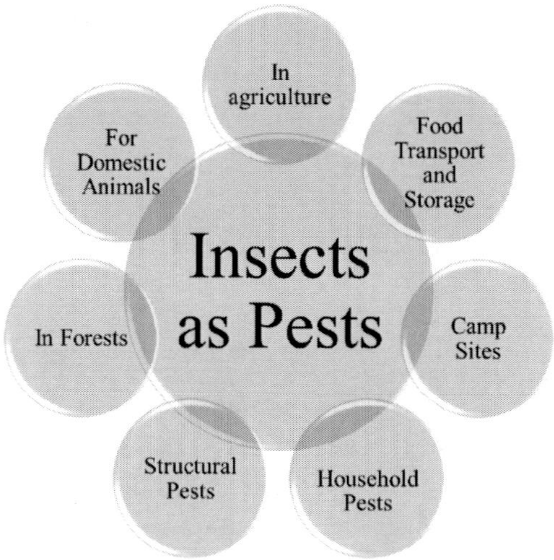

Figure 3. Different roles of insects as pests.

6. Insect-Borne Diseases

Bacterial and viral disorders spread by bug (insect) bites are known as insect-borne diseases. Ticks, fleas, mosquitoes, and sandflies are the most frequent insects that transmit disease. For instance, transmission of diseases like yellow fever, zika virus, and malaria is known to be caused by mosquitoes (Campbell, 2008). Diseases like Rocky Mountain Spotted Fever and Lyme disease are transmitted by ticks. Depending on the type of insect, the symptoms of diseases will vary. The most common insect-borne disease symptoms include chills, fever, headaches, nausea, sore muscles, skin rashes, and stomach pain. In severe cases, the symptoms may also include chest pain, the feeling of the throat closing, difficulty breathing, swelling in the lips, tongue, or face, vomiting, dizziness, and a racing heartbeat that lasts more than a few minutes. Most commonly, diseases are caused by the bites of blood-sucking insects, which infect humans. Bushes, sand, spots near still water, tall grasses, forests, and places around the world with heavy outbreaks are areas that pose the highest risk for humans to get infected by these insect-borne diseases. (Lehane, 2005). There are some insect-borne diseases caused by mosquitoes, as shown in Table 2.

Table 2. Diseases caused by mosquitoes in humans

S. No.	Disease name	Vector	Rate of infected people/year
1	Dengue fever	*Aedes aegypti* and *Aedes albopictus*	50 million
2	Malaria	*Anopheles*	500 million
3	Leishmaniasis	*Lutzomyia* and *Phlebotomus*	2 million
4	Bubonic plague	*Xenopsylla cheopis*	Several thousand per year
5	Sleeping sickness	Tsetse fly	Millions of people
6	Typhus	Mites, body lice and fleas	16 million
7	Wuchereria bancrofti	*Culex, Aedes, Anopheles, Mansonia*	120 million
8	Yellow fever	*A. aegypti, A. africanus* and *A. simpsoni, Haemagogus* and *Sabethes*	2 hundred thousand

There can also be long-term, life-affecting symptoms caused by some of the insect-borne diseases. For instance, the disease Lyme can affect you with long-term fatigue and sore muscles, as well as viral diseases like Zika virus, which can also be passed on to babies from mothers and cause babies' heads to be abnormally small, which is known as microcephaly and intellectual disability.

7. Insect Pests Control

Pests are organisms that cause severe economic damage to people, their goods, or anything else that irritates people in some way. Inferred from the aforementioned description are value judgements, which can differ depending on who makes them as well as the location and time of production. However, given a specific set of circumstances, there will be a threshold for economic harm (annoyance), calculated as the population density of a species. This is the point at which it is beneficial (profitable) to implement controls to reduce the species' density. Increasingly, there is a gap between the economic harm threshold and the population density. The profitability of control becomes more desirable. Therefore, pest control is primarily a sociological issue that involves politics, economics, and psychology. There are numerous techniques for eliminating insect infestations, as shown in Figure 4. Each of these techniques has benefits and drawbacks. These must be weighed against one another to decide which method (or combination of methods) is best in a particular situation. Even though some of these techniques are outstanding, they are just temporary. For instance, when large-scale pest outbreaks occur suddenly but are erratic and transient, others move more slowly yet have effects that are generally long-lasting. It can be applied to pests whose populations are rather steady while being more or less permanent (Gillott, 2005).

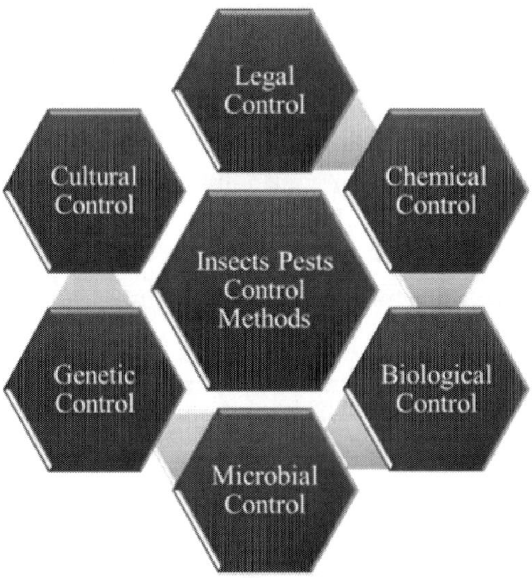

Figure 4. Insect pest control methods.

Conclusion

Since the beginning of soil cultivation, insects have competed with humans for the fruits of our labour. Undoubtedly, a large portion of the public, including homeowners, healthcare workers, agriculturalists, and those working with natural resources, view insects as ideal pests. Numerous species were introduced due to a significant increase in human geographic migration and a corresponding rise in trade, both plant and animal, into previously empty spaces. Some of these were successful in becoming established, and in the absence of the usual predators and parasitoids, in particular population regulators, they became significant pests and multiplied quickly. On occasion, as people settled in new places, for some of the endemic insect species in these regions, some of the introduced cultivated plants have proven to be a great source of food. Increasing demand for more affordable and plentiful food production through the mechanisation of agriculture refers to the practice of monoculture, which involves growing the same crop over an extended period on the same vast plot of land. Humans stimulate the growth of insect populations by reducing the complexity of the ecosystem. Additionally, because the rival insects are the main consumers, they typically have a rapid reproduction rate and a short generation time, i.e., they are found at the bottom of the food chain. Otherwise, there is a chance that populations of these species will grow quickly. Such efforts are essential because of several aspects of modern human evolution. Insects have long been valued by humans for their contribution to their health. For thousands of years, humans have consumed insects and/or their byproducts. Since over 5000 years ago, silk has been made from silkmoth pupae. The preservation of insects is still in its infancy, although public awareness of insect conservation is growing. Entomologists' job is to determine the most effective way to achieve this. Despite their ecological significance, insufficient attention has been paid to their conservation, but recent studies and reviews might be able to change that.

Acknowledgement

The authors are thankful to the Vice Chancellor, Banasthali Vidyapith and the Head of the Department of Bioscience and Biotechnology for providing the necessary facilities to facilitate the work. Facilities provided by DST-FIST to the University are highly acknowledged.

References

Allen, J. P. (2005). *The art of medicine in ancient Egypt*. Metropolitan Museum of Art.

Antonelli, A. L., & Ramsay, C. A. (2014). Livestock pests study guide.

Bhargava, M. C., & Kumawat, K. C. (2010). *Pests of stored grains and their management*. New India Publishing.

Biehler, D. D. (2013). *Pests in the city: Flies, bedbugs, cockroaches, and rats*. University of Washington Press.

Campbell, S. (2008). Insect-borne diseases and their prevention. *Primary Health Care (through 2013)*, *18*(6), 25.

Chand, B., Kumari, I., & Kumar, R. (2021). Evolution of Apiculture, History and Present Scenario. In *Honey* (pp. 1-22). CRC Press.

Considine, D. M. (2012). *Foods and food production encyclopedia*. Springer Science & Business Media.

Cooper, J., & Dobson, H. (2007). The benefits of pesticides to mankind and the environment. *Crop Protection*, *26*(9), 1337-1348.

Crouch, C. D., Grady, A. M., Wilhelmi, N. P., Hofstetter, R. W., DePinte, D. E., & Waring, K. M. (2021). Oystershell scale: an emerging invasive threat to aspen in the southwestern US. *Biological Invasions*, *23*(9), 2893-2912.

Erwin, T. L. (2004). The biodiversity question: How many species of terrestrial arthropods are there? *Forest Canopies*, *10*, 259-269.

Evans, A. V. (2008). *What's Bugging You?: A Fond Look at the Animals We Love to Hate*. University of Virginia Press.

Fei, M., Gols, R., & Harvey, J. A. (2023). The biology and ecology of parasitoid wasps of predatory arthropods. *Annual Review of Entomology*, *68*, 109-128.

Ferrell, G. T. (1996). The influence of insect pests and pathogens on Sierra forests.

Gillott, C. (2005). Insects and humans. *Entomology*, 725-782.

Govorushko, S. (2019). Economic and ecological importance of termites: A global review. *Entomological Science*, *22*(1), 21-35.

Grimaldi, D., Engel, M. S., Engel, M. S., & Engel, M. S. (2005). *Evolution of the Insects*. Cambridge University Press.

Hajek, A. E., & Eilenberg, J. (2018). *Natural enemies: an introduction to biological control*. Cambridge University Press.

Hammond, P. (1992). Species inventory. *Global biodiversity: status of the earth's living resources*, 17-39.

Heeps, J. (2016). *Insect management for food storage and processing*. Elsevier.

Hill, D. S. (2002). *Pests of stored foodstuffs and their control*. Springer Science & Business Media.

Hill, D. S., & Hill, D. S. (1997). Harmful Insects. *The Economic Importance of Insects*, 77-332.

Hogue, C. L. (1987). Cultural entomology. *Annual Review of Entomology*, *32*(1), 181-199.

Lehane, M. J. (2005). *The biology of blood-sucking in insects*. Cambridge University Press.

Mancini, S., Moruzzo, R., Riccioli, F., & Paci, G. (2019). European consumers' readiness to adopt insects as food. A review. *Food Research International*, *122*, 661-678.

Murefu, T. R., Macheka, L., Musundire, R., & Manditsera, F. A. (2019). Safety of wild harvested and reared edible insects: A review. *Food Control*, *101*, 209-224.

Nair, K. S. (2007). *Tropical forest insect pests: ecology, impact, and management*. Cambridge University Press.

Nayak, M. K., & Daglish, G. J. (2018). Importance of stored product insects. *Recent Advances in Stored Product Protection*, 1-17.

Ofuya, T. I., Okunlola, A. I., & Mbata, G. N. (2023). A Review of Insect Pest Management in Vegetable Crop Production in Nigeria. *Insects*, *14*(2), 111.

Payette, S., Bhiry, N., Delwaide, A., & Simard, M. (2000). Origin of the lichen woodland at its southern range limit in eastern Canada: the catastrophic impact of insect defoliators and fire on the spruce-moss forest. *Canadian Journal of Forest Research*, *30*(2), 288-305.

Pedigo, L. P., Rice, M. E., & Krell, R. K. (2021). *Entomology and pest management*. Waveland Press.

Pimentel, D. (Ed.). (2002). *Encyclopedia of pest management*. CRC press.

Pimentel, D., Acquay, H., Biltonen, M., Rice, P., Silva, M., Nelson, J., ... & D'amore, M. (1992). Environmental and economic costs of pesticide use. *BioScience*, *42*(10), 750-760.

Pimentel, D., Lach, L., Zuniga, R., & Morrison, D. (2000). Environmental and economic costs of nonindigenous species in the United States. *BioScience*, *50*(1), 53-65.

Pimentel, D., Wilson, C., McCullum, C., Huang, R., Dwen, P., Flack, J., ... & Cliff, B. (1997). Economic and environmental benefits of biodiversity. *BioScience*, *47*(11), 747-757.

Pontzer, H. (2017). Economy and endurance in human evolution. *Current Biology*, *27*(12), R613-R621.

Pratt, H. D. (1976). *Household and stored-food insects of public health importance and their control*. Center for Disease Control.

Raven, P. H., & Wagner, D. L. (2021). Agricultural intensification and climate change are rapidly decreasing insect biodiversity. *Proceedings of the National Academy of Sciences*, *118*(2), e2002548117.

Rawson, G. W. (1934). Arthropods and Their Relationship to Diseases of Domestic Animals. *The Scientific Monthly*, *39*(6), 497-510.

Robinson, W. (2020). *Urban entomology: insect and mite pests in the human environment*. Garland Science.

Sánchez-Bayo, F. (2021). Indirect effect of pesticides on insects and other arthropods. *Toxics*, *9*(8), 177.

Skendžić, S., Zovko, M., Živković, I. P., Lešić, V., & Lemić, D. (2021). The impact of climate change on agricultural insect pests. *Insects*, *12*(5), 440.

Southwood, T. R. E. (1977). Entomology and Mankind: Insects over the ages have greatly affected man's health and food supply and have played an important role as religious and cultural symbols. *American Scientist*, *65*(1), 30-39.

Stork, N. E. (1993). How many species are there? *Biodiversity & Conservation*, *2*, 215-232.

Van Huis, A. (2020). Insects as food and feed, a new emerging agricultural sector: a review. *Journal of Insects as Food and Feed*, *6*(1), 27-44.

Vogt, J. T., Gandhi, K. J., Bragg, D. C., Olatinwo, R., & Klepzig, K. D. (2020). *Interactions between weather-related disturbance and forest insects and diseases in the Southern United States*. United States Department of Agriculture, Forest Service, Southern Research Station.

Wermelinger, B. (2021). *Forest insects in Europe: diversity, functions and importance*. CRC Press.

Wright, L. (2005). Forest spirits, giant insects and world trees: The nature vision of Hayao Miyazaki. *The Journal of Religion and Popular Culture*, *10*(1), 3-3.

Yen, A. L. (2015). Insects as food and feed in the Asia Pacific region: current perspectives and future directions. *Journal of Insects as Food and Feed*, *1*(1), 33-55.

Zhang, C. X., Tang, X. D., & Cheng, J. A. (2008). The utilization and industrialization of insect resources in China. *Entomological research*, *38*, S38-S47.

Chapter 2

Major Insect Pests of Cereal Crops and Their Management by Pesticides

Souptik Bhattacharya*
Shaoli Das
and Sanjukta Banik
Department of Food Technology, Guru Nanak Institute of Technology, Kolkata, India

Abstract

Cereal crops play a significant role worldwide in terms of area, production, and diet composition. The current rate of population expansion would have been impossible without a comparable increase in the generation of food. Apart from humans, other animals also depend on cereals. The metamorphosis process of various insects is dependent on plants because of their abundant nutritional sources. The life cycle of a large group of insects is reliant on various crops. However, crops are also harmed by these insects because they form a parasitic relationship with the plant body. Thus, these pests are regarded as a significant productivity constraint agent. Insect abundance, distribution, and pest-related cereal losses are affected by global warming, abnormal weather, shifting agricultural patterns, and the adoption of technology throughout time and geography. To control and manage insect pests, pesticides and insecticides are used to control them, which results in a noteworthy increase in agricultural products. Today, many types of pesticides are crucial for reducing plant diseases and increasing agricultural production throughout the globe. Moreover, modern technology, including nanotechnology and genetically modified plants, is now being used as a good alternative to conventional pest management strategies. Thus, the present state of infestations of insects that affect important cereal crops is the primary subject matter of this chapter. Further key insect pest management technology by modern pesticides is discussed here. Additionally, this chapter also provides a summary of the use of advanced technologies like nanotechnology and genetically modified organisms to control cereal pest insects.

Keywords: cereal crops, pest control, pesticide, insecticide, food security, nanotechnology

---

* Corresponding Author's Email: souptikjduv@gmail.com; souptik.bhattacharya@gnit.ac.in.

In: Insects as Humans' Frenemies: From Friends to Foes
Editor: Istkhar Rao
ISBN: 979-8-89530-320-7

1. Introduction

Cereal grain crops are predominant crops, and they grow in larger quantities because they produce more food energy than any other type of crop. They consist of barley, rye, oats, and wheat. Pseudo-cereals are used in the same ways as edible grains and belong to different plant families, like buckwheat, quinoa, and chia. Cereal crops are vital to human civilisation and are grown from one end to the other of the world. The four main grains farmed for anthropological use in the tropics are maize, rice, sorghum, and pearl millet (Békés et al., 2017). Wheat and barley, which are wintertime cereals, are also cultivated in small numbers in tropical regions, particularly at higher altitudes. For example, wheat is cultivated in the African nation of Kenya, while barley is cultivated in the nation of Ethiopia. Despite being staple foods in the Indian subcontinent, wheat and barley are often produced at low altitudes throughout the winter months beyond the tropical regions. Today, tropical cereals are becoming more and more significant. The production of cereals in the tropics accounts for 30% of worldwide output; no fewer than 75 per cent of the grain produced across the world is made up of wheat, maize, and rice. In industrialised countries, more than 70% of the cereal crop is fed to livestock, while in emerging economies, 68% to 98% of the cereal crop feeds animals. Throughout the globe, cereal crops are threatened by pests, and different types of pesticides are used to manage them. Pest control agents are compounds known as pesticides. Herbicides are the most widely used pesticides, making up around 80% of all pesticide usage. The majority of pesticides are designed to act as crop protection agents, or crop protection products, which often shield plants from weeds, fungi, or insects. For instance, the aquatic weed *Salvinia* is controlled by the fungus *Alternaria solani* (Goussous et al., 2010). On the contrary, DDT was shown to be seriously threatening biodiversity in the 1960s because it was preventing many fish-eating birds from breeding. Despite the Stockholm Convention on Persistent Organic Pollutants' current ban on the use of DDT in agriculture, several countries continue to spray it on interior walls to kill or deter insects to avoid malaria and other tropical diseases. Over the past 50 years, conventional pesticides, notably carbamates, organophosphates, chlorinated hydrocarbons, and pyrethroids, have been effective in reducing losses in agricultural output by combating insect pests. Furthermore, a lot of these toxins disrupt the environment and are dangerous to both humans and healthy organisms. All these significant efforts have been made to reduce the harmful effects of pesticides on the environment, increase food production, and improve both human and domestic animal health. It is now crucial to study recent developments in insect control and create better crop security protocols to better understand the effects of agrochemicals. The "Green Revolution" era saw the arrival of high-yielding cultivars and agricultural commodities to India and other emerging countries in Asia, which boosted agricultural efficiency while concurrently reducing the amount of crop loss to pests like insects. Today, the desire for safer, more targeted insecticides that only target harmful pests while leaving other creatures and beneficial insect species unharmed is higher. As a constituent of 'integrated pest management' (IPM) programmes, research of novel alternatives and more environmentally friendly techniques of bug control is also encouraged by the quickly emerging resistance to conventional insecticides (Zhang et al., 2022). The development of new compounds that affect insect morphogenesis, such as 'chitin synthesis inhibitors' and 'juvenile hormone mimics', is one of these techniques that has attracted interest from all around the world. Public opinion surrounding the global colony collapse disorder situation is discouraging and delaying the evolution of insecticides. The use of the toxin produced by *Bacillus thuringiensis* for the control of lepidopteran, coleopteran, and dipteran pests has been introduced as an enhanced biocontrol agent (Gatehouse et al., 2011). A great deal of study has also gone into developing substances that target specific insect species by enhancing or suppressing the function of biochemical networks like metabolism (diafenthiuron), the 'nicotinyl acetyl chloride receptor' (imidacloprid and acetamiprid), and the glands that produce saliva from sucking pest infestations (pymetrozine).

2. The Relation Between Insects and Cereal Crops

Insects play an important role in our ecosystem. The ecological significance of insects sometimes goes unappreciated since they are typically viewed as pests or prospective problems (Kim, 1993). The four main ecological functions of insects in ecosystems are plant reproduction, food chain components, degradation of complex organics, and environmental rotation (Jankielsohn, 2018). Pests and insects in crop agriculture have been professed as rivals in the competition for existence. Over half of insect species are herbivorous, i.e., dependent on plants or related plant families for their survival. 18% of global agricultural productivity is damaged by herbivorous insects, which are primarily managed chemically (Cummins, 2014). Despite these harms, less than 0.5% of the known insects are classified as pests. Insects are not considered pests from an eco-friendly perspective, except for anthropocentric perception and social justice. As crops cannot be grown without the ecosystem services supplied by insects, insects are essential to human life. Insects are required to pollinate around 72% of all crops worldwide (Ndakidemi et al., 2016). Three-quarters of all crop kinds worldwide—or one-third of the volume of global agricultural production—have improved or stabilised yields thanks to pollinating insects. Many insect taxa have been connected to rising seed sets (Hoehn et al., 2008). The bugs that serve as insects that pollinate include plenty of distinct kinds of solitary honeybees, bumblebees, flies, beetles, and butterflies. The pollination of some crops depends more on natural varieties of bees than on honeybees (*Apis mellifera*) (Pardo & Borges, 2020). Insect pollination services are thought to provide 9.5% of the world's agricultural output yield (Robinson & Sutherland, 2002). Agriculture inevitably requires pest control. A shift in pest conditions has currently been seen in cereal crops over the past decade due to several causes, such as global warming, shifts in agricultural practices, weather, the implementation of novel agricultural methods, alteration of agronomic practices, and the launch of exceptional input-intensive superior cultivars and hybrid plants. Additionally, the unregulated application of pesticides to safeguard crops has led to issues including pest comeback, resistance to pesticides, and pollution of the total environment, all of which contribute to the issue of developing insect pests in certain areas. However, insect infestations from crops can be eliminated by predatory insects, which considerably benefit the functioning of the environment. Aphids, worms, root-feeding flies, and phytophagous beetles are only a few of the commercially significant crop insect pests that ground beetles effectively manage in large populations in crops that are planted (Kuotsu & Lalrinfeli, 2019). Furthermore, insects are essential for improving the soil in farming. Instead of chemical fertilisers, dung beetles significantly increase the levels of total proteins, calcium, magnesium, and nitrogen in the surrounding soil, thus increasing the yield of wheat crops (de Groot et al., 2002). Hence, insects have both good and bad effects on cereal crops.

3. Important Cereals Targeted by Insects and Their Management

3.1. Rice

The agroecosystem that supports rice cultivation is among the most diversified in the world. Over 65% of the global population depends on rice as their primary food source to survive. Insect infestations drastically limit the global cultivation of rice. Hundreds of insect species feed on paddy; twenty of these species are the most significant. Rice leaf folder (*Cnaphalocrocis medinalis*) and rice stem borers (*Scirpophaga incertulas* and *Scirpophaga innotata*) are regarded as the predominant insect pests of rice, causing significant damage throughout India's rice-growing regions. In India, seven states had a "severe" stem borer infestation, six states had a "moderate" infestation, and six states had a "low" infestation. Bandicoot rats, rice stem borers, white-backed planthoppers (*Sogatella furcifera*), rice hispa (*Dicladispa armigera*), brown planthoppers (*Nilaparvata lugens*), and other insect and vertebrate pests are the most prevalent in Bangladesh. Stem borers, which generally range from one to four species, are probably the most damaging group (Babendreier et al., 2022). Pest assessment studies frequently demonstrate that crops' responses to attacks by insects of similar densities vary greatly from site to site and year to year. *Sogatella furcifera* can reduce paddy yield by 40–57% (Subramanya et al., 2022). Aside from these insects, a few other insects and mites are regionally significant and only found in a few locations across the nation. For instance, low-lying and wet parts of eastern India are where caseworm infestations are found. Similar to this, only steep locations are home to the white grubs that infest rice. In Bihar (India), significant local insect pests of rice include the swarming caterpillar, green leafhopper, rice hispa, and mealybug' (*Maconellicoccus hirsutus*) (Tinzaara et al., 2002).

The dual goals of crop production and insect pest control are achieved by manipulating the cropping environment through cultural methods of insect control. The different integrated pest management strategies include clipping the tops of bundles of seedlings, maintaining appropriate spacing, field draining for one to two days, crop rotation, rice crop transplantation, etc. A thorough reconnaissance or surveillance of the infestation is essential for individualised pesticide treatment. Insects can be physically captured or squashed. The control of the yellow stem borer *Scirpophaga Incertulas* benefits greatly from the use of pheromone lures (Tinzaara et al., 2002).

3.2. Wheat

Among the cereal crop species, wheat is the most adaptable because it can be grown in a wide variety of environments in various regions of the world. More than 4.5 billion people in 94 developing nations depend on it for their daily sustenance (Amanullah et al., 2019). According to the US Department of Agriculture (USDA), India produces approximately 11.9% of the world's wheat. China is the world's largest wheat producer, with 16.9% of the world's wheat production. Canada, the United States of America, and the Russian Federation are the other major producers of wheat. Together, these five nations produce the majority of the world's wheat (Amanullah et al., 2019). However, sporadic pests like armyworm, cutworm, flea beetle, sugarcane leafhopper, and others have appeared in numerous outbreaks over the past three decades. Approximately eight insect pests are regarded as having a significant impact on wheat's economy. Nationally significant insects include the termite, wheat aphid, armyworm, American pod borer, pink stem borer, shoot fly, etc. Integrated pest management strategies have been suggested for the management of wheat pests in warmer, non-traditional wheat-growing areas due to the complexity of tropical pest problems. Anywhere wheat is grown, a good pest management system is important, but it may be even more important in the tropics.

In the months of May and June, extensive summer ploughing should be performed to expose the nematodes, rodents, and pupating larvae to radiation and pest control. The termite infestation can be reduced by applying well-rotted farmyard manure. Insecticides can be used, which will be discussed in detail in the later part of the chapter. The growing number of beneficial organisms reduced the use of insecticides in IPM-managed crops by 50% (Pecenka et al., 2021).

3.3. Maize

In developing and underdeveloped nations, maize is a major staple food crop that can be grown profitably in any season, including summer and spring. One of the most important biotic constraints to the production and storage of maize is the presence of insect pests, both in the storage and field. From sowing to harvest, about 220 arthropod species have been identified as maize pests. The most predominant of these are the stalk borers (*Chilo partellus*) and (*Sesamia inferens*), along with the shootfly (*Atherigona soccata*), thrips, climbing cutworm, and armyworm. These insects cause severe damage to maize crops. The most destructive maize pests are stemborers. Adult insects are most active at night when they mate and lay eggs that are laid on the underside of the leaf surfaces of plants, especially early harvests of maize and sorghum generated for feed. Following hatching, the juvenile larvae briefly eat the leaves. Then they make a hole in the central shaft of the plant and slide through it, producing a decayed plant (Deguine et al., 2021).

Shootflies, *Atherigona sp.*, severely harm spring-sown maize in the north of India. Additionally, termites are a major problem for tropical maize. *Microtermes obesi,* a type of termite, can attack maize shortly after sprouting and at any stage of growth. Dead plants in the row are an indication of early infestation. These plants and their chewed roots are simple to remove. Termite-infested mature plants wilt completely, lose strength in the roots, and frequently fall to the ground. The infested stem has earthen galleries on its outer surface, and the soil also fills the stem's inner hollow space (Deguine et al., 2021).

After harvest, stubble removal and destruction to reduce pest populations, crop residue destruction and the destruction of alternative host plants, and stem cutting to release larvae all have the potential to significantly reduce borer populations. Therefore, the first rain should be used to sow maize to prevent borer attacks. By disrupting the insects' access to a continuous food chain, crop rotation other than the host crops significantly reduces pest populations. It has been noticed that the population of plants significantly reduces insect populations. The incidence of insects rises in tandem with plant density. To reduce the number of insects that occur in the fields, the recommended plant population must be maintained.

3.4. Sorghum

As a major food crop, sorghum is essential to the survival of millions of people living in the tropics. Around 150 insects have been found to harm sorghum in various agroecosystems. Approximately twenty bug and pest species are considered to be relevant commercially. The nutritional value and production of sorghum are impacted by the stem borer and the shoot fly species, which are both common in tropical and subtropical areas. In India, insect pests account for more than 30% of the crop loss of sorghum, with the sorghum shoot fly accounting for 5% of that loss. No matter how much damage is done, a key issue is the diminished grain and fodder output caused by the pests' ability to seize vital nutrients. Because sorghum can flourish in arid, low-fertility regions, both of these crops are essential to the country's food security. There is a compelling need to offer workable solutions for managing insect pests due to their detrimental impacts on individuals and the ecosystem, as well as their exorbitant prices. Due to changes in cropping patterns, *Pyrilla perpusilla* (commonly known as planthopper) and grey weevil have lately become significant pests of sorghum. Among the insects that only occupy a few specific locations are the armyworm, cutworm, grasshopper, earhead caterpillar, shot bug, earhead bug, and aphid (Huang et al., 2013).

The different IPM strategies include ploughing the field immediately before crop harvest, collecting crop residues before monsoon to prevent pests passing from one season to the next, using freshly treated sorghum seed treated with carbo-furan, intercropping with legumes, etc. Sowing dates must be adjusted by the major pests' population dynamics to control the population. Shoot fly, midge, and head bug damage was reduced by early and undeviating sowing of the similar cultivar over large areas. Sorghum's pest management strategy ought to be built on the use of resistant cultivars. (Huang et al., 2013). It is necessary that we accelerate initiatives to create, disseminate, and distribute pest-resistant plant types. Sorghum can also benefit from its use as a component of integrated pest management.

3.5. Pearl Millet

Pearl millet (*Pennisetum typhoides*) is grown in low- to high-productivity ecological environments. Pearl millet outperforms other coarse cereals in these environments due to its adaptability to dry and low-fertility conditions. So, it is part of the country's food security. Pearl millet is not only a basic diet for billions of humans, but it also serves as animal fodder. Recent research has shown that the enormous economic losses brought on by possible pests have been underreported since pearl millet growers are unable to invest any expenditure on an agricultural product with low profit margins. At the national level, cutworms and white grub are considered significant insect pests. There are a few places in the country where other insects such as the "shoot fly, hairy caterpillar, grasshopper, grey weevil, earhead bug, stem borer, earhead worm, white ants, chaffer beetle, and blister beetle" can be found. Very few of the various methods of control have been tried out against the major pests in the real world. The development of high-yielding resistant varieties and hybrids is the most practical long-term insect management strategy for pearl millet. Late sown kharif crops suffer severe damage from shoot flies. Therefore, sowing should begin after a monsoon has begun or, at the latest, 10 to 15 days after the first rain of the monsoon. Because it increases the population of shot flies, staggered sowing in the area should be changed occasionally. Spraying crops with insecticides is another effective IPM strategy (Deguine et al., 2021).

4. Predominant Insecticide Groups and Their Effectiveness

4.1. Nicotinoids

Both the peripheral and central nervous systems are affected by nicotineyl insecticides, which cause excitement, paralysis, and ultimately fatality by activating nicotinic acetylcholine receptors (nAChR). Organophosphorous compounds that inhibit acetylcholine esterase cause these molecules to behave in a manner that is similar to that of acetylcholine. Two new nicotinoids, imidacloprid and acetamiprid, show potential for use in agriculture. Both demonstrate a strong affinity for the insect receptor with a relatively weak affinity for the vertebrate receptor and have a structure-activity relationship with the nAChR. That illustrates an increased predilection for insects over humans. The first neonicotinoids designed for commercial use were imidacloprid and acetamiprid, which are used to control pests in agriculture. The term "chloronicotinyl insecticides" also refers to them, highlighting the biological acceptance of the 'chlor' moiety in them (Figure 1). Imidacloprid is a crucial tool for managing aphids, grasshoppers, and whiteflies on attractive and vegetable plants. It is regarded as a comparatively polar substance with excellent 'xylem mobility', making it appropriate for soil amendments and seed treatment. Aphid feeding behaviour is strongly impacted by sub-lethal concentrations, which prevent honeydew excretion, roaming, and, ultimately, starving death. Acetamiprid was considerably more effective than imidacloprid when applied topically, whereas imidacloprid performed better when applied topically to suppress the whitefly *B. tabaci* (Sohrabi et al., 2011). These changes in biological activity might be the result of different xylem mobility when delivered through soil versus leaf immersion when applied to plants. Imidacloprid applied to soil has probably little to no impact on beneficial insects. Imidacloprid showed no impact on Diplopoda, spiders, or predatory mites following foliar application. The relatively high specificity of acetamiprid towards aphids, leafhoppers, and whiteflies, as well as the apparent harmless impact it has on certain natural adversaries, allow for its inclusion as an element in IPM regimes for eliminating sucking pests in field crops, vegetables, and decorative plants. (Khatun et al., 2023).

4.2. Pyrethroids

Evaluations regarding exposure intensification and permanence are carried out on agrochemicals to decide whether to add chemicals to prevent their discharge into the environment. To assess the possible danger to groups of species undergoing risk analysis, the valuation of exposure is set side by side with pertinent toxicity data. Pyrethroids are stereoisomeric mixes that naturally exist. Property evaluations have occasionally been done alongside the admixture and occasionally with the original form of stereoisomers. The data for a pyrethroid, independent of the stereoisomeric form utilised in an experiment, were pooled into a single grade rather than attempting to distinguish the data for one stereoisomeric state from another. As stereoisomers are typically combined to generate pyrethroids, it was anticipated that grouping all stereoisomeric data into one category would adequately describe pyrethroids (Protopopoff et al., 2018).

Instead of attempting to partition the data for each stereoisomeric state, the data for pyrethroids were unified into a single category regardless of the stereoisomeric form used in each experiment. It was believed that combining every stereoisomeric piece of data into one classification would accurately characterise pyrethroids in their entirety since stereoisomers are often mixed to produce pyrethroids. At both acidic and neutral pH levels, pyrethroids in an aqueous medium are only moderately stable, while alkaline situations soon cause them to hydrolyse. Bifenthrin, esfenvalerate, and permethrin are exceptions at higher pH levels because they hydrolyse slowly in an alkaline environment (Laskowski, 2002). The pyrethroids degrade at a fairly variable pace in aerobic soils.

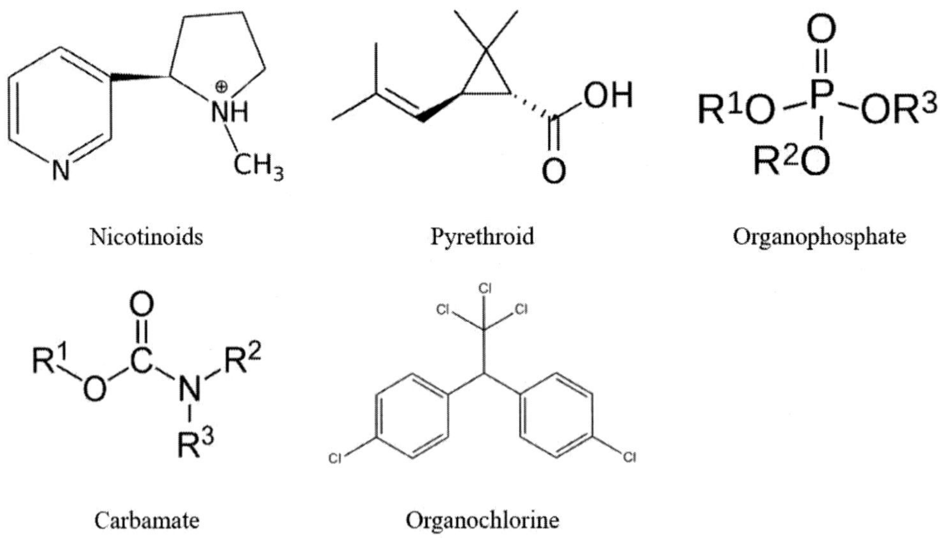

Figure 1. Structures of various types of insecticides.

4.3. Organophosphates

Organophosphates (OPs) are a broad category of synthetic chemicals with a wide range of applications. By reducing disease vectors like mosquitoes and ticks, OPs significantly contribute to safeguarding the health of the populace against illnesses including West Nile, Lyme disease, malaria, and many others. Nevertheless, these pesticides lack species selectivity, and because they contain very harmful compounds, they endanger not just the environment but also the well-being of the public and the whole environment altogether. Since they inhibit the acetylcholinesterase (AChE) enzyme and hence prevent the neurotransmitter acetylcholine (ACh) from functioning at brain terminals and at the neuromuscular junction, organophosphates are dangerous (NMJ). When AChE is blocked, ACh builds up, overexciting muscarinic and nicotinic ACh receptors, resulting in toxic hypercholinergic predominant symptoms that can extend from a couple of hours to a couple of days or weeks. Many OPs have been developed in laboratories throughout the world since the 1950s to make insecticides that are very harmful to insects yet safe for nontarget mammalian, avian, or aquatic species. Since its debut, malathion has served as the most extensively utilised insecticide worldwide for over 50 years, partially achieving these goals (Gupta et al., 2017).

The family of insecticides known as anticholinesterase OPs, which has a wide range of chemical structures but several characteristics, is widely used. A substantial cause of sickness and mortality is OP poisoning, especially in developing countries. Even if there are great antidotes accessible, the consequences of poisoning remain to be addressed. There is controversy around the long-term repercussions of OP poisoning, therefore, this problem has to be resolved (Marrs, 1993).

4.4. Carbamate

The active component of the insecticide TEMIK, aldicarb (carbamate), has been approved for use in over 100 countries to control a wide range of insect, mite, and nematode pests in agriculture. More than 280 animal studies on 12 different species of animals, two clinical human trials, and more than 20 human monitoring studies are all included in the toxicity database. The main cause of poisoning is the suppression of acetylcholinesterase in the peripheral somatic and autonomic nervous systems, which have nicotinic and parasympathetic effects. The neurotoxic effects of aldicarb, like those of other cholinergic N, N-dimethyl, N-methylcarbamate, and organophosphate insecticides, are caused by aldicarb's ability to block acetylcholinesterase at myoneural junctions and synapses. These insecticides work to inhibit acetylcholinesterase by first forming an enzyme-inhibitor complex, then reacting with the enzyme's active site to produce either a carbamoylated or phosphorylated enzyme. When a condition known as ageing has taken place, phosphorylated acetylcholinesterase is relatively persistent, making inhibition essentially irreversible. To recreate the original form of the active enzyme, carbamoylated acetylcholinesterase may be easily and spontaneously degraded (Baron, 1994).

Animals metabolise carbamate insecticides primarily through an oxidative pathway that is often mediated by MFO enzymes. As aldicarb sulfoxide is about 10-fold more effective as an anticholinesterase than aldicarb, it is thought that the metabolism of aldicarb to its sulfoxide involves an activation process. Yet, they are rather robust in neutral or acidic environments while being very vulnerable to alkaline hydrolysis. In general, methylcarbamate insecticides are relatively hazardous to animals. Since they are direct AChE inhibitors and do not require metabolic activation like many safe organophosphorus insecticides (such as malathion), they have a high acute toxicity. A number of these compounds have been changed into derivatives in hopes of enhancing the toxicological characteristics of methylcarbamate insecticides by swapping the hydrogen atom on the carbamate nitrogen for an appropriate functional group (Fukuto, 1990).

4.5. Organochlorine

Organochlorine (OC), a synthetically produced derivative of chlorinated hydrocarbon pesticides, is frequently used in farming operations, which are extensively used in agricultural sectors. These substances have significant toxic effects, slow deterioration, and high bioaccumulation. These substances fall under the category of organic contaminants with significant environmental persistence. Although OC pesticides were originally used successfully to treat malaria and typhus, they are now prohibited in almost all nations with industrialised economies. Statistics on the usage of various pesticides show that 40% of all pesticides are chemically representative of the organochlorine category. Organochlorine insecticides, including dichloro-diphenyl-trichloro-ethane (DDT), hexachloro-cyclo-hexane (HCH), aldrin, and dieldrin, have grown into some of the most prevalent pesticides due to their inexpensive price and their importance in helping to manage a variety of pests (Jayaraj et al., 2016).

5. Novel Methods to Reduce Use of Insecticides

5.1. Nanotechnology

A potential area of multidisciplinary study is nanotechnology. It brings up a wide range of prospects in several industries, including agriculture, medicines, electronics, and pesticides. Nanotechnology has a broad range of possible applications and advantages (Mazumder et al., 2020). Among them is the formulation of pesticides based on nanomaterials for the control of insect pests. Conventional approaches to pest management, such as integrated pest management (IPM), are insufficient, and the use of chemical pesticides has negative impacts on both people and animals, in addition to lowering soil fertility. As a result, nanotechnology would provide practical and ecologically friendly choices for the management of pest species in agriculture. This piece of art focuses on traditional approaches to insect pest management and the potential for nanoparticles to be used in cutting-edge nanotechnology applications. The uses and advantages of nanotechnology are countless. They include the improvement of agricultural production using nanoporous zeolites for delayed release and effective doses of water and fertiliser, nanocapsules for the delivery of herbicides and the management of vectors and pests, and nanosensors for the detection of pests. The modification of nanoparticles by atom-by-atom arrangement influences their size, shape, and orientation for response to the targeted tissues. It is now recognised that many insects have geomagnetic sensors located in their heads, thoraxes, and abdomens. The targeted nanoparticles frequently display unique properties, including extraordinary strength and increased chemical reactivity with these sensory organs. These nanoparticles can carry several pesticides, including fungicides, nematicides, insecticides, and herbicides. Pesticide-coated metal oxide nanoparticles with a UV photoprotection screen are extremely beneficial for eradicating dangerous insects (Deka et al., 2021).

Additionally, a lot of the pesticides used now are organic substances that are poorly soluble in water. To accomplish an efficient and consistent application of these materials in the field, large volumes of organic solvents are often used to dissolve them. Nevertheless, this causes the volatile solvents to contaminate the environment and raises the personal exposure of agricultural workers to harmful chemical compounds (Mazumder et al., 2020; Putatunda et al., 2019). Furthermore, organic diluents are very costly compared to water. The use of nanotechnology offers an enticing solution to these issues. The material's saturation solubility is increased by particle size reduction to the nanoscale (Bhattacharya et al., 2022). Additionally, nanoparticles are much more dynamic than agglomeration, enabling deeper absorption into insect tissues and enhancing the triggering of insecticide activity. This can be achieved by either quick absorption through proximity to the insect's exoskeleton or rapid penetration via the gastrointestinal system (Margulis-Goshen & Magdassi, 2013).

5.2. Genetically Modified Organisms

Transgenic crops having insect-resistant capability were first cultivated for sale in the middle of the 1990s, thanks to the introduction of genetically modified (GM) plants that expressed genes encoding the ‘ethmoidal endotoxin’ from *Bacillus thuringiensis* (Bt), a bacterium. Bt-based formulations have been used commercially to manage insect pests for approximately forty years, especially in Lepidoptera. The Bt genes, whose proteins function as insecticides, have been manipulated genetically to be expressed in several crops, including corn, cotton, and soybeans. The goal of Bt corn is to get rid of corn pests such as the southwestern corn borer, corn earworm, and European corn borer. The usage of traditional synthetic insecticides has decreased as a result of the use of Bt crops. Bt crops have not been shown to pose any major dangers to people's health, according to the EPA's analysis. In particular, tests by the EPA demonstrate that the Bt protein in genetically modified crops performs as one might expect for a dietary protein. It is not physically linked to a recognised food-allergic reaction or protein toxin. It does not exhibit cytotoxicity in humans when consumed (Gatehouse et al., 2011; Sheikh et al., 2017).

To maximise the efficiency of these Bt toxins against targets, it is essential to not only have a complete understanding of how they work but also to be aware of potential side effects. When Bt toxins are swallowed by an insect, they become soluble in the gastrointestinal tract, where they are proteolytically broken down (active form) to a 65–70 kDa molecule. The active molecule then attaches to a specific receptor before entering the membrane of the gastrointestinal epithelial cells, creating holes. Colloid osmotic lysis, caused by hole development in the intestinal epithelial lumen, kills the insect. These Bt toxins are expressed in their active form in the majority of commercial crop varieties (Gatehouse et al., 2011).

It is evident that Bt-expressing crops have significantly improved world agriculture in recent years, not least in terms of fewer pests and higher quality. Alternative tactics are, however, being developed because there is no efficient way to manage homopteran pests and because insect populations may evolve resistance. Some of them, like other insect pathogens or vegetative insecticidal proteins called VIPs, are derived from various *Bacillus* spp. Other approaches are based on using genes from animals or plants, such as those from insects that encode immunosuppressive proteins. The ability to locate and profit from endogenous resistance genes has lately been the focus of considerable advanced research using RNAi and functional genomics (Ehlers, 2011).

Concluding Remarks and Future Scope

As the green revolution began, the insect infestation patterns in cereal crops changed. Just a few of the small pests have become severe pests and led to substantial output losses both regionally and nationally. Many factors, including global warming, unusual weather, shifting agricultural patterns, and technological adoption, change insect population, distribution, and the losses brought on by pests. Activities related to the examination and monitoring of insect pests throughout the country should be frequently done to obtain trustworthy data on shifting trends in insect pests as well as associated damages to a variety of agricultural products, including food items such as cereal. The use of chemical pesticides is the most common strategy for controlling insect pests on crops. Although incredibly effective, chemical pesticides have negative consequences for both humans and other creatures. In addition, the ongoing use of chemical pesticides destroys beneficial soil bacteria and natural enemies, which results in insect resistance. Traditional methods, however, fall short when it comes to controlling pests in particularly sizable regions. Different new strategies have emerged for managing insect pests in agriculture without harming nature, including nanotechnology, genetically modified (GM) crops, biopesticides, etc.

Acknowledgements

The grant-in-aid authorised by the 'MoFPI', Government of India, and the 'MODROB-REG' programme of the 'AICTE, Ministry of Education, Government of India' a fully acknowledged.

Declaration of Competing Interest

None.

References

Amanullah, Khalid, S., Imran, Khan, H. A., Arif, M., Altawaha, A. R., Adnan, M., Fahad, S., & Parmar, B. (2019). *Organic Matter Management in Cereals Based System: Symbiosis for Improving Crop Productivity and Soil Health* (pp. 67–92). https://doi.org/10.1007/978-3-030-26265-5_3.

Babendreier, D., Tang, R., & Horgan, F. G. (2022). Prospects for Integrating Augmentative and Conservation Biological Control of Leaffolders and Stemborers in Rice. *Agronomy*, *12*(12), 2958. https://doi.org/10.3390/agronomy12122958.

Baron, R. L. (1994). A carbamate insecticide: a case study of aldicarb. *Environmental Health Perspectives*, *102*(suppl 11), 23–27. https://doi.org/10.1289/ehp.94102s1123.

Békés, F., Schoenlechner, R., & Tömösközi, S. (2017). Ancient Wheats and Pseudocereals for Possible use in Cereal-Grain Dietary Intolerances. In *Cereal Grains* (pp. 353–389). Elsevier. https://doi.org/10.1016/B978-0-08-100719-8.00014-0.

Bhattacharya, S., Mazumder, A., Sen, D., & Bhattacharjee, C. (2022). *Bioremediation of Dye Using Mesophilic Bacteria: Mechanism and Parametric Influence* (pp. 67–86). https://doi.org/10.1007/978-981-16-5932-4_3.

Cummins, T. N. (2014). The Economic Value of Public Goods. *Applied Mathematics*, *05*(18), 2862–2865. https://doi.org/10.4236/am.2014.518272.

de Groot, R. S., Wilson, M. A., & Boumans, R. M. . (2002). A typology for the classification, description and valuation of ecosystem functions, goods and services. *Ecological Economics*, *41*(3), 393–408. https://doi.org/10.1016/S0921-8009(02)00089-7.

Deguine, J.-P., Aubertot, J.-N., Flor, R. J., Lescourret, F., Wyckhuys, K. A. G., & Ratnadass, A. (2021). Integrated pest management: good intentions, hard realities. A review. *Agronomy for Sustainable Development*, *41*(3), 38. https://doi.org/10.1007/s13593-021-00689-w.

Deka, B., Babu, A., Baruah, C., & Barthakur, M. (2021). Nanopesticides: A Systematic Review of Their Prospects With Special Reference to Tea Pest Management. *Frontiers in Nutrition*, *8*. https://doi.org/10.3389/fnut.2021.686131.

Ehlers, U. (2011). The interplay between GMO regulation and pesticide regulation in the EU. *Journal Für Verbraucherschutz Und Lebensmittelsicherheit*, *6*(S1), 61–64. https://doi.org/10.1007/s00003-011-0672-9.

Fukuto, T. R. (1990). Mechanism of action of organophosphorus and carbamate insecticides. *Environmental Health Perspectives*, *87*, 245–254. https://doi.org/10.1289/ehp.9087245.

Gatehouse, A. M. R., Ferry, N., Edwards, M. G., & Bell, H. A. (2011). Insect-resistant biotech crops and their impacts on beneficial arthropods. *Philosophical Transactions of the Royal Society B: Biological Sciences*, *366*(1569), 1438–1452. https://doi.org/10.1098/rstb.2010.0330.

Goussous, S. J., Abu el-Samen, F. M., & Tahhan, R. A. (2010). Antifungal activity of several medicinal plant extracts against the early blight pathogen (Alternaria solani). *Archives Of Phytopathology And Plant Protection*, *43*(17), 1745–1757. https://doi.org/10.1080/03235401003633832.

Gupta, R. C., Miller Mukherjee, I. R., Doss, R. B., Malik, J. K., & Milatovic, D. (2017). Organophosphates and Carbamates. In *Reproductive and Developmental Toxicology* (pp. 609–631). Elsevier. https://doi.org/10.1016/B978-0-12-804239-7.00035-4.

Hoehn, P., Tscharntke, T., Tylianakis, J. M., & Steffan-Dewenter, I. (2008). Functional group diversity of bee pollinators increases crop yield. *Proceedings of the Royal Society B: Biological Sciences*, *275*(1648), 2283–2291. https://doi.org/10.1098/rspb.2008.0405.

Huang, Y., Sharma, H. C., & Dhillon, M. K. (2013). Bridging Conventional and Molecular Genetics of Sorghum Insect Resistance. In *Genomics of the Saccharinae* (pp. 367–389). Springer New York. https://doi.org/10.1007/978-1-4419-5947-8_16.

Jankielsohn, A. (2018). The Importance of Insects in Agricultural Ecosystems. *Advances in Entomology*, *06*(02), 62–73. https://doi.org/10.4236/ae.2018.62006.

Jayaraj, R., Megha, P., & Sreedev, P. (2016). Review Article. Organochlorine pesticides, their toxic effects on living organisms and their fate in the environment. *Interdisciplinary Toxicology*, *9*(3–4), 90–100. https://doi.org/10.1515/intox-2016-0012.

Khatun, M. F., Jahan, M., Das, K. R., Lee, K., & Kil, E. (2023). Population dynamics and rational management of sucking insect vectors on chilli (Capsicum annuum L.) in Bangladesh. *Archives of Insect Biochemistry and Physiology*, *112*(2). https://doi.org/10.1002/arch.21980.

Kim, K. C. (1993). Biodiversity, conservation and inventory: why insects matter. *Biodiversity and Conservation*, *2*(3), 191–214. https://doi.org/10.1007/BF00056668.

Kuotsu, K., & Lalrinfeli, R. (2019). Biodiversity of insect pests of major cereal crops in mid hills of Meghalaya. *The Pharma Innovation Journal*, *8*(8), 287–292.

Laskowski, D. A. (2002). *Physical and Chemical Properties of Pyrethroids* (pp. 49–170). https://doi.org/10.1007/978-1-4757-4260-2_3.

Margulis-Goshen, K., & Magdassi, S. (2013). Nanotechnology: An Advanced Approach to the Development of Potent Insecticides. In *Advanced Technologies for Managing Insect Pests* (pp. 295–314). Springer Netherlands. https://doi.org/10.1007/978-94-007-4497-4_15.

Marrs, T. C. (1993). Organophosphate poisoning. *Pharmacology & Therapeutics*, *58*(1), 51–66. https://doi.org/10.1016/0163-7258(93)90066-M.

Mazumder, A., Bhattacharya, S., & Bhattacharjee, C. (2020). *Role of Nano-photocatalysis in Heavy Metal Detoxification* (pp. 1–33). https://doi.org/10.1007/978-3-030-12619-3_1.

Ndakidemi, B., Mtei, K., & Ndakidemi, P. A. (2016). Impacts of Synthetic and Botanical Pesticides on Beneficial Insects. *Agricultural Sciences*, *07*(06), 364–372. https://doi.org/10.4236/as.2016.76038.

Pardo, A., & Borges, P. A. V. (2020). Worldwide importance of insect pollination in apple orchards: A review. *Agriculture, Ecosystems & Environment*, *293*, 106839. https://doi.org/10.1016/j.agee.2020.106839.

Pecenka, J. R., Ingwell, L. L., Foster, R. E., Krupke, C. H., & Kaplan, I. (2021). IPM reduces insecticide applications by 95% while maintaining or enhancing crop yields through wild pollinator conservation. *Proceedings of the National Academy of Sciences*, *118*(44). https://doi.org/10.1073/pnas.2108429118.

Protopopoff, N., Mosha, J. F., Lukole, E., Charlwood, J. D., Wright, A., Mwalimu, C. D., Manjurano, A., Mosha, F. W., Kisinza, W., Kleinschmidt, I., & Rowland, M. (2018). Effectiveness of a long-lasting piperonyl butoxide-treated insecticidal net and indoor residual spray interventions, separately and together, against malaria transmitted by pyrethroid-resistant mosquitoes: a cluster, randomised controlled, two-by-two fact. *The Lancet*, *391*(10130), 1577–1588. https://doi.org/10.1016/S0140-6736(18)30427-6.

Putatunda, S., Bhattacharya, S., Sen, D., & Bhattacharjee, C. (2019). A review on the application of different treatment processes for emulsified oily wastewater. In *International Journal of Environmental Science and Technology* (Vol. 16, Issue 5, pp. 2525–2536). Center for Environmental and Energy Research and Studies. https://doi.org/10.1007/s13762-018-2055-6.

Robinson, R. A., & Sutherland, W. J. (2002). Post-war changes in arable farming and biodiversity in Great Britain. *Journal of Applied Ecology*, *39*(1), 157–176. https://doi.org/10.1046/j.1365-2664.2002.00695.x.

Sheikh, A. A., Wani, M. A., Bano, P., Nabi, S., Bhat, T. A., Bhat, M. A., & Dar, M. S. (2017). An overview on resistance of insect pests against Bt Crops. *Journal of Entomology and Zoology Studies*, *5*(1), 941–948.

Sohrabi, F., Shishehbor, P., Saber, M., & Mosaddegh, M. S. (2011). Lethal and sublethal effects of buprofezin and imidacloprid on Bemisia tabaci (Hemiptera: Aleyrodidae). *Crop Protection*, *30*(9), 1190–1195. https://doi.org/10.1016/j.cropro.2011.05.004.

Subramanya, S., Sumanth, K., Gupta, P. K., Chayapathy, V., Keshamma, E., & Murugan, K. (2022). Formulation of green nanoemulsions for controlling agriculture insects. In *Bio-Based Nanoemulsions for Agri-Food Applications* (pp. 165–176). Elsevier. https://doi.org/10.1016/B978-0-323-89846-1.00002-4.

Tinzaara, W., Dicke, M., van Huis, A., & Gold, C. S. (2002). Use of infochemicals in Pest Management with Special Reference to the Banana Weevil, Cosmopolites sordidus (Germar) (Coleoptera: Curculionidae). *International Journal of Tropical Insect Science*, *22*(04), 241–261. https://doi.org/10.1017/S1742758400020877.

Zhang, H., Islam, T., & Liu, W. (2022). Integrated pest management programmeme for cereal blast fungus Magnaporthe oryzae. *Journal of Integrative Agriculture*, *21*(12), 3420–3433. https://doi.org/10.1016/j.jia.2022.08.056.

Chapter 3

An Introduction to Insects as Pests of Vegetable and Fruit Crops with a Note on Integrated Pest Management

Monika Chaudhary[1]
and Amar Prakash Garg[2,*]
[1]School of Biological Engineering & Life Sciences, Shobhit Institute of Engineering & Technology, Modipuram, Meerut, India
[2]Dean Academics and Director Research, Swami Vivekanand Subharti University, Subhartipuram, Meerut, India

Abstract

Given its diverse agro-climatic conditions, India has a distinct comparative advantage for cultivating the majority of horticulture crops (fruits and vegetables). The importance of horticulture in maintaining food and nutritional security, generating employment, and raising farm revenue has been maintained as an important subsector of agriculture. Horticulture is expected to have a global market value of USD 20.77 billion in 2021 and USD 40.24 billion in 2026. Horticulture production currently makes up 30.4% of agriculture's GDP and has become a significant driver of economic growth in several Indian states. India is the world's second-largest producer of fruits and vegetables and a host of other products, including mango, banana, coconut, cashew, papaya, and pomegranate. Similar to abiotic stress, biotic stress is a major factor in the global decline in agricultural productivity. One of the key reasons for decreased agricultural yield is the harm done by insect pests. Infestations of insects are the main biotic obstacle to the production of crops in India. Many of them not only directly injure people but also act as vectors for a variety of viral illnesses. There have been reports of 30 to 40 percent losses in the vegetable crop. Vegetable crops are greatly impacted by insect pests, which also reduces the output. Therefore, it is important to address these newly developing insect pests in the horticulture crop ecosystem, as well as the appropriate control strategies and management concerns, as well as sustainable development.

Keywords: horticulture crop ecosystem, insect pests, management, sustainable development

---

* Corresponding Author's Email: amarprakashgarg@yahoo.com.

In: Insects as Humans' Frenemies: From Friends to Foes
Editor: Istkhar Rao
ISBN: 979-8-89530-320-7

1. Introduction

Due to their importance as a source of nutrients, including vitamins, minerals, and carbohydrates, vegetables are an essential part of the human diet. Some vegetables are still only cultivated in specific civilisations, while others have spread across several continents and gained widespread popularity. Most of the natural resources are rapidly decreasing and declining as a result of rising urbanisation and deforestation. The world must feed its growing population without upsetting the natural balance due to its limited land and diminishing water supplies. Considering the economic value of the produce, the production of vegetables is one of the most active areas of agriculture. For rural people, vegetables are quickly becoming a significant source of revenue. India is lucky to have a variety of agro-climatic conditions spread over the whole nation, allowing for the production of both tropical and temperate crops. With an estimated production of around 125.89 million metric tonnes from 7.80 million hectares, India is presently the second-largest producer of vegetables after China. Vegetable production has expanded significantly over the previous 50 years (Indian Horticulture Data Base, 2008). 13.38% of the world's total vegetable production comes from India. India ranks first globally for cauliflower, second globally for onions, and third globally for cabbage. West Bengal has the most land planted in a variety of vegetables, followed by the U.P., MP, Bihar, and Gujarat, according to the state-by-state distribution of the nation's vegetable output (Indian Horticulture Data Base, 2020). India produces more fruits and vegetables than any other country in the world, coming in at number two. There are various reasons why India's production is so poor when compared to other developing nations, including its horticulture crops. Poor crop growth and production are caused by factors such as high temperatures, high potential transpiration, unpredictable rainfall, low soil fertility, and very low-quality groundwater. In addition to environmental restrictions, biotic issues like pests and diseases also provide a barrier to agricultural production in India. Pest-related production losses account for almost 30% of the nation's entire economic output. However, research on pests and their control is crucial to the development of horticultural crops. To obtain food, insects harm plants, either directly or indirectly. Almost all of the plant's components—the roots, stem, bark, leaves, buds, flowers, and fruits—are impacted. The study of signs and symptoms that different plant parts exhibit as a result of injury from insect pests is known as symptomatology.

2. Losses in Vegetable Crops

One of the main biotic factors that impact production and productivity is insect pests that harm vegetables. Even though the use of pesticides has been somewhat effective in controlling the pest population, it has been established that indiscriminate use of these dangerous chemicals damages soil quality, pollutes the environment, and harms animal health (Dash et al., 2022). Insect pests are the primary cause of agricultural production losses all over the world, and pest control is crucial to guaranteeing both food security and income for farmers. Insect pests may be responsible for 8.7% of the anticipated worldwide loss in vegetable production, which, if unchecked, might result in considerably larger losses. According to some reports, the projected yearly vegetable losses worldwide due to insects alone are 18–20% during storage and 15-20% throughout field production (Zhang et al., 2018, Nigel., 1980, FAO, 1997). Pests and illnesses are projected to cause a 30–35% gross loss in agricultural output in India, or 50,000 crore (U$500 billion) annually (Chakraborty, 2017). It has been calculated from time to time how much Indian agriculture has lost because of insect infestations. Losses have usually grown in the post-green revolution era compared to the pre-green revolution era. From 7.2% in the early 1960s to 23.3% in the early 2000s, losses as a whole grew. The biggest increase in loss was seen in cotton (18.0–50%), then other crops, which include sorghum and millets (3.5–30%), maize (5.0–25%), and oilseeds (other than groundnut) (5.0–25%). (Dhaliwal & Arora, 1996, 2002; Dhaliwal et al., 2003, 2004).

Table 1. Yield losses due to major insect pests in crops

S. No	Crop	Pests	Yield Loss (%)	References
1	Tomato	Fruit borer (*Helicoverpa armigera*)	24-73	Shivalingaswamy et al. (2002)
2	Brinjal	Fruit and shoot borer (*Leucinodes orbonalis*)	11-93	Dhillon et al. (2005)
3	Chillies	Thrips (*Scirtothrips dorsalis*) Mites (*Polyphagotarsonemus latus*)	12-90 34	Satpathy et al. (2005)
4	Okra	Fruit borer (*H. armigera*) Leafhopper (*Amrasca biguttula biguttula*) Whitefly (*Bemisia tabaci*) Shoot and fruit borer (*Earias vittella*)	22 54-66 54 23-54	Raju et al. (2007)
5	Cabbage	Cabbage leaf webber (*Crocidolomia binotalis*) Cabbage borer (*Hellula undalis*) Diamondback moth (*Plutella xylostella*) Cabbage caterpillar (*Pieris brassicae*)	28-51 30-58 17-99 69	Singh et al. (2007)
6	Cucurbits	Fruit fly (*Bactrocera cucurbitae*) Bitter gourd Cucumber Ivy gourd Musk melon Snake gourd Sponge gourd	 60-80 20-39 63 76-100 63 50	Ghosal et al. (2012)

Within the agricultural economy, vegetable crops hold a significant position, and they are a vital part of human nutrition. Since potatoes are propagated by vegetative methods, all pest groups that have an economic presence in this crop cause damage to it. Around the world, insect pests have been blamed for a real loss of 39% in potato production; without crop protection, this number could rise to 71%. Actual overall losses were estimated to have varied from 24% across Europe to more than 50% in Africa (Oerke & Dehne, 2004; Oerke, 2006). Vegetable crop losses in India have ranged from 30 to 40 per cent (Rai et al. 2014). *H. armigera*, a fruit borer, can reduce tomato yields by 73%. Additionally, substantial occurrences of the fruit flies *Dacus dorsalis* (Hendel) and *Plutella xylostella* (Linnaeus) in Cole and Cucurbit crops, respectively, have been associated with crop failure. In Uttar Pradesh, India, *P. xylostella* caused a cauliflower pandemic that caused a 100% loss of the crop. According to Ahmed et al. (2009), *P. xylostella* is one of the most serious pests of cruciferous crops globally. Fruit borer can dramatically lower the production of brinjal. *Leucinodes orbonalis* Guenee, also known as the shoot and fruit borer, is believed to reduce Brinjal yields by 95%, while other researchers have noted losses ranging from 20 to 92% due to this pest (Naresh et al., 1986). Additionally, a loss of 54.04% has been reported in cases where okra was infested with *B. tabaci* whiteflies. Various researchers have estimated the yield loss on chilli crops caused by the thrips,

Scirtothrips dorsalis Hood, ranging from 11.8 to more than 90%. Yield losses to important vegetable crops due to different pests are summarised in Table 1.

Changes in cropping practices, climatic conditions, and the introduction of exceptionally input-intensive, high-yielding cultivars and hybrids have all contributed to a change in pest status in recent years. The majority of plant protection guidelines for vegetables until date suggested applying pesticides according to the calendar. The majority of farmers in the nation who cultivate vegetables now do this regularly. Numerous pests have adapted to new hosts, gained pesticide resistance, and frequently experienced secondary outbreaks.

3. Major Vegetable Crop Pests and Their Control Measures

3.1. Brinjal

Some significant pests of brinjal include the shoot and fruit borer *Leucinodes orbonalis*, the hadda beetle or leaf-eating beetle *Henosepilachna vigintioctopunctata*, the leaf hopper A*mrasca bigutula biguttula*, the aphid *Aphis gossypii*, and the red spider mite *Tetranychus cinnabarinus* (Figure 1). Due to a weak natural enemy complex and considerable pesticide usage in brinjal, the shoot and fruit borer has continued to be a significant pest for the past 20 years. Shoot and fruit borer cause crop losses in brinjal that vary from 25.82 to 92.50% and yield reductions of 20 to 60%. In certain areas, hadda bugs skeletonise the brinjal plant by feeding on the leaves. Both adult beetles and grubs cause this devastation. Another serious pest of brinjal is the stem borer, which burrows into the stem and kills the plant (Kumar et al., 2022).

Control: Effective treatments include using metacid 50 EC 1 ml/l, carbaryl 0.2% (sevin 50% WP 4 g/l), or phosphamidon 0.03% (dimecron 100 EC, 0.3 ml/l of water) every two weeks. Phosphamidon and dimethoate are needed to be sprayed three times in alternate months at a 0.05% concentration.

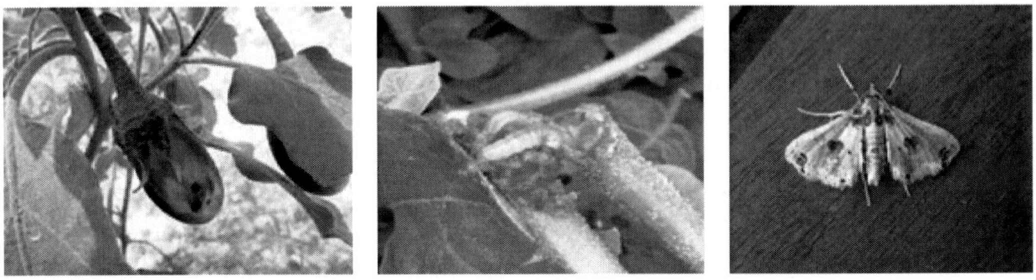

Figure 1. Shoot and fruit borer, stem borer and brown leaf hopper pests of brinjal.

3.2. Tomato

The cotton leafworm (*S. littoralis*), which predominantly harms summer harvests, and the cotton aphid (*A. gossypii*) are sucking insects that can infest this plant (Figure 2). In addition to causing defoliation, they can penetrate fruits' interiors and feed there. Tomato leaf attacks by the leaf miner (*L. trifolii*) result in a variety of losses. The main pest in tomatoes, *H. armigera*, feeds on buds, flowers, and fruits and reduces productivity by an average of 46% (Dhandpani et al., 2003).

Control: Picking the fruits and using contact pesticides like Carbaryl 0.2%, Inchlorophos 0.25%, 81% Malathion + 0.1% Sandovi 4 sprays every 15 days, or Metasystox @ 1 ml/litre of water + Sandovit 0.1% are two methods for controlling pests.

Figure 2. Fruit borer, white fly and, thrips pests of tomato.

3.3. Chilli

Chilli is one of the most significant spice crops and an essential ingredient in many Indian curries and chutneys. The primary pests of chillies include thrips, *S. dorsalis*, mites, *P. latus,* fruit borer, *S. litura, and H. armigera.*

Control: Effective treatments include Rogor or Metasystox 1.5ml/litre of water, followed after 10 to 15 days by Malathion 1 ml/litre of water. 0.1% of people benefit by spraying Metasys ox at 1.5 ml/litre with stickers like Triton, Sandrif, or Sandast.

3.4. Okra

This category includes the commercially significant vegetable okra *(Abelmoschus esculantus (L.) Moench*). Okra plants are preyed upon by white flies *(B. tabaci*), jassids (*A. bigutalla bigutulla*), and aphids (*Aphis gossypii*). The shoot and fruit borer (*Earias sp.*) mostly targets flowers and pods, but the American bollworm (*H. armigera*) larva prefers the reproductive parts of the plant, such as buds, flowers, and fruits (Figure 3).

Figure 3. Fruit borer and white fly pests of okra.

Control: Once or twice, spray 2nd Malathion or 1 ml Thiodan per litre of water, avoiding harvesting for up to 5 days following. Effective spraying includes cupping infected portions and burning them in the soil before the blossoming stage. Spraying the local crop is advised. Clean cultivation and summer ploughing both aid in lowering the bug population.

3.5. Cabbage and Cauliflower

Around the world, crops of cruciferous vegetables, including cabbage, cauliflower, and knolkhol, are grown. Cruciferous vegetables produce little and are thus attacked by a wide range of insect pests. The most significant are as follows: DBM (*Plutella xylostella Linn.*) (Figure 4), leaf webber (*Crocidolomia binotalis Zeller*), stem borer (*Hellula undalis Zeller*), aphid (*Brevicoryne brassicae Linn, Hyadaphis erysimi Kaltenbach*), stink bug (*Bagrada cruciferarum Kirkaldy*), striped flea beetle (*Phyllotreta striolata Fabr.*), and the mustard sawfly (*Athalia lugens proxima Klug*). To keep cabbage plants free of bug infestations, farmers use cow dung and ash-soil combinations (Lal & Verma, 2006).

Control: As soon as an infestation starts, spray 0.20% malathion or thiodon at intervals of 10- 15 days to combat cabbage butterflies, diamondback moths, tobacco caterpillars, and aphids. Instead of malathion when the cabbage heads develop, spray 0.1% pyrethrins.

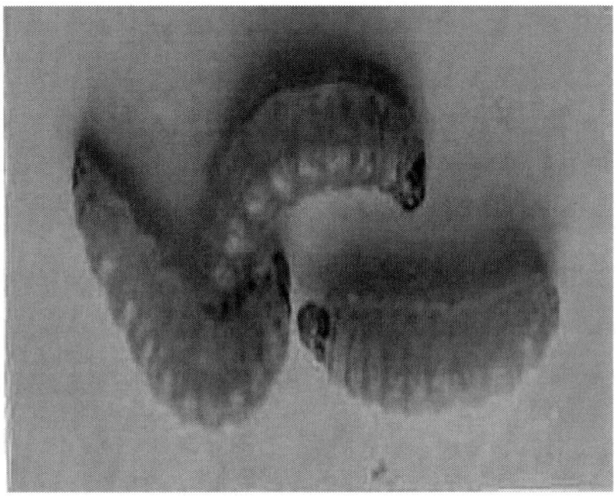

Figure 4. Cabbage borer pest of cole crops.

4. Fruit Crops: Losses in Fruit Crops, Insect Pests and Their Management

Due to their high vitamin, mineral, and antioxidant content, fruits are regarded as preventive meals that shield people against a wide range of illnesses. On a surface of 55.08 million hectares, 599.3 million metric tonnes of fruit are produced worldwide during this time. According to area and output, the top three fruit-growing nations are China, India, and Brazil (Anonymous 2012). In the category of biotic stressors, insect pests—aside from diseases—cause significant yield losses. According to estimates from 1996, insects damage fruit crops by 6% even when pesticides are used and by 23% when protection from insecticides is not there.

Insects not only directly affect fruit crop yields by destroying diverse fruit tree elements, including foliage, twigs, blooms, and fruits, but they also act as carriers of numerous infections that cause illness (Krattiger 1997). A list of some of the important insect pests of fruits are mentioned in Table 2.

Table 2. Insect pests of fruits

S. No	Fruit	Pests	References
1	Mango	Stem borer Bark borer Shoot borer Mango hoppers Inflorescence midge	Anonymous (2013)
2	Banana	Rhizome weevil (*Cosmopolites sordidus)* Pseudostem borer (*Odoiporus longicollis)* Banana aphid (*Pentalonia nigronervosa f. typica)* Tingid or Lacewing bug (Stephanitis typicus*)*	Moore (2004)
3	Citrus	Citrus aphids, Black aphid (*Toxoptera aurantii, Toxoptera citricida)* Citrus psyllid (*Diaphorina citri)* Mealy bug (*Planoccus citri)* Citrus butterfly (*Papilio demolious, P. Polytes, P. helenus)*	Dadmal & Pawar (2001)
4	Pomegranate	Fruit butterfly (*Dorsophila melanogaster*) Shoot and fruit borer (*Deudorix Isocrates*)	Cai & Geng (1997)
5	Guava	Fruit fly, bark borer and fruit borer	Haseeb & Sharma (2007)
6	Sapota	Leaf webber, Budworm, Hairy caterpillar	Tanwar et al. (2007)
7	Ber	Fruit fly (*Carpomyia vesuviana*)	(Jayachandran et al., 2022)

4.1. Mango

Tropical and subtropical fruits like the mango, or *Mangifera indica L.* (Family: *Anacardiaceae*), are native to the Indo-Burma region. With over a thousand distinct varieties of mangoes, India has the widest selection. Major mango-producing countries include India, China, Thailand, Mexico, Pakistan, the Philippines, Indonesia, Brazil, Nigeria, and Egypt. India occupies the top rank and produces about 50% of the world's mangoes, with 2.5 million hectares generating 18.0 million metric tonnes each year. The ability of mango cultivars to produce their maximum yield is severely constrained by insect pests (Mukherjee, 1997). Leafhoppers (Hemiptera: *Cicadellidae),* thrips (Thysanoptera: *Thripidae*), Mango Leaf Webbers (Family: *Pyralidae*), Mealybugs (Hemiptera: *Pseudococcidae*), Midges (Diptera: *Cecidomyiidae*), Scale Insects (Hemiptera: *Coccidae*), Shoot Gall Psylla (Hemiptera: *Psyllidae*), Mites (Acari: *Tetranychidae and Eriophyidae*), Mango Shoot Borers (Lepidoptera: Noctuidae), Mango Stem Borers (Coleoptera: *Cerambycidae*), Fruit Borers (Family*: Pyralidae*) are certain insect pest of mango.

Mango Hoppers: Flowers and flower buds wither and shed due to nymphs and adults. Tiny honeydew drops on the bottom leaves, then sooty mould begins to form. It's not uncommon for jassids to roam about among the leaves while making a clicking noise.

Idioscopus nitidulus: white band across the wing and three dots on the scutellum.

Idioscopus clypealis: dark patches and two spots on the scutellum are present.

Amritodus atkinsoni: The scutellum has two spots on it.

Control: Spraying malathion (0.15%), diazinon (0.02%), endrin (0.04%), and carbaryl (0.15%) will control it. Phosphamidon (0.05%) or nuvacron (0.04%), twice—once after the emergence of the panicle and once again during the stage of fruit set, spraying with 0.025 per cent methyl parathion will be effective.

Gall Midges:

Procystiphora mangiferae: deforms flowers and sheds droppings. Both the adult and the maggot are orange.

Dasineura amaramanjarae: destroys floral buds and causes buds to fall.

Erosomyia mangiferae: stunting and distortion of the inflorescence are the results. The maggots are yellow.

Control: As of yet, no firm control measure has been developed. However, four weekly sprays of 2 to 3 per cent tar oil throughout the egg-laying cycle may help lessen the occurrence. Sprays containing endrin (0.04%) are also effective against this pest. While the CVs are not particularly susceptible, Tommy Atkins and Heidi are. Sensation, Irwin, Keitt, and Kent seem unaffected by galling (Schoeman, 1995).

Shoot Webber: Defoliation and webbing of terminal leaves are caused by larvae.

Larvae have a prothoracic shield and a brown head. The adult moth is dark and has wavy lines on its forewings.

Control: Spraying 1200 gallons of water per acre with 500 cc of malathion (50 EC) or 40 kg of carbaryl (50 WP) works effectively.

Shoot Borer: Upward from the developing point to a depth of 5 to 6 inches, larvae enter the tissue, the entire seedling remains short-lived, and each twig has an odd, bunchy look at the terminal end (Figure 5). Dark pink larva with a noticeable dark brown prothoracic shield. The adult moth is greyish, with wavy patterns on its dark grey wings.

Control: Attacking shoots need to be severed and eliminated. From the beginning of a new flush, spraying carbaryl (0.2%) or quinalphos (0.05%) at intervals of two weeks can successfully control the pest. It may be necessary to spray twice at intervals of three weeks, starting with the onset of the fresh flush of leaves.

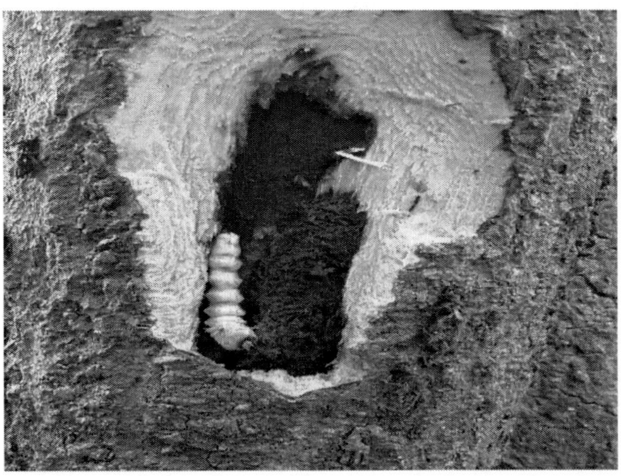

Figure 5. Stem borer pest of mango.

Whitefly: White flies are present on the ventral side of plants and create areas of yellowing caused by nymphs and adults. Nymphs are seen in groups and are greyish-white. Adults have a drab white colouration.

Control: Spray 1200 litres of water per acre with 40 kg of carbaryl or 500 cc of malathion (50 EC).

Fruit Fly: Fruit droppings and decaying patches are seen on semi-ripe fruits. Maggots seem yellow. The wings of an adult fly are translucent and light brown.

Control: The key to control is the quick collection and disposal of the damaged fruits, either by burying them deeply in the ground or water. To reduce the prevalence of fruit flies, 10 wide-mouthed containers with poisoned bait are placed on each acre. The formulas that have been demonstrated to be effective are (i) sodium fluosilicate (1 part) and molasses or syrup (6 parts) in water (100 parts), (ii) lead arsenate (1 part) and molasses (24 parts), and (iii) a 100 ml emulsion of methyl eugenol (0.1%) and malathion (0.1%). Starting at the egg-laying stage, sprays of carbaryl (0.2%) combined with protein hydrolysate or molasses (0.1%) can be used to control flies.

4.2. Banana

Rhizome Weevil: The grub kills opening pipes and causes outer leaves to wither. Plants are killed by grubs that drill into the rhizome. The grub has a redhead and is apodous and yellowish-white. Adult weevils are black in hue.

Control: Clean cultivation is crucial. Make sure you only plant healthy suckers. The planting companion that has been afflicted spreads the borer. Therefore, *in vitro,* planting materials are an option. Additionally, quarantine procedures are required to stop the borer from spreading to unaffected areas. It is shielded from weevil assault by phorate pit application or by dipping suckers in monocrotophos for 30 minutes. The most effective control of weevil damage to bananas was achieved by applying phorate (20 g) to the soil during planting and three months later (Charles et al., 1996). Biological pest control techniques like entomopathogenic nematodes (*Steinernema* and *Heterorhabditis sp.*) and fungal pathogens (*Beauveria bassiana*).

Pseudostem Borer: The grub penetrates and bores into the pseudostem, causing the plant to wilt. Grub has a black head and is apodous and creamy white. Strong reddish brown and black colours are shown by weevils in adult form.

Banana Aphid: Bunchy top disease is spread by nymphs and adults. On pseudostems and leaf axils, they grow in colonies. Adults and nymphs are both black in colouration. Adults with wings have dark veins.

Control: Spray dimethoate at a rate of 2 ml/ltr. of water or 1.5 ml/ltr of cypermethrin.

Thrips: Yellowing of the leaves is caused by leaf thrips. Adults have wings that have fringes. Fruit rust thrips: These pests cause rusty development on fruit and yellowing of the leaves. Adult wings are tinted and yellowish-white. Corky scab is caused by flower thrips on fruits and flowers.

Control: At the time of shooting, securely wrap the bunch in a 100-gauge polythene sleeve with 6–10% perforations or polypropylene bunch leaves so that there is no space between the cover and the bunch. Remove the male flower buds after opening all the hands. Imidacloprid 17.8EC @ 2 ml chemical solution (0.2 ml of chemical diluted in 500 ml of water) should be injected into each bud while it is upright using a disposable syringe. Spray a mixture of sticking agent (at 0.5 ml/litre of water) and chlorpyriphos (at 2.5 ml/litre of water/acetamiprid 20SP (0.0025%)) twice on the bunch: once when the hands open for the first time and again after all the hands have opened.

Banana Aphid (*Pentalonia nigronervosa*): Due to the fact that it serves as the virus's vector, bunchy top sickness, it is of special importance.

Control: Malathion and metasystox can be sprayed to suppress the bug. It was discovered that applying granules at the time of planting to the rhizome, base of the sucker, or base of the leaf axils did not substantially vary from each other in their trial to control banana aphids using granular pesticides. Furthermore, it was discovered that, unlike carbofuran, the effects of disulfoton and phorate lasted for at least five months.

4.3. Citrus

Citrus Bark Borer (*Inderbela tetraonis, Inderbela quadrinotata*): Citrus bark and branch borers are important pests, particularly in abandoned and ageing plantations. Caterpillars that consume bark eat away at the bark's relocating tissues. Since they feed at night, they frequently go unnoticed, but the presence of a loose mass of small bits of wood and pellets of excrement combined with silky sticky materials on afflicted trees serves as a telltale sign of their existence.

Control: Several researchers have recommended the use of sprays containing the chemicals malathion (0.05%), monocrotophos (0.025%), carbaryl (0.1%), phosphamidon (0.025%), and parathion (0.025%) as efficient chemical controls for the insect.

Neem (*Azadirachta indica*) oil and azadirachtin (0.03%) provided a high level (90%) of control over citrus psylla.

Citrus Leaf Miner: The leaf lamina is distorted and covered in serpentine mines as a result of the larval infestation. The larva is apodous, tiny, and reddish or yellow. The adult is a little moth with a black spot towards the tip of the forewing.

Control: Insect pests can be effectively controlled by using phosphamidon (0.03% EC) or monocrotophos (0,035%) every week during the flushing season. The greatest results were achieved when le Spam, tebufenozide, was combined with summer mineral oil at rates of 14.4 and 18 g/hectoliter. Tebufenozide showed outstanding selectivity in tests against other beneficial insects. 90% of *P. citrella* was controlled by spraying abamectin (1.8%, 20 cm3/100 l water) and oridacloprid (35%, 25 cm3/100 L water).

Citrus Butterfly: Tender leaves are defoliated by the larva. Early stages of larva mimic bird droppings. The mature larva is robust, cylindrical, green, and has brown lateral oblique streaks. Adult swallowtail butterflies are dark brown with many yellow patterns.

Control: Spray dimethoate at a rate of 2 ml/ltr. of water or 1.5 ml/ltr of cypermethrin.

Fruit-Sucking moths: Fruit drops and rots at the feeding location as a result of adult moths penetrating it and sucking the liquid. The velvety, black body of the semi-looper larva, which feeds on the weed host, is marked with orange, blue, and yellow spots. The adult moth is a strong insect with orange and grey wings. There are three black spots on the forewings.

Control: Spraying with carbaryl (0.1%), phosphamidon (0.025%), monocrotophos (0.05%), and malathion (0.05%).

Citrus Whitefly: The sap that causes the leaves to curl and fall off is consumed by both adults and nymphs. The nymph has purple eyes and a pale-yellow body.

Control: Avoid planting too closely and being flooded. Avoid using too much nitrogen fertiliser and irrigation. Avoid growing bugs' secondary hosts, such as guava, sapota, and pomegranate. At the 50% egg hatching stage, foliar applications of imidacloprid 17.8 SL at 0.5 ml, acephate 75 WP at 1.25 g, phosalone 35 EC at 1.5 ml, dimethoate 30 EC at 2 ml, or novaluron 10 EC at 0.79 ml/l of water are made to the undersides of leaves covering the whole tree canopy. A second application of any of the aforementioned insecticides or 10 ml/l of neem oil should be made after 15 days.

4.4. Guava

Mealy Bug (*Ferrisia virgata*): Mango mealy insect and citrus mealybug have also been found in guava.

Control: The pest was effectively managed by banding the tree trunks with 30 cm wide cage polythene mixed with grease on a 5 cm region on its lower end and sealing the space between the band and the trunk. Spraying with a variety of insecticides, including chlorpyrifos (0.08%), acephate (0.08%), permethrin (0,02%), methyl demeton (0.08%), monocrotophos (0.08%), quinalphos (0.08%), dimethoate (0.08%), and phenthoate (0.08%), has reportedly been just as successful at reducing mango mealy bug larvae that are infesting guavas (Ghule & Dhumal, 1992).

Bark-Eating Caterpillar (*Indarbela quadrinotata, I. tetraonis*): On hatching, larvae consume the bark and dig into the trunk. More infestations are seen in older, abandoned orchards. Apple and Seedless varieties are more susceptible than Allahabad Strawberry, Cattleya Guava, and Safeda (Butani, 1974).

Control: The orchard may be kept clean to lessen the infestation. Wire can be inserted in the openings to kill the larvae.

Fruit Borer: Boreholes in infected fruits have been filled with a larval anal segment. Fruit that is severely infested starts to decay and drop off. Larvae have a small, robust build, a filthy dark brown colour, and are coated with short hair. The butterfly's mature colour is bluish-brown. The female's forewing bears a "V"-shaped patch. Fruits have holes in them. The adult butterfly is tinted metallic red.

Control: Use a light trap at 1/ha to keep an eye on adult activities. Malathion 50 EC 0.1% has two cycles of insecticides, one at flower development and the second at fruit set.

Tea Mosquito Bug: fruit damage manifests as corky scab development. The inflorescence blight, terminal drying of young shoots, and water-soaked lesions that turn into brownish patches at feeding sites are some of the infestations caused by nymphs and adults. Long, reddish brown beetles with a black head, a red thorax, and a black and white abdomen make up adults and nymphs.

Control: Malathion 50 EC 0.2%, applied biweekly during the period of blooming, is an insecticide.

4.5. Sapota

Chickoo Moth or Leaf Webber: The chlorophyll on the leaves is scraped off by the larva once they are webbed together in a clump. A group of dry leaves are dangling from the webbed stalks. As a result, flower buds and delicate fruits become dull, wither, and lose their petals. The larva is pinkish and has three dorsolateral brown stripes on each side. The forewings of the adult moth are hairy brown or have black patches, while the hindwings are semi-hyaline.

Control: Malathion application at 0.5% is extremely efficient.

Budworm: Webbed together and shed are flower and bud buds. The larva is small, slender, and pinkish brown, with a black head and a yellowish-brown prothoracic shield. Adult moths are grey in hue and have a black spot on their wings.

Hairy Caterpillar: The irregular feeding of larvae on leaves results in defoliation. Larvae are robust, hairy, and greyish brown. The adult is a robust, brownish-grey moth. The centre of the male's forewings has a chocolate brown band and pectinate antennae. The female has wavy transverse stripes on her wings and is larger than the male.

Control: Dimethoate (30 ml in 18 litres of water) is sprayed on this bug one to two times to suppress it.

Fruit Fly: Fruit that is just semi-ripe displays decaying patches and drops later.

Maggots seem yellow. The wings of an adult fly are translucent and light brown.

Control: Infested fruits that have fallen to the ground and undersized fruits that have been left on the tree should be collected and disposed of. If the trees are less in number, another option is to bag the fruit in cloth or paper bags.

Chlorpyriphos 20 EC @ 5 ml/ltr should be applied to the soil beneath the tree after it has been scraped up. Ten methyl eugenol traps per acre are used to monitor the adult fruit fly population. This aids in mass trapping as well, and when carried out on a large scale in a neighbourhood, it may significantly lower insect concentrations.

4.6. Ber

Fruit Fly (*Carpomyia vesuviana*): In Ber, fruit flies are the most problematic pest. The fruit set begins in September, which coincides with the commencement of the fruit fly infestation. The mature female places her ovipositor within the developing fruit to lay her eggs. Larvae emerge after 2 to 5 days and eat the pulp. Fruits with infestations may decay, become misshapen, and drop to the ground in vast numbers.

Control: Utilised resistant cultivars to test for fruit fly resistance and discovered that Tikadi was the most resistant, with Ilaichi being moderately resistant. Dandan, Gola, Gola Cungaon No. 3, Kaithli, and Kakrola Gola were found to be the most vulnerable. Chinese, Sanaur-1, Safeda Ilaichi, Mirchia, ZG 3, and Umran cultivars were found to be resistant to fruit flies, followed by endosulfan at 0.07%, also sprayed three times, according to Mann and Bindra (1976). The largest yields were likewise produced by fenthion, which was followed by endosulfan and 0.04% malathion.

4.7. Pomegranate

Fruit Fly (*Drosophila melanogaster*): Fruit rotting is a sign of an infestation. The fruits' pulp is where the maggots eat, and damage to the fruit includes brown resinous fluid coming from it, twisted and misshapen fruits, premature fruit falling, and making it unsuitable for food. Maggots are apodous and whitish. Adults have hyaline wings, or wings that are brown with a faint yellow stripe on the third tergite.

Control: Spray dimethoate at a rate of 2 ml/ltr. of water or 1.5 ml/ltr of cypermethrin.

Shoot and Fruit Borer (*Deudorix isocrates*): Fruits have holes made by larvae. The larva has a black head and prothoracic shield, is pale green with a pinkish tint, and has fine hairs. Adult pale golden moth with medium size and little black spots and wing patches of darkness (Figure 6).

Control: Application of metacid 50 EC 1 ml/l, carbaryl 0.2% (sevin 50% WP 4 g/l), or phosphamidon 0.03% (dimecron 100 EC, 0.3 ml/l of water) at biweekly intervals is beneficial for controlling fruit butterflies or shoot and fruit borer.

Figure 6. Whitefly pest of pomegranate.

5. Integrated Pest Management

Integrated Pest Management (IPM) is a thorough approach to controlling pests, including herbivores, diseases, and weeds. Utilising artificial pesticides only when essential, it combines therapeutic and preventive methods. To create efficient and dependable crop protection measures that do not lead to the development of resistance in herbivores, diseases, and weeds, it was proposed that an evolutionary framework be incorporated into IPM. This is comparable to the emerging realisation that the understanding and prediction of interactions between hosts, illnesses, and medicinal interventions might benefit from an evolutionary viewpoint in medicine. Such a framework would enhance each element of pest control and strengthen their interdependencies, in addition to reducing the evolution of pest resistance. It is important to highlight important components of IPM that would particularly benefit from a solid grasp of evolution. The benefits and difficulties of establishing communication between scientific groups established in diverse biological fields, as well as between scientists and society, were also discussed. Furthermore, suggestions were also provided that might improve the use of evolutionary concepts in IPM and support the development of sustainable agriculture that is immune to both present and future pests (Karlsson et al., 2020).

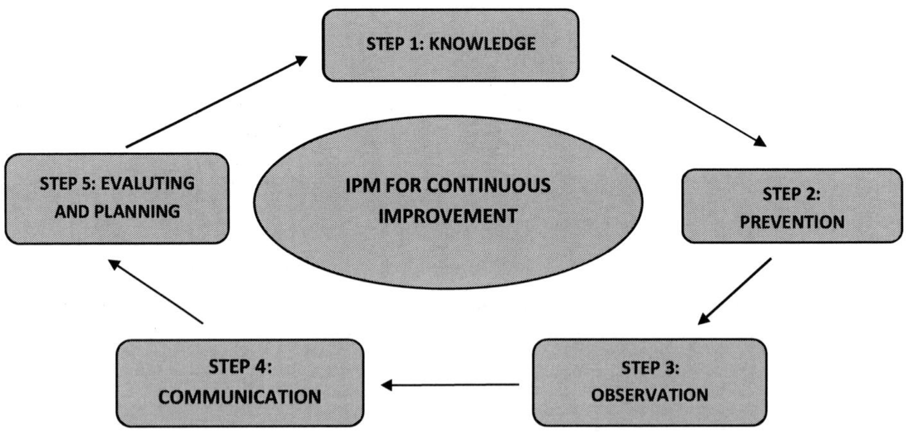

Figure 7. Scheme for integrated pest management.

IPM is a systems-based strategy for pest management that integrates the responsible use of pesticides with non-chemical methods such as biological, cultural, and others. IPM minimises the detrimental effects of pest control on human health and natural resources while maintaining pest populations below those that are economically detrimental. IPM is a constantly changing system where all appropriate control measures, readily available surveillance data, and forecasting information are combined into an extensive management programme and provided to the farmer as an aspect of sustainable crop production technologies at suitable intervals. IPM strives to combine all possible approaches or instruments of insect pest management in a methodical way that minimises the use of insecticides and disruption to the environment while yet becoming a multidisciplinary one (Kumar et al., 2022). IPM, often known as integrated pest management, is a concept of pest management. It entails combining several management strategies (such as bin cleaning, cleanliness, facility maintenance, temperature manipulation, etc.) to bring pest infestation levels down to economically viable levels. A specified level of insect infestation that justifies using a pesticide as a prophylactic strategy is known as an economic threshold.

5.1. Components in IPM

For successful implementation of IPM, certain steps need to be followed (Figure 7). The new IPM model is divided into four major components that address various pest management strategies: the grower's knowledge and resources to address the pest issue; organising and preparing information to take necessary actions; and upholding good communication to acquire information about pests and their management (Karlsson et al., 2020).

1) Pest management
   The phrase "pest control" has changed through time to "pest management" with the knowledge that, save for recently imported invasive pests, a balanced approach to regulating populations of pests to levels that do not cause financial losses is preferable to destroying or eradicating them. It includes-
   - Resistance in Host Plants
   - Cultural Control
   - Biological Control
   - Behavioural Control
   - Physical or Mechanical Control
   - Chemical control
   - Microbial control
2) Knowledge and resources
   The farmer can choose the best course of action for their scenario by knowing the different management strategies, pest biology and harm potential, and the applicability of available resources. It includes-
   - Pests
   - Available Control Options
   - Tools and technology
3) Planning and organisation
   This part of the new IPM model focuses on the administration of data gathering, organisation, and real pest control measures. It includes-
   - Pest Monitoring
   - Managing Information
   - Corrective Actions
4) Communication
   The final element of the new IPM model is effective communication to impart individual or collective knowledge for the benefit of everyone. As researchers create knowledge about endemic and invasive pests, emerging threats, and new control strategies, outreach can use both traditional and modern communication tools. It includes-
   - Staying Informed.
   - Communication- within the Group.

- Communication- among Growers.
- Communication- with the Public.

5.2. Challenges with IPM

1) It is unknown how evolutionary-based management affects yield.
2) Information is lacking about the possibility of resistance emerging to integrated control approaches.
3) There is a paucity of understanding in several areas that may be exploited to create new management techniques, such as plant tolerance.
4) Farmers may not be very keen on using evolutionary-based management.

Conclusion and Future Prospects

IPM is a strategy for ensuring environmental safety via the prudent application of all available pest control tools, societal acceptance due to the low cost and high safety of IPM-based food, and economic viability at both the consumer and producer levels (sellers are always expected to turn a profit). Despite the use of pesticides, illnesses and insect pests result in significant losses for crops. In addition, many insect pests have become resistant to the pesticides used to treat them, necessitating more insecticide applications, rising insurance costs, subsequent outbreaks, and pest comebacks. After evaluating current practices and, if necessary, adapting them to meet various ecological needs, it is urgently necessary to popularise new technologies.

After evaluating current practices and, if necessary, adapting them to meet various ecological needs, it is urgently necessary to popularise new technologies. Other effective control methods, such as cultural practices, host plant resistance, biocontrol, and the use of biopathogens and botanicals, should be prioritised in order to limit the use of pesticides. The adoption of integrated pest control for food quality and high-yield production for domestic and worldwide markets of vegetable crops can better defend against insect pests.

Given the current climate change crisis, the sowing of various agricultural and horticultural crops needs to be re-planned. For the preservation of microbial diversity, the use of systemic harsh agrochemicals needs to be prohibited for a suitable environment and development. The development and progression of agriculture shortly will be determined by the employment of bio-control agents and an integrated pest management system. Integrated pest management is necessary for sustainable development, even when it is not cost-effective.

Acknowledgements

The authors are thankful to Dr. Rajendra K. Singh, Department of Entomology, SVP University of Agriculture & Technology, Meerut, for permitting them to take photographs from their specimen collection. We are also thankful to Professor Jamal Ahemad Siddiqui, Librarian, C.C.S. University, Meerut, for the plagiarism check.

Conflict of Interests

The authors declare that they have no conflict of interest.

References

Ahmed T, Ansari M, & Ali H. (2009). Outbreak of diamondback moth, Plutella xylostella in Aligarh, India. *Trends in Biosciences*, *2*,10–12.

Anonymous. (2012). Major fruit producing countries in the world. Indian Horticulture database 2011:239.

Anonymous. Indian Horticulture Database, 2008.

Anonymous. Indian Horticulture Database, 2020.

Butani, D. K. (1974). Insect Pests of Fruit Crops and Their Control. 7. Mango. *Pesticides*, *8*, 37-41.

Cai, H., & Geng, Z. T. (1997.) Occurrence and control of *Ophideres fullonica* Linnaeus. *Plant Protection*, 23, 33–34.

Chakraborty, P. K. (2017). Pests eat away 35% of the total crop yield. The Hindu Kolkata 25 February, 2017.

Charles, J. F, Nielsen-LeRoux, C., Delecluse, A. (1996). *Bacillus sphaericus* toxins: molecular biology and mode of action. *Annual Review of Entomology*, *41*,451–472.

Dadmal, S. M., & Pawar, N. P. (2001). The fruit sucking moth, Eudocima (=Othreis) fullonica on Nagpur mandarin in Vidarbha Region. *Insect Environment*, *6*:167.

Dash, L., Padhy, D., Manish, M. K., & Ramalakshmi, V. (2022). Sources of pest resistance in vegetable crops: A review. *Journal of Entomological Research*, *7(4)*,190-192.

Dhaliwal, G. S. & Arora, R. (2002). Estimation of losses due to insect pests in field crops In: B. Sarath Babu, K. S. Varaprasad, K. Anitha, R. D. V. J. Prasada Rao, S. K. Chakrabarty and P. S. Chandukar (eds.) Resources Management in Plant Protection. Vol. 1. Plant Protection Association of India, Hyderabad, pp. 11-23.

Dhaliwal, G. S., Arora, R. & Dhawan, A. K. (2004). Crop losses due to insect pests in Indian agriculture: An update. *Indian Journal of Ecology, 31(1)*,1-7.

Dhaliwal, G. S., Arora, R. & Dhawan, A. K. (2003). Crop losses due to insect pests and determination of economic threshold levels. In: A. Singh, T. P. Trivedi, H. R. Sardana, O. P. Sharma and N. Sabir (eds.) Recent Advances in Integrated Pest Management. National Centre for Integrated Pest Management, New Delhi, pp.12-20.

Dhaliwal, G. S. & Arora, R. (1996). An estimate of yield losses due to insect pests in Indian agriculture. *Indian Journal of Ecology*, *23*, 70-73.

Dhandapani N., Umeshchandra R. S., & Murugan M. (2003). Biointensive pest management (BIPM) in major vegetable crops: an Indian perspective. *Food, Agriculture and Environment*, *1(2)*, 333-339.

Dhillon, M. K., Singh, R., Naresh, J. S. & Sharma, H. C. (2005). The melon fruit fly, Bactrocera cucurbitae: A review of its biology and management. *Journal of Insect Science*, 10.1673/1536-2442(2005)5[1: TMFFBC]2.0.CO;2.

FAO. (1997). Quarterly Bulletin of Statistics; Food and Agriculture of United Nations: Rome, Italy, 65p.

Ghosal, A., Chatterjee, M. L. and Manna, D. 2012. Studies on some insecticides with novel mode of action for the management of tomato fruit borer (Helicoverpa armigera Hub.). *Journal of Crop and Weed*, *8(2)*,126-129.

Ghule, B. D., & Dhumal, V. S. (1992). Chemical control of scale insects on pomegranate. Journal-Maharashtra Agricultural Universities, 17, 322-322.f *Maharashtra Agricultural Universities 17(2)*,322-323.

Haseeb M, & Sharma S (2007) Studies on incidence and crop losses by fruit borer, Deudorix isocrates (Lep: Lycaenidae) on guava. In: Singh G, Kishan R, Chandra R (eds) Ist International guaua symposium. *Acta Horticulturae*, 735, doi: 10.17660/ActaHortic.2007.735.66.

Karlsson, G. K., Stenberg, J. A., Lankinen, Å. (2020). Making sense of Integrated Pest Management (IPM) in the light of evolution. *Evolutionary Applications*, *13*,1791-1805. https://doi.org/10.1111/eva.13067.

Kumar, A., Tiwari, G. & Singh, A. K. (2022). IPM practices for insect pests of major vegetable crops: An overview. *The Pharma Innovation Journal*, *11(3S)*, 1728-1734.

Lal, C., & Verma, L. R. (2006). Use of certain bio-products for insect pest control. *Indian Journal of Traditional Knowledge, 5(1)*,79-82.

Mann, G. S. & Bindra, O. S. (1976). Evaluation of different jujube (Ziziphus mauritiana) cultivars for resistance to the ber fruitfly. *The Punjab horticultural journal*, 16(1/2),64-67.

Moore, D. (2004). Biological control of Rastrococcus invadens. Review article. *Biocontrol News and Information*, *25*,17–27.

Mukherjee, S. K. (1997). Introduction: botany and importance. In: Litz RE (ed) The mango: botany, production and uses. CAB International, Wallingford, pp 1–19.

Naresh, J. S., Malik, V., Balam, J. S. et al. (1986). A new record of Trathala sp. A larval endoparasite attacking brinjal fruit borer, Leucinodes orbonalis Guen. *Bulletin of Entomological Research*, *27*,74.

Nigel, S. (1980). Pest and Disease Control Handbook; Lovenham-Press Limited: Lavenhannem, UK, 1980; pp. 163–172.

Oerke E. C. (2006). Crop losses to pests. *The Journal of Agricultural Science*, *144*,31–43.

Oerke, E. C., Dehne, H. W. (2004). Safeguarding production – losses in major crops and the role of crop protection. *Crop Protection*, *23*, 275–285.

Rai, A. B., Halder, J., & Kodandaram, M. H. (2014). Emerging insect pest problems in vegetable crops and their management in India: An appraisal. *Pest Management in Horticultural Ecosystems*, *20*,113–122.

Raju, S. V. S., Bar, U. K., Shankar, U, & Kumar, S. (2007). Scenario of Infestation and Management of Eggplant Shoot and Fruit Borer, Leucinodes orbonalis Guen., in India. *Resistant Pest Management Newsletter*, *16(2)*, 14-16.

Satpathy, S., Kumar, A., Singh, A. K. & Pandey, P. K. (2005). Chlorfenapyr: A new molecule for diamondback moth (Plutella xylostella L.) management in cabbage. *Annals of Plant Protection Sciences*, *13*, 88-90.

Schoeman, M. H. (1995). Effect of temperature on guava disease development. ITSC.

Shivalingaswamy, T. M. S. Satpathy, S. & Banerjee, M. K. (2002). Estimation of crop losses due to insect pests in vegetables, In: Resource Management in plant protection, Vol-I (Sarat Babu, B., Vara Prasad, K. S., Anita, K. Rao R. D., Chakraborty, S. K. and Chandukar, P. S. eds.) Plant Projection association of India, Hyderabad, pp-24-31.

Shukla R. & V. Prasad. (1985). Population Fluctuations of the Oriental Fruit Fly, Dacus Dorsalis Hendel, in Relation to Hosts and Abiotic Factors. *Tropical Pest Management*, *31(4)*, 1985, 273-275. doi:10.1080/09670878509370999.

Singh, S. K., Singh, A. K. & Singh, H. M. (2007). Relative resistance of okra germplasm to shoot and fruit borer, Earias vittella Fabr. under field conditions. *Journal of Applied Zoological Researches*, *18(2)*, 121-123.

Tanwar, R. K., Jeyakumar, P., & Monga, D. (2007). Mealybugs and their management, Technical Bulletin 19, August, 2007, National Centre for Integrated Pest Management, New Delhi.

Visalakshi A., Nair G. M., Beevi S. N. & Amma A. M. K. (1989) Occurance of Odoiporus longicollis Olivier (Coleoptera: Curculionidae) as a pest of banana in Kerala. *Entomon*, *14*, 367-368.

Zhang, W., Kato, E., Bianchi, F., Bhandary, P., Gort, G., van der Werf, W. (2018). Farmers' perceptions of crop pest severity in Nigeria are associated with landscape, agronomic and socio-economic factors. *Agriculture, Ecosystems & Environment*, *259*, 159–167.

Chapter 4

Common Insect Pests of Vegetable Crops and Their Management Through Integrated Pest Management in India: A Review

Himani Sharma[1,*]
Istkhar Rao[2]
and Ashok Kumar Chaubey[1]
[1]Nematology Laboratory, Department of Zoology, Chaudhary Charan Singh University, Meerut, U.P., India
[2]Department of Bioscience and Biotechnology, Banasthali Vidyapith, Banasthali, Rajasthan, India

Abstract

Pest insects have narrowed agriculture's ever-expanding frontiers. After China, India is one of the world's top producers of vegetables, although significant losses have been linked to insect infestations. These insect pests belong to the orders Coleoptera, Diptera, Lepidoptera, Hemiptera, Homoptera, Hymenoptera, Isoptera, Mantodean, Mecopteran, Neuropteran, Odonatan, Orthoptera, Phasmida, Plecopteran, Psocoptera, Siphonaptera, Thysanoptera, Thysanura, and Trichoptera. Among all the orders, Lepidoptera is the major group of insect pests of vegetables, such as *Helicoverpa armigera* on tomatoes and okra, thrips (*Scirtothrips dorsalis*) and mites (*Polyphagotarsonemus latus*) on chillies, *Plutella xylostella* on cabbage, etc. Furthermore, dynamic changes in the environment due to various activities such as tillage, the development of HYV (high-yielding variety) seeds, and much more lead to differences in the current pest status. The changes in the host range, resistance development, and expanding geographical range have also become major constraints for the agriculture industry. The broadening of the host range is a threat to vegetables. For example, *Liriomyza trifolii* is a major pest of tomatoes, but it also affects brinjal, cowpeas, french beans, squash, leafy vegetables, and cucumbers. These shifts were seen in different areas of India, i.e., Andhra Pradesh, Karnataka, Chhattisgarh, etc. To control these outbreaks, chemical pesticides became a boon for controlling them, however, for a shorter period. To overcome this, biocontrol is a promising area for restricting the population. Biocontrol can be achieved using strains of efficient and pathogenic entomopathogenic nematodes, fungi, and predators against various pests. The application of Integrated Pest Management (IPM) in India is in its dormant stages; however, various proposals and native strains are present for the proper execution of this method against pests. This chapter will emphasise the current status of the vegetable pest along with its present and shifted host, its control mechanisms, and challenges in the application of biocontrol agents.

---

* Corresponding Author's Email: sharmahimani1995hariom@gmail.com.

In: Insects as Humans' Frenemies: From Friends to Foes
Editor: Istkhar Rao
ISBN: 979-8-89530-320-7

Keywords: Lepidoptera, shifted hosts, vegetable pests, biocontrol, IPM

1. Introduction

Insect pests pose the biggest threat to the agriculture sector and are getting harder to manage every day. Because the generally used control measures are not keeping up with the population of these insect pests, the agriculture industry has reported damage that is growing daily. India is known for its biodiversity and agricultural fields; however, the agroecosystems are found to be affected by pests and major diseases, which lead to yield losses ranging from 33 to 40% (Rai et al., 2014; Pati et al., 2021). There are varieties of pests that affect vegetable crop yield, viz., *Spodoptera litura*, *Daphnia indica*, *Earias vittella*, etc. (Halder & Rai, 2021). The damage caused by the insect pest was found to be increasing per year due to the continuous disturbances in the agroecosystem. The indiscriminate use of pesticides in agricultural fields to control the pest population shows side effects, mainly the development of resistance in the pest population, infertility, pollutant contamination of the soil ecosystem, biomagnification, etc. The biggest problem with chemical pesticides is biomagnification, which leads to diseases in the human population due to the consumption of polluted vegetables (Allahyari et al., 2016; Karkanis & Athanassiou, 2021). Scientists are now looking for sustainable solutions to reduce the harmful effects of pesticides, which include natural insecticides based on secondary metabolites secreted from plants and biocontrol. Biocontrol could become a promising alternative to all the artificial pesticides applied in the field to regulate pest populations. Though biocontrol is a safer solution to control pests and pest-related diseases, scientists and researchers are more focused on integrated pest management programmes (IPM). IPM exploits the natural potential of various biocontrol agents, viz., entomopathogenic nematodes (EPN), bacteria (EPB), entomopathogenic fungi (EPF), protozoa, and predatory insects, to restrict the pest outbreak (Walia et al. 2017). The potential of these biocontrol agents needs to be commercialised so that farmers can use them instead of pesticides in their fields. For commercialisation purposes, several formulations of EPN are available in the world that have achieved similar or better control over chemical pesticides (Pati et al., 2021). However, developing countries such as India are behind in formulation development. For a successful biocontrol programme, the formulated product should be cost-efficient, adapted to the specific localities, and regulate the pest population. The adaptation of the formulated product is very important for controlling the pest population, due to which indigenous strains of these biocontrol agents should be used. The trapping of insects harmful to vegetable crops and their status in the field play a significant role in the application of insecticides and biocontrol agents. Besides this, there are potent plants that serve as natural insecticides and can be used for the target pest. Previous studies revealed the uses of plants in insecticide production, and their efficacy was tested against insect pests in laboratory conditions (Siegwart et al., 2015; Pavela, 2016; Grieneisen & Isman, 2018; Isman, 2020). These plant-derived insecticides showed a broad range of bioactivities when tested under laboratory conditions. These activities restrict population growth, control reproductive cycles, produce organic compounds that have deleterious effects on the host, etc. The major challenge in the regulation of pest populations using insecticides is the development of resistance. To overcome this, post-insecticide effects and combinations of two or more insecticides for a particular pest are suggested. However, some pest species have already developed resistance to pesticides, and their regulation with other methods is a little difficult, especially with insecticides. In these cases, the variable sites of the pest should be targeted with plant-derived insecticides. All the insecticides don't work at every developmental stage, which further suggests laboratory and field trials against the different stages of the pests.

2. Vegetable Crops Grown in India

India is known for its agricultural land area, and different vegetable crops are grown along with other cash crops. Several examples of vegetables can be given that are grown in India and have high nutritional qualities. *Lagenaria siceraria* (Bottle Gourd; Louki), has a high moisture content as compared to phosphorus, iron, vitamin C, A, B, calcium, and other minerals. *Momordica charantia* L. (bitter gourd; karela) comprises high moisture, calcium, and vitamin A content concerning other macro- and micronutrients. *Momordica dioica* Roxb. (spine gourd; kakrol) also contains a high amount of moisture, calcium, phosphorus, vitamin A, and C. *Luffa acutangula* (L.) Roxb. (ribbed gourd; kali tori) contains a high amount of moisture and phosphorus; however, it lacks vitamin C micronutrients. Likewise, *Luffa cylindrica* (L.) Roem (sponge gourd; ghia tori) is found to be deficient in vitamin C content but contains a high amount of moisture and vitamin A. *Benincasa hispida* (Thunb.) Cogn. (wax or ash gourd; petha) is found to be deficient in vitamin A and also contains vitamins B, and C, phosphorus, calcium, carbohydrate, protein, iron, and a very low concentration of fat. *Trichosanthes cucumerina* (snake gourd; chinchinda) contains enormous moisture content but lacks vitamin C. *Trichosanthes dioica* Roxb. (pointed gourd; parwal) contains all described macro- and micronutrients with the highest moisture and phosphorus content and low levels of fat. *Cucumis sativus* L. (cucumber; khira) has a high moisture content in the absence of vitamin A. *Cucumis melo var. momordica* (snapmelon; phoot) also contains moisture content but is found to be deficient in vitamin B, phosphorus, calcium, and iron. *Cucumis melo* var. *utilissimus* (long or serpentmelon; kakri) also contains high moisture content. *Praecitrullus fistulosus* Pang. (round melon; tinda) contains moisture and all the vitamins A, B, and C with iron, calcium, and phosphorus content, with low levels of carbohydrates and fats. *Coccinia grandis* (L.) Voigt (ivy gourd; kundru) comprises water content with carbohydrates, protein, vitamins A, B, C, iron, phosphorus, calcium, and fats (Mondal et al., 2020).

India also has high cultivation of potatoes (*Solanum tuberosum* L.) in different climatic and agricultural conditions (Rana, 2011; Pandey & Kang, 2003). There are six described zones of potato in India due to the varying agroecosystems. The six zones described are the North-Western Plains, West-Central Plains, North-Eastern Plains, Plateau Region, North-Western and Central Hills, North-Eastern Hills, and Southern Hills, which cover all the states of India suitable for potato cultivation. The North-Western plains covered the cultivated land of Punjab, Haryana, and Rajasthan, whereas the West-Central plains covered major agricultural lands of Madhya Pradesh, North-Western Gujarat, and West-Central Uttar Pradesh, which provide a major proportion to potato cultivation with other states, viz., West Bengal and Bihar. The north-eastern plains covered West Bengal, Bihar, Orissa, Jharkhand, Uttar Pradesh, Assam, Pradesh, Madhya Pradesh, and Chhattisgarh. Some parts of Orissa, Gujarat, and Madhya Pradesh are categorised under the Plateau region, along with Maharashtra. The North-Western and Central Hills consist of colder regions such as Uttarakhand, Jammu and Kashmir, and Himachal

Pradesh. The northeastern and southern hills covered the left-over state with less production of potatoes due to improper irrigation as precipitation is not so good (Chandel et al., 2022).

Another known and globally consumed vegetable is okra which has conventional importance in India. The okra, (*Abelmoschus esculentus* Linn.) contains various macro- and micronutrients, viz., vitamin C, calcium, potassium, and other properties, which make it consumable as a dietary source (Halder et al., 2021). The other most consumed vegetable is tomato (*Solanum lycopersicum* L.), classified under the Solanaceae family, which also includes potato, tobacco, and eggplant (Knapp, 2002). The tomato cultivation began in the Andes, where Peru and Mexico were found to be important regions for the diversification of this plant (Peralta & Spooner, 2007; Larry & Joanne, 2007). The original size of the tomato was similar to that of the peas; however, advancements in genetics yielded today's fruit size (Soyk, 2017). The tomato is very important in terms of nutrients, as it comprises ascorbic acid, vitamin A, high moisture content, low fat and carbohydrates, β-carotene, and many more minerals (Wilcox et al., 2003). India is one of the five countries with the highest cultivation of this vegetable (Melomey et al., 2019). The other important vegetable from the family Solanaceae grown in India is eggplant, which is commonly named brinjal (*Solanum melongena* L.). Apart from India, this is found to be grown in various other countries, such as Africa, Central America, Asia, and the Mediterranean, due to its dietary importance (Mat-Sulaiman et al., 2020). The production of this plant started in the areas of Africa, some parts of Asia, and Europe and was also referred to as an old-world crop in contrast to a new-world crop, which is tomatoes and potatoes (Chapman, 2019). The eggplant also comprises a high moisture content of approximately 92%, along with iron, carbohydrates, protein, vitamins A, and C, and a low amount of fat of approximately 0.3%. The eggplant is also rich in calcium, phosphorus, vitamin B, and other nutrients (Oladosu et al., 2021).

3. Pests Associated with Vegetable Crops

As already mentioned, a variety of insect pests belonging to different insect orders affect the growth and yield of vegetables. A few important ones are listed below.

3.1. Coleopteran Pests

3.1.1. White Grubs

The vegetable crops, viz., cucurbits and potatoes, found to be highly infected with the white grubs belong to the family Scarabaeidae. They are harmful and notorious pests of vegetables, especially potatoes, and pose a major threat to their cultivation. These soil-inhabited pests are known to damage potato crops as they reach up to 20 cm in the soil. The different life stages of beetles, including the larval forms (grub stages), tend to feed on various parts of the plant (Chandla, 1985). The adult stages are found in the soil approximately 20 cm underground, which makes it convenient for the larval stages to feed on the underground parts of the plant (Musthak Ali, 2001). The other crops, viz., sugarcane, pigeon pea, garden grass, rice, ginger, maize, etc., were also found to be destroyed by these white grubs (Mishra, 2001). If we look at the potato crops only, the larvae are found to be harmful as they feed on the underground parts.

The grub's first stage after hatching is to feed on the tubers, roots, and other parts; however, due to their polyphagous nature, they can also feed on the available nutrients present in the soil (Mehta et al., 2010). Potatoes are found to be damaged by the 2nd and 3rd larval stages (grub), which leads to the production of holes in the vegetable. The infection at these stages could be very high, and this could lead to the development of a tunnel in the tuber as the number of holes increases. The biggest challenge faced by the farmers is that the severity of infection can't be determined by the symptoms, as there are no specific warning signals detected from the plant, which leads to the development of infection and great losses at the time of harvesting. Till now, 20 species of this pest have been identified with their effect on the potato crop. The *Brahmina coriacea*, *B. flavoserica* Brenske, *Melolontha indica* Blanch, *Holotrichia longipennis*, and *H. seticollis* Moser belong to the subfamily Melolonthinae; *Anomala dimidiata* Hope, *A. polita*, and *A. rugosa* belong to subfamily Rutelinae; and *Phyllognathus dionysius* belongs to subfamily Dynastinae, known to affect the potato crop in hilly areas (Misra & Chandla, 1989; Chandel et al., 1997; Butani & Jotwani, 1984; Mehta et al., 2008; Bhalla & Pawar, 1977; Haq, 1962; Rai & Joshi, 1988; Sushil et al., 2006). Furthermore, the other species of beetles, viz., *Holotrichia repetita*, *H. rustica*, *H. serrata*, and *Anomala* sp., are known to damage the potato crop in southern India, especially Karnataka (Lingappa & Giraddi, 1995; Veersh et al., 1991). The Tamil Nadu agricultural fields of potato were found to be affected by *H. excisa* Moser, *H. nototiocollis*, *A. communis* Brenske, and *A. nathani* Frey grubs (Regupathy et al., 1997; Butani & Jotwani, 1984).

3.1.2. Wireworms

Wireworms are the larval stages of beetles, which cause damage to the agricultural industry. The beetles, viz., *Drasterius, Agronichis*, and Lacon, belonging to the family Elateridae, possess these larval stages in their life cycle and are voracious feeders of crops, especially potatoes. Like other pests, they also attacked the potato tuber and formed cylindrical holes, either inviting other infections or reducing the crop yield. The population of wireworms at the initial stages is found to be controllable; however, if left unseen, they can cause potent destruction to crops (Chandel & Chandla, 2003; Chandel et al., 2008a).

3.1.3. Pumpkin Beetles

The beetle belongs to the family Chrysomelidae and is a serious pest of vegetables, especially the *Aulacophora* genus, which comprises several species. Among all species, *Aulacophora foveicollis* causes the most damage to the cucurbits in comparison to other pests. Though they are harmful pests, their life stages can bear only a 27–32°C limit. Slight changes in this prescribed temperature led to the mortality of the pest (Pavlakos, 1943; Al-Ali et al., 1982). This temperature barrier becomes the reason for recording its highest activity in February. The larvae and the beetles were found to be responsible for damaging the crops. In India, they are the pests of cucumber, muskmelon, bottle gourd, and watermelon, and as the name suggests, they also feed on pumpkins. The threatening population of the pest was found in Madhya Pradesh and Tripura with seventy-five adults per five sweepings and one hundred four beetles per thirty leaves, respectively (Shinde et al., 1994; Roy & Mukhopadhyay, 1980).

3.1.4. Hadda Beetle

These hadda beetles belong to the family Coccinellidae and the genus *Henosepilachna*, also known as *Epilachna*, which are the potent pests of cucurbits and Solanaceae. They feed on various crops, as reported from the hilly areas of Himachal Pradesh and Jammu & Kashmir. Furthermore, the pest population was also reported in the agricultural fields of Uttar Pradesh, Punjab, and Bengal. They are the major pests of *Momordica charantia* L. (Karela) in India. The other families of vegetables attacked by this pest are Malvaceae, Fabaceae, and Convolvulaceae (Mondal et al., 2020). *E. vigintioctopunctata* and *E. dodecastigma* were found to be responsible for the damage of 13%–88% in the leaves and 42% in the pod of cowpea (Halder & Rai, 2021).

3.1.5. Flea Beetles

The pest population belongs to the Chrysomelidae family and is an agricultural pest that causes damage to crops. *Psyllodes plana* Maulik attacks the plant's leaves after emerging from their cocoon. The infestation of these beetles can be identified by the presence of round holes (Chandla, 1985). The pest population is naturally regulated during the winter, as they hide themselves and cause less damage. The favourable period of the beetles becomes a bane for crops as they can destroy whole fields. The white larval stage also feeds on the leaves (Chandel et al., 2011).

3.1.6. Blister Beetle

The pest belongs to the family Meloidae and feeds on various crops belonging to the families Asteraceae, Leguminosae, Solanaceae, Cucurbitacae, and many others. *Epicauta hirticornis* is commonly known as blister beetles due to their harmful effects on humans as well. The beetle's release of poisonous substances is responsible for causing human skin swelling via direct contact of the skin with the beetles. The pest is restricted to two geographical locations: one on the African continent and the other on the Asian continent. The adult stages were found to be responsible for the destruction of the agricultural fields, as were the larval stages found in association with the grasshoppers (Chandel et al., 2022; Mondal et al., 2020).

3.2. Lepidopteran Pests

3.2.1. Cutworms

There are many species of cutworms, which pose a threat to the food industry as a pest found worldwide and possess a tendency to feed on various fruits and vegetables. They are the most serious potato pest in India. There are various species of the genus *Agrotis*; however, *Agrotis segetum* and *Agrotis ipsilon* are found to damage crops in comparison to other known species. *A. segetum* inhabits hilly areas and becomes active in the summers (Singh, 1987). *A. ipsilon* is found in the agricultural areas of Bihar, Punjab, and other states and is found to be active during the late winters, especially in March and April (Singh, 1987). The management of this pest becomes challenging as it shows nocturnal feeding and does considerable damage to the roots and shoots. The caterpillars destroy the potato crop by making holes in the tuber, which can be attacked by other infections due to the open environment, ultimately leading to crop loss (Das & Ram, 1988; Kishore & Misra, 1988).

3.2.2. Bottle Gourd Plume Moth

These moths, as the name suggests, are the major pests of *Lagenaria siceraria* (Bottle Gourd), belonging to the family Pterophoridae. This is a major pest of cucurbits, pigeon peas, luffas, and many others. *Sphenarches caffer* larval stages and cause damage to the plants, especially leaves, fruits, and flowers, reducing the yield (Mondal et al., 2020).

3.2.3. Snake Gourd Semilooper

This is a major pest of *Trichosanthes cucumerina* (snake gourd), belonging to the family Noctuidae. *Anadevidia peponis* larval stages were found to be destructive for the cucumber, bitter gourd, snake gourd, *Cucurbita moschata*, and *Citrullus lanatus*. The larvae or caterpillars have a green-to-white colour, which helps them feed on the leaves of the above-described vegetables. The life cycle includes egg, larva, pupa, and adult stages that last for 6 days. Due to its short life cycle and high fecundity, early management of the pest is necessary (Mondal et al., 2020).

3.2.4. Pumpkin Caterpillar or Cucumber Moth

Diaphania indica, the pumpkin caterpillar or cucumber moth, is a common pest of the Cucurbitaceae family, especially targeting *Cucumis sativus* in India. The population of this pest is naturally regulated in the winter; however, it has enough potential to affect yields in warmer climates. The life cycle consists of four stages, viz., egg, larval, pupal, and adult stages, whereas larval stages, greenish with two longitudinal stripes, are the voracious feeders of the mentioned vegetables (Mondal et al., 2020). The attacking period of this pest differs for different crops, as they attack *Cucumis sativus* at the time of flowering (Vanisree et al., 2005). In contrast to this, they attack squash in the summer (Zehnder 2011). Moisture played a key role in the abundance of this pest population (Mondal et al., 2020).

3.2.5. Americal Bollworm

American Bollworm *Helicoverpa armigera* (Hübner) is one of the serious pests in the agricultural industry that has been reported in 181 plant species, including vegetables, in India (Manjunath et al., 1989). The life cycle of this insect comprises adults, eggs, six larval stages, and pupa (Figure 1). The vegetable crops, mainly tomato (Yadav et al., 1985; Pawar et al., 1986a), potato (Yadav et al., 1985), pigeon pea (Pawar et al., 1986; Sithanantham et al., 1982), and okra (Thontadarya et al., 1978; Sithanantham et al., 1982), showed high levels of infections. The average loss caused by this pest on tomato plants was found to be in the range of 7%–68% (Halder & Rai, 2021).

3.2.6. Tobacco Cutworm

Spodoptera litura (Fab.), commonly known as the tobacco cutworm, showed a cosmopolitan distribution. The pest belongs to the Noctuidae family and is responsible for damaging field crops. The life cycle consists of four stages, i.e., egg, larval/caterpillar, pupal, and adult stages (Figure 2). The larval stages feed voraciously on the leaves, and one more characteristic possessed by these larvae is nocturnal feeding, which makes the crops more vulnerable to pests and their management more difficult in comparison to other pests (Chandel et al., 2022). There are other species of *Spodoptera* like *S. exigua* reported from *Solanum tuberosum* L. and *S. furgiperda* from *Zea mays* plants; however, *S. furgiperda* is one of the invasive species that has caused great damage in Chikkaballapur, South India (Butani & Jotwani, 1984; Shylesha et al., 2018).

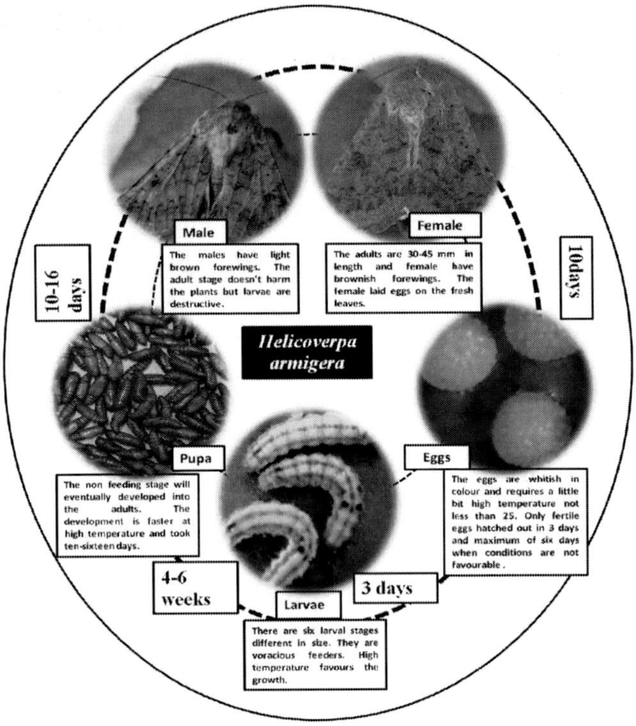

Figure 1. Different stages present in the life cycle of *Helicoverpa armigera*.

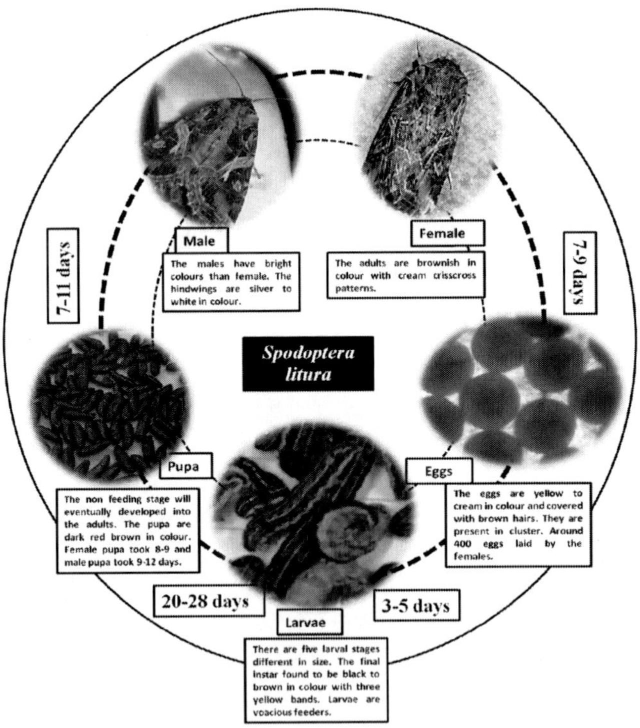

Figure 2. Different stages present in the life cycle of *Spodoptera litura*.

3.2.7. Cabbage Semi Looper

This is a pest of radish, pea, and many other vegetable crops in India. The pest *Plusia orichalcea* (Fab.) population belongs to the Noctuidae family and causes great damage to agricultural fields as they feed on the leaves, leading to the defoliation of the whole plant (Misra et al., 1995). The pest population clusters on the plant and can be spotted by their colours, which generally range from green to brown (Chandel et al., 2022). The adult stage is found on the flowers of the plant. The larval stages are responsible for the destruction of the plants as they feed voraciously on the leaves (Misra et al., 1995).

3.2.8. Oriental Armyworm

This is a serious pest with a broad geographical range covering India, destroying the agricultural fields as these were fed on the leaves. This pest, *Mythimna separata* (Walker), belongs to the Noctuidae family, where the larval stages are dependent on the leaves for their growth. The feeding of pests causes damage to the tissues along with disruption of the leaf framework. Furthermore, the late instars were found to be very quick at damaging the leaves, as they were not restricted to one leaf or one branch. The severity of infection can be determined by the condition of the leaves, and low to moderate levels of infection lead to the destruction of the whole field. The larval stages can be identified by their green to brownish colour and the presence of clear longitudinal bands that are blackish to greenish and four in number. The larval stages transform into pupa, and from the pupa, adults emerge (Chandel et al., 2011; 2020).

3.2.9. Bihar Hairy Caterpillar

This is a pest of crops that causes damage to potatoes, fruit trees, etc. The pest *Spilosoma obliqua* (Walker), which belongs to the family Arctiidae, attacked the crops either in the November–March period or in the April–June months. The pest follows the basic life cycle of holometabolous insects, which consists of four developmental stages: egg, larva, pupa, and adult. The larval stages target the leaves, causing defoliation of the agricultural fields if the infestation is high. It's noteworthy to describe the nature of the larval stages as they target mature leaves of the mature plants in comparison to the younger leaves when choices are available in the fields; otherwise, they can feed on any leaf (Chandel et al., 2011; 2022). The fecundity of females found was higher, which becomes stressful when restricting the pest population in the field. The larval stages include 7 instars possessing 7 wide transverse stripes. Besides this, a bunch of hairs were found attached to the stripes (Srivastava & Butani, 1998). The peak of the winter inhibits the growth of this pest as they become inactive during this period (Chandel et al., 2011; 2022).

3.2.10. Hairy Caterpillar

This is an agricultural pest with a broad host range ranging from coffee to cauliflower. *Dasychira mendosa* (Hübner) belongs to the Lymantriidae family and consists of a grey-brownish larval stage that feeds voraciously on the leaves. Heavy infestations of this pest lead to the defoliation of the fields (Chandel et al., 2011; 2022).

3.2.11. Eggplant Borer

This pest, belonging to the family Crambidae, is an agricultural pest for many vegetable and non-vegetable crops. *Leucinodes orbonalis* Guenee is a major pest of eggplant but is found infecting the *Solanum tuberosum* (potato) crops in India. The larval stage targets the stem part, leading to the wilting of the leaves and weakening of the plants. The larval stage can be identified by its pinkish colour, which transforms into pupa after a sufficient amount of feeding (Chandel et al., 2011; 2022).

3.2.12. Cabbage Butterfly

Plutella xylostella belongs to the family Plutellidae and is responsible for the destruction of $1 billion in losses worldwide (Talekar & Shelton, 1993; Sarfraz et al., 2006). The pest was first reported from China; however, there are some contradictions, as some scientists believed that it was reported from South Africa (Liu et al., 2000; Kfir, 1998). The pest was first reported from India in 1914 (Fletcher, 1914). The management of this pest becomes challenging due to its wide host range and cosmopolitan distribution. To manage the pests in India through integrated pest management, the costs come to approximately $168 million per year (Sandur, 2004).

3.3. Dipteran Pests

3.3.1. Fruit Flies

These pests are classified under the Tephritidae family and are found to be responsible for the great losses to cucurbit yields. This family comprises 4000 species, of which some are very important from an agricultural point of view, mainly *Bactrocera cucurbitae*, *B. diversa*, *B. tau*, *B. latifrons*, *B. ciliates*, *B. correcta*, and *B. tryoni* (Enkerlin 2003). Though the genus *Bactrocera* is found to be the most destructive pest of vegetable crops, the maximum damage done by *Bactrocera cucurbitae* itself to the agriculture industry The flies were found to affect the cucurbit crops mainly the bitter, snake, ivy, pointed, and wax gourd; watermelon, snap melon, cucumber, and many other crops (Clausen et al., 1965; Kapoor & Agarwal, 1983; Doharey, 1983). The losses reported by these flies for the cucumber vegetables were 10–100%; the bitter gourd and watermelon affected by these flies led to a loss of 30–80%; and other crops such as cucumber, snake gourd, and sponge gourd had an estimated loss of 20–100% (Shooker et al., 2006). The life cycle consists of four stages, i.e., egg, larva/ maggot, pupa, and adult; however, the maggot stage consists of 2-3 larval stages, which enter the fruit through the holes and feed there until it has moulted into the pupal stage in the soil.

3.3.2. Serpentine Leaf Miner

The pest, which belongs to the family Agromyzidae, was first reported from the USA; however, it wasn't reported under this genus. The pest was classified under the Chloropidae family but shifted to this family in 1925 (de Meijere, 1925). *Liriomyza trifolii* (Burgess) is a serious pest of *Solanum lycopersicum*, but it also damages the crops of capsicum, members of Cucurbitaceae, and many more. This pest was first reported in 1990 from India, as its invasion occurred due to *Chrysanthemum* from the USA (Singh, 2004). Maggots and adults are both responsible for damaging crops, as adults target the leaves for reproduction and egg-laying purposes, whereas maggots attack the leaves, damaging the crops (Mondal et al., 2020).

3.4. Hemipteran Pests

3.4.1. Leafhopper

Amrasca biguttala biguttala is a major insect pest of okra in India, belonging to the family Cicadellidae (Jayasimha et al., 2012). The life cycle consists of four stages, i.e., egg, nymph, and adult (male and female), where the nymph and adult are found to be responsible for the decreasing plant sap leading to the weakening of the okra plant. Five nymphal stages differ from their previous ones in size and colour. The colour of the first instar is yellow; the second instar has a yellow and red colour; the third instar possesses a yellow and greenish colour; and the fourth instar is similar to the third one, with differences in size as they are bigger than the previous one. The final instar has nine abdomen segments, green in colour, an average length of 2.28 mm, and pads at the wings (Jayasimha et al., 2012). Temperature plays an important role in the pest population, as it's been reported from studies that high temperatures favour the growth of their populations (Srinivasan et al., 1988; Jayasimha et al., 2012). The *Empoasca devastans* are found to feed on the potato plant and are responsible for its destruction by transmitting various diseases (Butani & Jotwani, 1984; Misra, 1995). The 'hopper burn' is the characteristic feature of this pest and occurs due to the presence of scratches on the leaves (Trivedi & Rajagopal, 1999).

3.4.2. Whiteflies

The white flies are classified under the Aleyrodidae family, and the *Bemisia* genus has a cosmopolitan distribution. This is a major pest of okra, potato, and tomato vegetable crops in India. The pest becomes active during the autumn, especially in the agricultural fields near the Ganges (Kumar et al., 2003). The pest *Bemisia tabaci* is found to feed on multiple crops, which is damaging to crop yield and makes its management difficult (Puri et al., 1995). The pest feeds on the potato crop, specifically in the winter season, causing the dropping of the leaves, discolouration, and the occurrence of spots on the leaves (Broad & Puri, 1993; Chandel et al., 2010). The growth of the plants becomes retarded, and the dropping of the leaves increases in cases of high infection by pests in the fields (Malik et al., 2005). This pest was found to be a vector of potato apical leaf curl disease caused by the geminiviruses in potato crops (Chandel et al., 2010). The spiralling whitefly (*Aleurodicus disperses* Russell) and silver leaf whitefly (*Bemisia argentifolii* Bellows) were found to be invasive species that were introduced in India during 1993 and 1999, respectively (Palaniswami et al., 1995; Singh, 2004).

3.4.3. Sap-Sucking Bugs

Nezara viridula bugs belong to the family Pentatomidae and are distributed throughout the world. These are the minor pests on vegetable crops in India. This bug is a major pest of coriander; however, it also feeds on other vegetables like tomatoes, potatoes, etc. due to its nature of feeding on multiple hosts. Both larval and adult stages are responsible for the damage, as they derive their nutrition from the leaf and shoot sap (Butani & Jotwani, 1984; Chandel et al., 2022). The life cycle includes eggs, 4-5 nymphal instars, and adult stages where nymphs are found to be brown-reddish, possessing spotting of different colours (Butani & Jotwani, 1984). *Piezodorus hybneri* (legume stink bug) is also reported from vegetable crops and is considered to be less harmful than other pests. They were found on the leaves of the potato crop.

There are other bugs belonging to the family Miridae that feed on a broad range of vegetable and non-vegetable crops. *Creontiades pallidifer* (Walker) follows the basic life cycle, which consists of eggs, larval stages (nymphs), and adults, where both larval and adult stages feed on the plant leaves of vegetable plants, mainly eggplant, melon, potato, etc. The pest was found to survive in the capital (Delhi), where it attacked eggplant, potatoes, and other crops in the winter areas, but other vegetable plants specifically belonging to the Cucurbitaceae family were attacked by this pest in the summers (Butani & Jotwani, 1984).

3.5. Homopteran Pests

3.5.1. Aphids

They belong to the family Aphididae and are considered to be harmful pests in the agricultural industry. They feed on the sap of the plant, causing it to weaken. Furthermore, they serve as vectors for many microorganisms, destroying the fields. The potato crop was found to be infested by *Macrosiphum euphorbiae* Thomas, which also possesses the ability to damage cucurbit crops. The distribution of the pest is worldwide, except in the winter region, where extreme cold climatic conditions inhibit the pest population. Both larvae and adults feed on the plants, and they feed voraciously, leading to a decline in yield (Mondal et al., 2020; Chandel et al., 2022).

3.6. Isopteran pests

3.6.1. Termites

The pest belongs to the Termitidae family and is a serious pest in the agriculture industry due to the high fecundity of the queen (female termite), which is about 70–80 thousand eggs per day. These are eusocial insects that live and feed on a specific area, and after damaging the area, the whole colony flies to new places. The rainfall and precipitation play an essential role in the growth of the termite colony. It's reported from previous studies that after rainfall or precipitation, agricultural fields become more susceptible to termite infection. The workers feed on the leaves of the plant; however, their jaws can cut the wood, due to which affects the stem of the plant. The species of termites reported from India are *Eromotermes* sp., *Odontotermes obesus* (Rambur), and *Microtermes obesi* (Holmgren) (Srivastava & Butani, 1998; Butani & Jotwani, 1984; Chandel et al., 2022).

3.7. Thysanopteran Pests

3.7.1. Thrips

These pests belong to the families Thripidae and Phlaeothripidae and are found to affect vegetable crops in India. Besides this, they are the vectors of the Tospo virus, which is found in the association of the first larval stage during its voracious feeding on the leaves. The infection was spread by adults as the virus multiplied inside the host (Chandel et al., 2008c; 2022). The common species reported from India that are serious pests of vegetables are

Scirtothrips dorsalis Hood, which is a major pest of chillies and is also reported from the *Solanum tuberosum* crops; *Caliothrips collaris* (Bagnall); and *Haplothrips* sp., also cause damage to the *Solanum tuberosum* crops (Khurana et al., 2001).

3.8. Mites

3.8.1. Red Spider Mite

The pest belongs to the family Tetranychidae, has a broad host range, and causes great damage to various crops, including vegetable crops. There are various species reported from other families as well, viz., spider and false spider mites, gall mites, and yellow mites, which cause huge-scale destruction. The mites also target the leaves of the plants, which leads to the defoliation of the fields. The red spider mite, *Tetranychus urticae*, has a broad geographical range, as it's found worldwide and has been found to damage cucurbit crops. The high-temperature conditions favour the growth of this pest (Mondal et al., 2020).

4. Integrated Pest Management

4.1. Biological Control

The management of insect pests using biological control agents has gained the attention of scientists working in this field, as pesticides are harmful to human health and soil fauna, and many insect pests have developed resistance to them. The application of insecticides, pesticides, and other chemicals is much easier in agricultural fields. The pesticide and insecticide applications don't specifically target the pest, but other related organisms in the vicinity were also affected. All the insects are not harmful, as there are beneficial insects or control agents present that can naturally regulate the pest population. Vegetable crops are important from a food and nutrition perspective, as damage caused by insect pests leads to a decrease in yield and quality. To maintain a healthy environment, it's necessary to exploit the natural characteristics of biocontrol agents in the regulation of pest populations (Altieri et al., 2005; Mahr et al., 2008; Halder et al., 2011; Pati et al., 2021). The biodiversity of India is rich in biological agents that can be used for controlling purposes (Table 1).

4.1.1. Predators

Predators are those insects that can derive their nutrition from other insects without damaging the plants. This characteristic of predators was found to be beneficial for regulating the pest population. The predatory population belongs to spiders, mantises, bugs, birds, etc., which feed on the common insect pests protecting agricultural fields. These predators showed a broad host range, which helps in the regulation of different pest populations at the same time. The specialised mouth apparatus of predators helps attack prey (Sampaio et al., 2010). Some predators can feed on other food types; however, some feed on the pest population completely. The predators have well-developed and strong body parts for capturing, teasing, and piercing the prey population.

Table 1. List of common pests of vegetables along with their natural enemies for their management in the fields

Insect Pest	Insecticides/Chemicals	Biocontrol agents (Predators, Parasitoids)	Entomopathogenic nematodes (EPN)	Reference
White grubs	Methyl Parathion, Carbaryl, Or Monocrotophos	-	*Steinernema feltiae, Heterorhabditis indica, Steinernema glaseri*	Chandel et al., 1994b; Chandla et al., 1988; m Shapiro-Ilan & Gaugler (2010)
Wireworms	Phorate	-	-	Chandel & Chandla, 2003
Pumpkin Beetles	Indoxacarb, Chlorpyrifos, Carbaryl	*Chauliognathus pensylvanicus*	-	Mondal et al., 2020; Pati et al., 2021
Hadda Beetle	Malathion, Dichlorvos, Endosulfan, Chlorpyriphos, Or Carbaryl	-	-	Pandey, 2002; Mondal et al., 2020
Flea beetles	Malathion, Endosulfan, Or Chlorpyriphos	-	-	Chandel et al., 2022
Blister beetle	Chlorpyriphos Or Carbaryl, Beta-Cyfluthrin, Azadirachtin, Spinosad, Phosalone	-	-	Chandel et al., 2022
Cutworms	Oxydemeton Methyl	-	*Heterorhabditis indica, Steinernema brazilense*	Trivedi & Rajagopal, 1999; Shapiro-Ilan & Gaugler (2010)
Bottle Gourd Plume Moth	-	Apanteles *ruidus* and *A. paludicola, Bacillus thuringiensis kurstaki*	-	Mondal et al., 2020
Snake Gourd Semilooper	Carbaryl	Apanteles paragame, *A. plusiae* and *Mesochorus plusiaephilus.*	-	Mondal et al., 2020
Diaphania indica	Carbaryl, Malathion, Dimethoate	*Apanteles* spp., Common Myna (*Acridotheres tristis*) and cattle egret (*Bubulcus ibis*)	-	Mondal et al., 2020
Helicoverpa armigera	Carbaryl, Endosulfan, Monocrotophos, Quinalphos, Chlorpyriphos, Or Malathion	-	*Heterorhabditis indica, Steinernema abbasi*	Pandey 2002, Srivastava et al., 2022
Spodoptera litura	Carbaryl, Endosulfan, Monocrotophos, Quinalphos, Chlorpyriphos, Or Malathion	-	*Heterorhabditis indica, Steinernema abbasi*	Pandey 2002, Bhat et al., 2021; Shapiro-Ilan & Gaugler (2010)
Cabbage semi looper	Carbaryl, Endosulfan, Monocrotophos, Quinalphos, Chlorpyriphos, Or Malathion	-	-	Pandey, 2002
Oriental armyworm	Carbaryl, Endosulfan, Monocrotophos, Quinalphos, Chlorpyriphos, Or Malathion	-	-	Pandey, 2002
Bihar hairy caterpillar	Carbaryl, Endosulfan, Monocrotophos, Quinalphos, Chlorpyriphos, Or Malathion	-	-	Pandey, 2002
Hairy caterpillar	Carbaryl, Endosulfan, Monocrotophos, Quinalphos, Chlorpyriphos, Or Malathion	-	-	Pandey, 2002

Insect Pest	Insecticides/Chemicals	Biocontrol agents (Predators, Parasitoids)	Entomopathogenic nematodes (EPN)	Reference
Eggplant borer	Carbaryl, Endosulfan, Monocrotophos, Quinalphos, Chlorpyriphos, Or Malathion	-	-	Pandey, 2002
Plutella xylostella		*Chrysoperla carnea*, coccinellids	*Heterorhabditis indica*, *Steinernema carpocapsae*	Pati et al., 2021; Shapiro-Ilan & Gaugler (2010)
Fruit flies	Malathion, Carbaryl	*Opius fletcheri*, *Fopius arisanus*,	*Steinernema carpocapsae*	Mondal et al., 2020
Serpentine Leaf Miner	Abamectin, Cyromazine	Lacewings, lady beetle, spiders, fire ants, dragonfly, robber fly, *Tetrastichus ovularum* (egg), *Gronotoma micromorpha*	*Steinernema carpocapsae*, *Steinernema feltiae*	Mondal et al., 2020; Pati et al., 2021
Leafhopper	Dimethoate And Methyl Demeton, Phorate	Coccinellids, *Distina albino*, *Chrysoperla* spp., mired bug	-	Pandey, 2002; Pati et al., 2021
Whiteflies	Imidacloprid	Mirid bug (*Dicyphus hesperus*), (*Geocoris* sp), *Encarsia* sp, *Eretmocerus* sp, *Chrysocharis pentheus*	-	Malik et al., 2005; Pati et al., 2021
Aphids	Dimethoate, Metasystox, Phorate, Imidacloprid and Thiamethoxam	-	-	Chandla et al., 2004; Chandel et al., 2008c
Termites	Chlorpyriphos	-	-	Raj & Parihar, 1993
Thrips	Imidacloprid	*Aelothrips* spp. Coccinellids *Galendromus occidentalis*, lacewings	-	Singh et al., 2000; Pati et al., 2021
Red Spider Mite	Clofentezine, Hexythiazox, Etoxazole, Chlorfenapyr, Pyridazinone, Bifenazate	*Galendromus occidentalis* and *Phytoseiulus* mite, *Scolothrips sexmaculatus*, *Stethorus* picipes, *Feltiella acarivora*	-	Mondal et al., 2020

The common pests of okra, viz., white fly and borer, were found to be attacked by *Chrysoperla carnea*, *Geocoris* spp., *Eocanthecona furcellata*, *Coccinellid beetles*, and mired bugs. Likewise, *Solanum lycopersicum* is damaged by various pest populations, mainly leafminers, whiteflies, and many more, which are also killed by their natural enemies, viz., *Macrolophus pygmaeus*, *Nesidiocoris tenuis*, *Macrolophus pygmaeus*, *Nabis pseudoferus*, lady beetles, spiders, etc. The cabbage crop has been majorly destroyed by the *Plutella xylostella* pest, which can be controlled by *Chrysoperla carnea* and coccinellids. Furthermore, eggplant is attacked by the eggplant borer and aphids, which can be regulated by the introduction of lacewigs, mired bugs, ladybird beetles, predatory coccinellids, and many more. The Solanum tuberosum crop is destroyed by moths, leaf minors, white flies, and other minor pests, which can be controlled by the red ant, dragonfly, spider, hoverfly, etc.

The Cucurbitaceae family was damaged by various major and minor pests that have been killed by their natural enemies, viz., *Orius* spp., *Mallada basalis*, *Chrysoperla carnea*, *Stethorus punctillum*, *Oligota* spp., *Aphidoletis aphidimyza*, *Feltiella minuta*, etc. (Pati et al., 2021).

4.1.2. Parasitoids

These are the organisms that derive their nutrition from other organisms. They are quite different from parasites, as parasites don't want the death of their host, but parasitoids feed on the other organism to the point where the host dies due to a lack of sufficient nutrients. This quality of parasitoids can be exploited for the natural management of insects without disturbing the natural flora and fauna (Sanda & Sunusi, 2014). The classification of these parasitoids primarily depends on the presence of the stages attached to the hosts. For example, some species lay eggs with other species' eggs; however, the whole growth stage will be finished within the larvae of each species. They are commonly referred to as egg parasitoids. It's been reported from previous studies that any stage of the life cycle of an insect or bird acts as a parasitoid (Sampaio et al., 2010). The commonly occurring parasitoids of insect pests are T*richogramm*a sp., *Bracon gelechiae*, *Mesochorus* sp., *Eriborus* sp., *Cotesia plutellae*, *Diadromus* sp., *Aphidius colemani*, *Diaeretiella* sp., *Aphelinus* spp., etc. (Pati et al., 2021).

4.1.3. Entomopathogenic Nematodes

The entomopathogenic nematodes showed a symbiotic association with the commonly known entomopathogenic bacteria. The frequency of Entomopathogenic nematodes was found to be 0.7–50% in the land ecosystem (Benseddiq et al. 2020). Entomopathogenic nematodes, as their name suggests, target the insect pest in the field and are responsible for the mortality of the pest within 48–72 hours. The adults of EPN feed inside and reproduce inside the host, giving rise to the first juvenile stage, which moults to the second, and finally to the third, which is a free-living stage and is referred to as an infective juvenile. The infective juvenile showed association with the bacteria, and this is the only stage showing mutual interaction with the bacteria (Grewal et al. 1994). The free-living third infective juvenile enters the insect pest, simultaneously acting as a vector for bacteria. The bacteria, after coming out of the infective juvenile, have produced secondary metabolites, which lead to the mortality of the pest (Dillman et al. 2012). Mutualism benefits both the bacteria and the nematode, and their association becomes toxic for the insect pest (Gaugler & Campbell, 1993; Sharma et al., 2021). The current status reveals the efficacy of Heterorhabditis indica against various pests. This native isolate was found efficacious against Helicoverpa armigera, Spodoptera litura, thrips pest, and many more (Kajol et al., 2020; Srivastava et al., 2022; Istkhar & Chaubey, 2017, 2016). Likewise, the virulence of Steinernema abbasi was shown against the pests Helicoverpa armigera, Spodoptera litura, and thrips (Bhat et al., 2021; Pati et al., 2021). These are the native strains of the Indian subcontinent, which can be isolated from the soil very easily, and a cadaver-based formulation using Galleria mellonella larvae may be used so that it could be available to farmers (Figure 3).

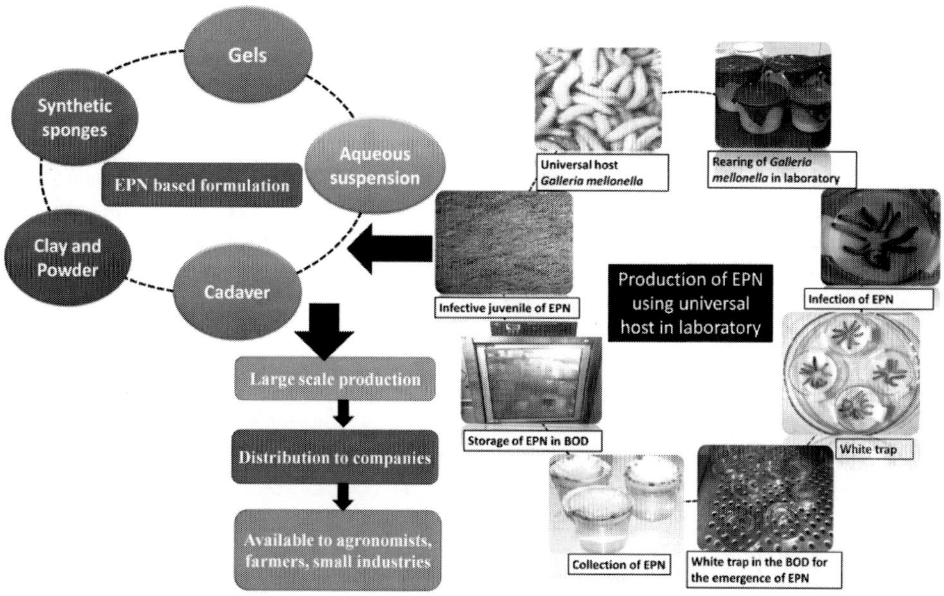

Figure 3. Schematic diagram of production of EPN using the universal host *Galleria mellonella*.

Conclusion

The management of these insect pests' populations is very necessary as their host and geographical ranges increase along with the development of resistance against the chemicals. The exploitation of natural enemies is a safer alternative; however, there are some challenges, as the cost of mass production of a pesticide is very difficult. This problem can be overcome by using native EPN species, which are safer to use, and their formulation using alginate gels, aqueous suspensions, cadavers, etc. increases the shelf life of EPN. In India, this work is in its infancy, but developed countries have already adopted this method for the management of pest populations.

References

Al-Ali, A. S., Al-Neamy, I. K., & Alwan, M. S. (1982). On the biology and host preference of Aulacophora foveicollis Lucas (Coleoptera, Galerucidae). *Zeitschrift für Angewandte Entomologie*, *94*(1-5), 82–86. https://doi.org/10.1111/j.1439–0418.1982.tb02548.x.

Ali, A. T. M. (2001). Biosystematics of phytophagous Scarabaeidae–an Indian overview. *Indian Phytophagous Scarabs and their Management. Present Status and Future Strategies (G. Sharma, YS Mathur and RBL Gupta, eds) Agrobios (India), Jodhpur*, 5–37.

Allahyari, M. S., Damalas, C. A., & Ebadattalab, M. (2016). Determinants of integrated pest management adoption for olive fruit fly (Bactrocera oleae) in Roudbar, Iran. *Crop Protection*, *84*, 113–120.

Altieri, M. A., Nicholls, C. I., & Fritz, M. A. (2005). *Manage insects on your farm: a guide to ecological strategies* (No. Handbook Series Book 7). *Sustainable Agriculture Research and Education (SARE).*

Benseddik, Y., Boutaleb Joutei, A., Blenzar, A., Ezrari, S., Molina, C. M., Radouane, N., ... & Dababat, A. A. (2020). Occurrence and distribution of entomopathogenic nematodes (Steinernematidae and Heterorhabditidae) in Morocco. *Biocontrol Science and Technology*, *30*(10), 1060–1072.

Bhalla, O. P., & Pawar, A. D. (1977). A survey study of insect and non–insect pests of economic importance in Himachal Pradesh. *A survey study of insect and non–insect pests of economic importance in Himachal Pradesh.*

Bhat, A. H., Rana, A., Chaubey, A. K., Shokoohi, E., & Machado, R. A. (2021). Characterisation of Steinernema abbasi (Rhabditida: Steinernematidae) isolated from Indian agricultural soils and their efficacy against insect pests. *Biocontrol Science and Technology*, *31*(10), 1027–1051.

Broad, V. K., & Puri, S. N. (1993). Some studies on the behaviour of whitefly. *Journal of Maharashtra University, 18,* 101–103.

Butani, D. K., & Jotwani, M. G. (1984). Insects in Vegetables. *Colour Publications, Mumbai, India.*

Campbell, J. F., & Gaugler, R. (1993). Nictation behaviour and its ecological implications in the host search strategies of entomopathogenic nematodes (Heterorhabditidae and Steinernematidae). *Behaviour*, 155–169.

Chandel, R. S., & Chandla, V. K. (2003). Managing tuber damaging pests of potato. *Indian Horticulture (India)*, *48*(2).

Chandel, R. S., & Kashyap, N. P. (1997). About White Grubs and Their Management. *Farmer and Parliament XXXVII*, pp. 29–30.

Chandel, R. S., Banyal, D. K., Singh, B. P., Malik, K., & Lakra, B. S. (2010). Integrated management of whitefly, Bemisia tabaci (Gennadius) and potato apical leaf curl virus in India. *Potato Research*, *53*, 129–139.

Chandel, R. S., Chandla, V. K., Verma, K. S., & Pathania, M. (2022). Insect pests of potato in India: biology and management. In *Insect Pests of Potato* (pp. 371–400). Academic Press.

Chandel, R. S., Dhiman, K. R., Chandla, V. K., & Desh, R. (2008a). Insect pests of potato: root and tuber eating pests. *Pestology*, 32, 39–46.

Chandel, R. S., Dhiman, K. R., Garg, I. D., & Desh, R. (2008c). Insect pests of potato e II: sap feeding pests. *Pestology* 32, 47–53.

Chandel, R. S., Sharma, P. C., Verma, K. S., Mehta, P. K., & Vinod, K. (2011). Insect pests of potato e III: leaf eating and defoliating insects. *Pestology*, 35, 60–66.

Chandla, V. K. (1985). Potato pests and their management.34, 31–32.

Chapman, M. A. (2019). Introduction: The importance of eggplant. *The Eggplant Genome*, 1–10.

Clausen, C. P., Clancy, D. W., & Chock, Q. C. (1965). *Biological control of the oriental fruit fly (Dacus dorsalis Hendel) and other fruit flies in Hawaii* (No. 1322). Agricultural Research Service. US Department of Agriculture.

Das, B. B., & Ram, G. (1988). Incidence, damage and carry–over of cutworm (Agrotis ypsilon) attacking potato (Solanum tuberosum) crop in Bihar. *Indian Journal of Agricultural Sciences*, *58*(8), 650–651.

de Meijere, J. C. (1925). *Die larven der Agromyzinen*. Nederlandsche Entomolog. *Vereeniging*.

Dillman, A. R., Chaston, J. M., Adams, B. J., Ciche, T. A., Goodrich–Blair, H., Stock, S. P., & Sternberg, P. W. (2012). An entomopathogenic nematode by any other name. *PLoS Pathogens*, *8*(3), e1002527.

Doharey, K. L. (1983). Efficacy of some insecticides against fruit flies. *Indian Journal of Entomology,* 45, 465–469.

Enkerlin, W. (2003). Economics of area–wide SIT control programs. In: Recent trends on sterile insect technique and area–wide integrated pest management–economic feasibility, control projects,farmer organization and bactrocera dorsalis complex control study. *Research Institute for Subtropics, Naha, Japan*, pp 1–10.

Fletcher, T. B. (1914). Some South Indian Insects. *Superintendent Government Press, Madras.* pp: 565.

Grewal, P. S., Selvan, S., & Gaugler, R. (1994). Nematodes: niche breadth for infection, establishment, and reproduction. *Journal of Thermal Biology 19*, 245–253.

Grieneisen, M. L., & Isman, M. B. (2018). Recent developments in the registration and usage of botanical pesticides in California. *Managing and analyzing pesticide use data for pest management, environmental monitoring, public health, and public policy*, 149–169.

Haider, J., & Rai, A. B. (2021). Emergence of new insect pests on vegetables during the last decade: a case study. *Current Horticulture*, *9*(1), 20–26.

Halder, J., Divekar, P. A., & Rani, A. T. (2021). Compatibility of entomopathogenic fungi and botanicals against sucking pests of okra: an ecofriendly approach. *Egyptian Journal of Biological Pest Control*, *31*(1), 1–7.

Halder, J., Rai, A. B., Kodandaram, M. H., & Dey, D. (2011). Pediobius foveolatus (Crowford): A promising bioagents against Epilachna vigitionctopunctata. In *National conference on Horti–business linking farmers with market Held at Dehradun, 28–31May, 2011.* (2011) Pp–133.

Haq, A. (1962). Notes on the bionomics of Lachnosterna longipennis B1. (Melolonthinae: Coleoptera). *Indian Journal of Entomology*, 24, 220–221.

Isman, M. B. (2020). Commercial development of plant essential oils and their constituents as active ingredients in bioinsecticides. *Phytochemistry reviews*, *19*, 235–241.

Jayasimha, G. T., Rachana, R. R., Manjunatha, M., & Rajkumar, V. B. (2012). Biology and seasonal incidence of leafhopper, Amrasca biguttula biguttula (Ishida)(Hemiptera: Cicadellidae) on okra. *Pest Management in Horticultural Ecosystems*, *18*(2), 149–153.

Kajol, A. H., Bhat, A., & Chaubey, A. K. (2020). Biochemical and molecular characterization of Photorhabdus akhurstii associated with Heterorhabditis indica from Meerut, India. *Pakistan Journal of Nematology*, *38*(1).

Kapoor, V. C., & Agarwal, M. L. (1983). Fruit flies and their natural enemies in India. In *Fruit flies of economic importance. Proceedings of the CEC/IOBC International Symposium, Athens, Greece, Balkema Publications, Rotterdam*, (1983) pp 104–105.

Karkanis, A. C., & Athanassiou, C. G. (2021). Natural insecticides from native plants of the Mediterranean basin and their activity for the control of major insect pests in vegetable crops: shifting from the past to the future. *Journal of Pest Science*, *94*(2), 187–202.

Kfir, R. (1998). Origin of the diamondback moth (Lepidoptera: Plutellidae). *Annals of the Entomological Society of America*, *91*(2), 164–167.

Khurana, S. P., Bhale, U., & Garg, I. D. (2001). Stem necrosis disease of potato. *Technical Bulletin*, *54*.

Kishore, R., & Misra, S. S. (1988). Impact of different planting dates on the incidence of cut worm on potato crop. *Bulletin of Entomological Research* 29, 223–225.

Knapp, S. (2002). Tobacco to tomatoes: a phylogenetic perspective on fruit diversity in the Solanaceae. *Journal of Experimental Botany*, *53*(377), 2001–2022.

Kumar, S., Singh, P. H., Garg, I. D., & Khurana, S. M. P. (2003). Integrated management of potato diseases. *Indian Horticulture (India)*, *48*(2).

Lingappa S, Giraddi RS. (1995). Insect pests of potato and their management. In: Research Highlights on Potato. *Division of Horticulture UAS Dharwad, Karnataka, India*, pp. 19–22.

Liu, S., Wang, X., Guo, S., He, J., & Shi, Z. (2000). Seasonal abundance of the parasitoid complex associated with the diamondback moth, *Plutella xylostella* (Lepidoptera:Plutellidae) in Hangzhou, China. *Bulletin of Entomological Research* 90, 221–231.

Mahr, D. L., Whitaker, P., & Ridgway, N. M. (2008). Biological control of insects and mites: An introduction to beneficial natural enemies and their use in pest management. *University of Wisconsin Cooperative Extension, No. A3842.*

Malik, K., Chandel, R. S., Singh, B. P., & Chandla, V. K. (2005). Studies on potato apical leaf curl virus disease and its whitefly vector Bemisia tabaci. In: Proceedings, *Annual Meeting of Indian Society of Plant Pathologists and Centenary Symposium on Plant Pathology, 7–8 April 2005, India. CPRI, Shimla, India.*

Manjunath, T. M., Bhatnagar, V. S., Pawar, C. S., & Sithanantham, S. (1989). Economic importance of Heliothis spp. in India and an assessment of their natural enemies and host plants. *In Proceedings of the*

Workshop on Biological Control of Heliothis: increasing the effectiveness of natural enemies New Delhi, India, pp. 197–228.

Mat Sulaiman, N. N., Rafii, M. Y., Duangjit, J., Ramlee, S. I., Phumichai, C., Oladosu, Y. et al. (2020). Genetic variability of eggplant germplasm evaluated under open field and glasshouse cropping conditions. *Agronomy* 10(3), 436.

Mehta, P. K., Chandel, R. S., & Mathur, Y. S. ((2008). Phytophagous White Grubs of Himachal Pradesh. *Technical Bulletin: Department of Entomology* CSK HPKV, *Palampur, India.*

Mehta, P. K., Chandel, R. S., & Mathur, Y. S. (2010). Status of white grubs in north-western Himalaya. *Journal of Insect Science (Ludhiana)*, *23*(1), 1–14.

Melomey, L. D., Danquah, A., Offei, S. K., Ofori, K., Danquah, E., & Osei, M. (2019). Review on tomato (Solanum lycopersicum, L.) improvement programmes in Ghana. *Recent advances in tomato breeding and production*, *49*.

Mishra, P. N. (2001). Scarab fauna of Himalayan region and their management. *Indian Phytophagous Scarabs and their Management: Present Status and Future Strategies (G Sharma, YS Mathur and RBL Gupta, eds) Agrobios (India), Jodhpur*, 74–85.

Misra, S. S., & Chandla, V. K. (1989). White grubs infesting potatoes and their management. *Journal of Indian Plant Association*, *16*, 29–33.

Misra, S. S., Chandla, V. K., & Singh, A. K. (1995). Potato Pests and Their Management. Technical Bulletin. 45. CPRI, Shimla, India.

Mondal, B., Mondal, C. K., & Mondal, P. (2020). *Stresses of cucurbits: current status and management* . Springer Singapore, Imprint: Springer. (pp. 115–222).

Oladosu, Y., Rafii, M. Y., Arolu, F., Chukwu, S. C., Salisu, M. A., Olaniyan, B. A., ... & Muftaudeen, T. K. (2021). Genetic diversity and utilization of cultivated eggplant germplasm in varietal improvement. *Plants*, *10*(8), 1714.

Palaniswami, M. S., Pillai, K. S., Nair, R. R., & Mohandas, C. (1995). A new cassava pest in India. *Cassava Newsletter*, *19*(1), 6–7.

Pandey, S. K., & Kang, G. S. (2003). Ecological zones and varietal improvement. *The Potato Production and Utilization in Sub Tropics." (Eds SM Khurana Paul, JS Minhas and SK Pandey)*, 48–60.

Pati, P., Behera, S. K., Raghu, S., & Annamalai, M. (2021). Biological control of insect pests in vegetable crops: An eco–friendly approach. *International Journal of Current Microbiology and Applied Sciences*, *10*, 1358–1373. doi: https://doi.org/10.20546/ijcmas.2021.1001.162.

Pavela, R. (2016). History, presence and perspective of using plant extracts as commercial botanical insecticides and farm products for protection against insects—a review. *Plant Protection Science* 52, 229–241.

Pavlakos, J. G. (1943). The Biology and Control of Aulacophora foveicollis, Luc. in Greece. *Zeitschrift fur Angewandte Entomologie*, *30*(pt. 1).

Pawar, C. S., Bhatnagar, V. S., & Jadhav, D. R. (1986). Host Plants and Natural Enemies of *Heliothis* spp. in India: A Compendium. *International Crops Research Institute for the Semi–arid Tropics, Cropping Entomology, progress report 13, Patancheru, Andhra Pradesh.*

Peralta, I. E., & Spooner, D. M. (2006). History, origin and early cultivation of tomato (Solanaceae). *Genetic improvement of solanaceous crops*, *2*, 1–27.

Puri, S. N., Baranwal, V. K., & Kumar, S. (1995). Cotton leaf curl disease and its management. *NCIPM Extension Folder*, *1*, *National Centre for Integrated Pest Management, New Delhi, India*, (1995).

Rai, A. B., Halder, J., & Kodandaram, M. H. (2014). Emerging insect pest problems in vegetable crops and their management in India: An appraisal. *Pest Management in Horticultural Ecosystems*, *20*(2), 113–122.

Rai, K. M., & Joshi, R. (1988). Control of white grubs (Kurmula) damaging potato crop in U.P. Hills Progress. *Horticulture,* 20, 333–334.

Rana, R. K. (2011). The Indian potato–processing industry: Global comparison and business prospects. *Outlook on agriculture*, *40*(3), 237–243. Regupathy A., Palanisamy S., Chandramohan N., Gunathilagaraj, K. (1997). A Guide on Crop Pests. Sooriya Desktop Publishers, Coimbatore, India.

Robertson, L. D., & Labate, J. A. (2006). Genetic resources of tomato (Lycopersicon esculentum Mill.) and wild relatives. *Genetic improvement of Solanaceous crops*, *2*, 25–75.

Roy, S., & Mukhopadhyay, S. (1980). Epidemiology of a virus disease of pumpkin in West Bengal. Indian *Plant Pathology,* 32(2), 207–209.

Sampaio, M. V., Bueno, V. H. P., Silveira, L. C. P. et al. (2010). Tropical biology and conservation management – Vol. III – *Biological Control of Insect Pests In The Tropics.*

Sanda, N. B., & Sunusi, M. (2014). Fundamentals of biological control of pests. *International Journal of Chemical and Biochemical Science,* 1(6).

Sandur, S. (2004). Implications of diamondback moth control for Indian farmers. Consultant report for the Centre for Environmental Stress and Adaptation Research. *La Trobe University, Victoria, Australia.*

Sarfraz, M., Dosdall, L. M., & Keddie, B. A. (2006). Diamondback moth–host plant interactions: implications for pest management. *Crop Protection, 25*(7), 625–639.

Sharma, H., Rana, A., Bhat, A. H., & Chaubey, A. K. (2021). Entomopathogenic Nematodes: Their Characterization, Bio–Control Properties and New Perspectives. *Nematodes–Recent Advances, Management and New Perspectives*, 187–210.

Shinde, H. J., Desai, U. T., Masalkar, S. D., & Choudhari, S. M. (1994). Efficacy of plant growth regulators to control vine length in watermelon. *Journal of Maharashtra Agricultural University,* 19(1), 150–151.

Shooker, P., Khayrattee, F., & Permalloo, S. (2006). Use of maize as a trap crops for the control of melon fly, B. cucurbitae (Diptera: Tephritidae) with GF–120. *Bio–control and other Control Methods.*

Shylesha, A. N., Jalali, S. K., Gupta, A., Varshney, R., Venkatesan, T., Shetty, P.,... & Raghavendra, A. (2018). Studies on new invasive pest Spodoptera frugiperda (JE Smith) (Lepidoptera: Noctuidae) and its natural enemies. *Journal of Biological control, 32*(3), 1–7.

Siegwart, M., Graillot, B., Blachere Lopez, C., Besse, S., Bardin, M., Nicot, P. C., & Lopez–Ferber, M. (2015). Resistance to bio–insecticides or how to enhance their sustainability: a review. *Frontiers in plant science, 6*, 381. https ://doi.org/10.3389/fpls.2015.00381.

Singh, S. P. (1987). Studies on some aspects of the bioecology of potato cutworms in India. *Journal of Soil Biology and Ecology, 7*, 135–143.

Singh, S. P. (2004). Some Success Stories in Classical Biological Control of Agriculture pest in India. *APAARI, Bangkok Thailand.*

Sithanantham, S., Bhatnagar, V. S., & Jadhav, D. R. (1982). Some aspects of Trichogramma spp. in eggs of *Heliothis armigera* (Hb.), (Lepidoptera: Noctuidae).

Soyk, S., Müller, N. A., Park, S. J., Schmalenbach, I., Jiang, K., Hayama, R., ... & Lippman, Z. B. (2017). Variation in the flowering gene SELF PRUNING 5G promotes day–neutrality and early yield in tomato. *Nature Genetics, 49*(1), 162–168.

Srivastava, K. P., & Butani, D. K. (1998). Pest Management in Vegetables. *Part I Periodicals and Book Publishing House, El Paso, Texas.*

Srivastava, S., Heena, & Chaubey, A. K. (2022). Studies on biological control potential of Steinernema abbasi CS–39 against Helicoverpa armigera (Hub.) (Lepidoptera: Noctuidae) by adopting the regression probit model. *Egyptian Journal of Biological Pest Control, 32*(1), 9.

Sushil, S. N., Mohan, M., Selvakumar, G., & Bhatt, J. C. (2006). Relative abundance and host preference of white grubs (Coleoptera: Scarabaeidae) in Kumaon hills of Indian Himalayas. *Indian Journal of Agricultural Sciences, 76*(5), 338–339.

Talekar, N. S., & Shelton, A. M. (1993). Biology, ecology, and management of the diamondback moth. *Annual Review of Entomology, 38*(1), 275–301.

Thontadarya, T. S., Hanumanna, M., & Rao, K. J. (1978). Record of Trichogramma australicum Girault on the eggs of Heliothis armigera (Hubner) infesting bhendi (Abelmoschus esculentus {H.} Moench). *Current research journal of biological sciences,* 7(12):206.

Trivedi, T. P., & Rajagopal, D. (1999). Integrated pest management in potato. In: Upadhayay, R. K., Mukeriji, K. G., Dubey, O. P. (Eds.), IPM System in Agriculture, Cash Crops, vol. 6. *Aditya Books Pvt. Ltd., New Delhi, India*, pp. 299–313.

Vanisree, K., Rajasekhar, P., Rama Subba Rao, V., & Srinivasa Rao, V. (2005). Seasonal incidence of pumpkin caterpillar, Diaphania indica (Saunders) on cucumber in Krishna–Godavari Zone. *Journal of Plant Protection Environment,* 2, 127–129.

Veersh, G. K., Kumar, A. R. V., & Musthak Ali, T. M. (1991). Biogeography of pest species of white grubs of Karnataka. In: Veersh, G. K., Rajagopal, D., Viraktamath, C. A. (Eds.), Advances in Management and

Conservation of Soil Fauna Oxford and IBP. *Publishing Company Pvt Ltd., Bangalore, India*, pp. 191–198.

Walia, S., Saha, S., Tripathi, V., & Sharma, K. K. (2017). Phytochemical biopesticides: some recent developments. *Phytochemistry reviews*, *16*, 989–1007.

Willcox, J. K., Catignani, G. L., & Lazarus, S. (2003). Tomatoes and cardiovascular health.1–18.

Yadav, D. N., Patel, R. C., & Patel, D. S. (1985). Impact of inundative releases of Trichogramma chilonis Ishii against Heliothis armigera (Hbn.) in Gujarat (India). *Journal of Entomological Research,* 9, 153–159.

Zehnder, G. (2011). Biology and management of pickleworm and melonworm in organic curcurbit production systems. *Electronic.*

Chapter 5

Major Insect Pests of Orchard Crops and Their Management

Nisha Rana[1]
Aasha Rana[2,*]
and Barkat Hussain[3]
[1]Department of Zoology, KVSCOS; Research & Development Cell, Swami Vivekanand Subharti University, Meerut, India
[2]Department of Zoology, Faculty of Basic & Applied Sciences Madhav University, Pindwara (Sirohi), Rajasthan, India
[3]Division of Entomology, Sher-e-Kashmir University of Agricultural Sciences of Technology of Kashmir, Jammu and Kashmir, India

Abstract

Numerous insect pests damage orchard crops, resulting in significant losses every year. Among these pests, several are well-known and cause economic harm. These insects significantly reduce orchard productivity, and to manage these insect pests, orchard agricultural practices are mainly managed by chemical pesticides and fertilisers, which lead to serious and long-term environmental impacts, contamination of soil, environmental health hazards, and biodiversity loss. Reducing the number of pesticides used during orchard culturing is a global environmental concern since it helps to preserve and increase beneficial microbial diversity in the orchard ecosystem. Orchard systems offer favourable conditions to maintain food webs for microbial communities within the agrosystem. In apple, apricot, ber, pomegranate, mango, sapota, citrus, and peach orchards, researchers have examined the significant economic insect pests and their impacts on plant manipulation growth and their components, as well as the chemicals and bio-control techniques of control. For better biological control, they feel that further study is necessary to identify the processes involved in various metrics. The orchard system could be restructured to optimise ecosystem benefits and maintain biodiversity by exploring soft management strategies.

Keywords: orchard crops, fruits, insect pests, microbial diversity, biological control

---

* Corresponding Author's Email: aasha.aasharana@ymail.com; aasha.rana@madhavuniversity.edu.in.

In: Insects as Humans' Frenemies: From Friends to Foes
Editor: Istkhar Rao
ISBN: 979-8-89530-320-7

1. Introduction

In addition to being essential for enhancing nutrition, agricultural production serves as many people's primary source of income. Raising crop yields is essential for both social and economic advancement as well as the eradication of hunger. Crop production has drastically transformed in the last few decades on a global scale. In many nations across the world, there has also been an increase in the diversity of regimes. In the past, the main agricultural products were grains, roots, and other basic crops. Legumes, fruits, vegetables, nuts, seeds, and other foods have all been included in this. International trade has also greatly increased agricultural production. Food production was traditionally localised worldwide, with farmers producing food for their households or communities. The food trade was primarily restricted to domestic markets, even as it expanded. India is a nation based on agriculture, with the agricultural sector employing the bulk of the population and contributing to approximately 16.1% of India's GDP (Agricultural Department of India, CMIE reports, Cooperative Sugar, 2019). Farmers cultivate a variety of crops, including food crops (wheat, maize, rice, millets, and pulses, for example); commercial crops (sugarcane, tobacco, cotton, jute, and oilseeds, for example); vegetable crops (tomato, potato, okra, cabbages, and brinjal, for example); plantation crops (coffee, coconut, tea, and rubber, for example); and orchards (apple, ber, sapota, apricot, pomegranate, citrus, mango, etc.). Several pests affect the growth, yield, and quality of these food crops. As per the definition, a pest may be "any organism that occurs in large numbers, is noxious, destructive, troublesome to humans, and conflicts with his welfare or interests, convenience, profit, or is likely to do so." Pests include insects, nematodes, mites, snails, slugs, fungi, bacteria, rodents, birds, and others that cause economic damage by harming or terminating crops and spoiling stored items. Insects are the main rivals of humans for food and fibre among them. They lower the amount and quality of food, which results in enormous losses for the entire world. Insects affect all the parts of the plant, including the stem, root, shoot, leaves, bark, fruit colour, size, weight, etc. They have been found to have increased growth problems and have to bear tremendous economic losses (Khursheed et al., 2021). Each year, harmful insect pests drastically lower agricultural output; thus, control of their population is required. It is predicted that 78% of the fruit, 54% of vegetables, and 32% of cereal crops will vanish if not managed properly and timely (Lamichhane, 2017; Tudi et al., 2021). Agrochemicals play a big role in agricultural productivity and disease management. The agriculture industry uses 3 million metric tonnes of technical-grade pesticides per year, costing $40 billion, out of which only 0.1% of the pesticide volume is consumed by the target pest (Bernardes et al., 2015). 99.9% of these remain as pesticide residues, which build up in the environment and pollute it, thereby causing considerable damage to the biosphere (Carriger et al., 2006). Large amounts of pesticides are absorbed by the soil, causing severe soil pollution that degrades some soil properties where pesticides adhere to soil particles due to the application of chemical pesticides and fertilisers (Bošković et al., 2020). Food quality has become a major concern, and global demand for fruits and associated by-products has surged as a result of growing awareness of the nutritious value of horticultural produce in the human diet. Today advances in horticulture crop production technologies have considerably contributed to addressing the increased demand for orchard crops due to the growing human population and animals.

2. Horticulture Crops

One of the biggest groups of blooming plants, orchards, are found all over the world. 29,199 species have been recognised as of this writing, and approximately 31,000 species are thought to exist overall, with several hundred new species names being published annually (Hinsley et al., 2017). Orchids have long been marketed as beautiful plants in the horticultural and floricultural industries, appealing to a wide range of markets and consumer demographics. For many generations, orchards have also been an integral part of communities, serving as a hub, a place for people to congregate, and a place where humans and the natural world collaborate to produce bountiful harvests that supply fresh fruit well in advance of the global freight era. Because they offer a plethora of alternatives for crop diversification, horticulture crops have been instrumental in helping the farming community increase their revenue over the past 20 years. India, with 2.5% of the global land area and 16% of the global population, has also documented the relevance of horticulture crops as a nutritional and national food security engine. India ranked second in the world in fruit output, producing 88.98 million metric tonnes from 7.21 million hectares (Anonymous, 2017). Horticulture is one of the main sources of revenue generation in India (Kimothi et al., 1997). It is much higher than food grain production (Anonymous, 2017). India produces a wide range of fruits, the most important of which are bananas, mangoes, citrus, papaya, apple, guava, pomegranate, grape, sapota, pineapple, and litchi. Maharashtra is leading in India's fruit basket, contributing 15.12%, followed by Andhra Pradesh (11.81%), Gujarat (8.99%), Tamil Nadu (8.28%), Uttar Pradesh (7.74%), Karnataka (7.47%), Madhya Pradesh (6.40%), Telangana (4.99%), and Bihar (4.51%), respectively. It is expected that field functionaries and forward-thinking farmers will raise awareness about correct insect pest identification and prompt pest management solutions. Horticulture has an important role in increasing land productivity, creating jobs, increasing exports, and improving the economic conditions of farmers and entrepreneurs, hence providing adequate nutrition and food to the population (Usha & Singh, 2013). The expenditures of cultivating fruit have surpassed the benefits throughout the last 30 years (Clayton, 1989). One of the most significant temperate fruit crops in the world is apples (*Malus domestica*). Apple manufacturing contributes significantly to rural economies by directly or indirectly creating jobs and raising living standards for millions of people (Abdessemed et al., 2022).

3. Insect Pests in Orchards

The main culprits behind crop damage and productivity loss are pests, including insects. They might devour a substantial amount of grain and destroy the entire crop. If left untreated, they can lower crop output by 30–50% (%) per year. For instance, production losses of up to 5% are common for perennial crops like apples and stone fruits in the Netherlands, but losses of up to 45% might happen in a coffee crop in Brazil. One of the special strategies to avoid getting sick is to control pests and insects (Manosathiyadevan et al., 2017). Many main food crops can have yield losses of up to 70% in the absence of crop protection; weeds cause the most losses at 30%, followed by animal pests and diseases at 23% and 17%, respectively (Gordon, 2023). According to the FAO, pests cause 20–40% of output losses globally on average, costing the economy $290 billion in total (FAO, 2024).

Table 1. Insect pests of orchards, their effects on plant parts, and their chemical and biological control measure

S. No.	Insect Pest Name	Scientific Name	Infective stage	Infected orchard	Infected part	Chemical Control	Bio-management Measures
1	Apple aphid	*Aphis pomi,*	Nymph & adult	Apple and other pome fruits	Mostly terminal branches, leaves, young branches	Thioclorpid	-
2	Codling Moth	*Cydia pomonella*	larva	Apple and other pome fruits	Fruit, seed	Dimethoate, Imidacloprid, Cypermethrin, chlorpyriphos	-
3	Comstock (San Jose Scale)	*Quadraspidiotus perniciosus*	Crawlers	pome fruits and Apple		Chlorpyriphos, Qunalphos, H.M.O., Cypermethrin	-
4	Woolly apple aphid	*Eriosoma lanigerum*	Nymphs and adults	Apple	Mostly terminal branches, leaves, young branches and roots	Dimethoate, Imidacloprid, Cypermethrin, chlorpyriphos	-
5	European red mite	*Panonychus ulmi*	Immatures & adults	Apple, pear and other fruit crops	leaves	Oberon, Fluvalinate	-
6	Oriental Fruit Moth	*Grapholita molesta*	larva	Apple, Cherry	Shoot, fruit	Dimethoate, Imidacloprid, Cypermethrin, chlorpyriphos	green lacewings (Chrysopidae)
7	Ber fruit fly	*Carpomyia vesuviana*	larva	Ber	Fruit	-	-
8	Ber stone weevil	*Aubeus himalayanus*	-	-	-	-	-
9	Ber butterfly	*Tarucus theophrastus*	-	-	-	-	-
10	Ber fruit borer	*Meridarchis scyrodes*	-	-	-	-	-
11	Ber weevil	*Myllocerus dentifer, M. blandus & Amblyrrhinus poricollis*	-	-	-	-	-
12	Tea Mosquito bug	*Helopeltis antonii*	nymphs and adults	Guvava	fruit	Malathion	-
13	Citrus leaf miner	*Phyllocnistis citrella*	larval	Citrus	leaves	Imidacloprid Fenvalerate	Neem oil, Coccinellids, Chrysopids predators and eulophid parasitoids
14	Citrus psylla	*Diaphorina citri*	nymphs and adults	Citrus	Leaves and flowers	quinalphos or novaluron or thiamethoxam	Coccinellids, Chrysopids, Eulophids, *Mallada desjardensi*

S. No.	Insect Pest Name	Scientific Name	Infective stage	Infected orchard	Infected part	Chemical Control	Bio-management Measures
15	Citrus blackfly	*Aleurocanthus woglumi*	adults and nymphs	Citrus	Effect entire plant	imidacloprid or acephate or phosalone or dimethoate or novaluron	Neem oil
16	Citrus whitefly	*Dialeurodes citri*	adults and nymphs	Citrus	Effect entire plant	imidacloprid or acephate or phosalone or dimethoate or novaluron	Neem oil
17	Citrus aphid	*Toxoptera aurantii*	adults and nymphs	Citrus	Leaves and young shoots	Dimethoate or acephate or imidacloprid	-
18	Fruit sucking moth	*Eudocima maternal*	Adults	Citrus	Fruit	Malathion + jaggary + mandarin juice	Neem oil
19	Citrus rust mite	*Phyllocoptruta oleivora*	Adult	Citrus, Grapes,	Fruit	wettable sulfur or dicofol or propargite or ethion	-
20	Mango hopper	*Idioscopus clypealis, I.nitidulus and Amritodus atkinsoni*	Nymphs, adults	Mango	fruits	cypermethrin or fenvelerate or profenophos or chlorpyriphos or carbaryl or profenophos imidachloprid or thiamethoxam diamethoate or Lambda cyhalothrin	*Mallada boninensis, Chrysopa lacciperda*, egg parasite, *Polynema* spp. *Gonatocerus* sp. *Tetrastichus* sp. and fungus, *Verticillium lecanii*
21	Mealybug	*Drosicha mangiferae*	Nymphs and adults	Mango	Fruits and leaves	Carbaryl	*Menochilus sexmaculatus, Rodolia fumida, Sumnius Renardi*
22	Thrips	*Caliothrips indicus, Rhipiphorothrips cruentatus and Scirtothrips dorsalis*	larva	Mango	Inflorescence, leaves, fruits	Dimethoate or spinosad or thiamithoxam	predatory mites (e.g., *Amblyseius* spp.) pirate bugs (*Orius* spp.), ground beetles, lacewings, hoverflies, and spiders
23	Stem borer	*Batocera rufomaculata*	grub	Mango	stem	DDVP or kerosene/ petrol or Carbaryl	-
24	Shoot borer	*Chlumetia transversa*	larva	Mango	Shoot and leaves	Quinalphos	-
25	Fruit fly	*Bactocera dorsalis*	Maggot & adults	Mango	Fruit	imidacloprid dimethoate	-

Table 1. (Continued)

S. No.	Insect Pest Name	Scientific Name	Infective stage	Infected orchard	Infected part	Chemical Control	Bio-management Measures
26	Termites	*Odontotermes* sp	Workers termites	Mango	stem	Malathion Chlorpyriphos	Ground mahua cake
27	Shoot gall psylla	*Apsylla cistellata*	Shoot and nymph	mango	Shoo, leaves	thiomithoxam or quinalphos	-
28	Inflorescence midge	*Erosomyia indica*	Larva	mango	Inflorescence	thiomithoxam dimethoate	-
29	Bark-eating caterpillar	*Indarbela quadrinotata*	Larva	mango	Stem	monocrotophos or DDVP	-
30	Fruit fly	*Bactrocera dorsalis*	Larva	Mango, peach, guava	fruits	Chlorpyriphos, methyl eugenol traps	-
31	Fruit borer	*Deudorix Isocrates*	Larva	Pomegranate	fruits	Deltamethrin, malathion, methomyl	neem formulations
32	Thrips	*Scirtothrips dorsalis* and *Rhipiphorothrips cruentatus*	Nymph and adults	Pomegranate	Leaves and fruit	Acetamiprid, chlorpyrifos or imidacloprid	-
33	Stem borer	*Coelosterna spinator* and *Zeuzera* sp.	Grub	pomegranate	Stem	fenvalerate or dichlorvos or quinalphos or chlorpyrifos	-
34	Bark-eating caterpillar	*Indarbela* sp.	Larva	pomegranate	Stem	quinalphos or fenvalerate or carbaryl or dichlorovos	-
35	Fruit piercing/sucking moth	*Othreis fullonia, O. materna* and *O. homaena*	adults	pomegranate	Fruit	95% molasses/jaggery + 5% malathion	Egg parasitoids, *Telenomus* sp., *Ooencyrtus* sp. and *Trichogramma* sp
36	Bud borer	*Anarsia achrasella*	Larva	Sapota	Bud and flower, fruit	polytrin C44 (profenphos and cypermethrin) or Emamectin benzoate	azadirachtin
37	Seed borer	*Trymalitis margarias*	Larva	Sapota	Fruit	Deltamethrin, Lambda cyhalothrin or Profenofos or Indoxacarb or Novaluron	-
38	Leaf webber	*Nephopteryx eugraphella*	Larva	Sapota	Buds, leaves	Quinolphos	
39	Fruit flies	*Bactrocera dorsalis*	Larva	Sapota, Guava	Fruit	chlorpyriphos	

S. No.	Insect Pest Name	Scientific Name	Infective stage	Infected orchard	Infected part	Chemical Control	Bio-management Measures
40	Leaf miner	*Acrocercops gemoniella*		Sapota	leaves	methyl demeton	
41	Mealy bug	*Rastrococcus iceryoides*	Nymph, adults	Sapota	leaves, inflorescence, branches, fruits	malathion or buprofezin	predators, *Rodolia fumida,Menochilus sexmaculatus, Sumnius renardi* and Australian lady bird beetle, *Cryptolaemus montrouzieri*
42	Hairy caterpillar	*Metanastria hyrtaca*	Larva	Sapota	Tree	cypermethrin	

Because of environmental and technological changes, insect pest problems in agriculture have shifted significantly over the first decade of the twenty-first century (Dhaliwal, 2010). The growing global trade in agriculture increases the possibility of the introduction of exotic pests. There is an impending shift in pest status due to changes in cropping patterns, indiscriminate pesticide usage, climate change, and the introduction of input-intensive, high-yielding types, hybrids, and cultivars. Many pests have spread their host range, developed pesticide resistance, and emerged as important pests (Rathee, 2018). In the Indian states of Jammu & Kashmir and Himachal Pradesh, the San Jose scale has a diverse host range that includes apples, forest trees, and ornamental plants. Table 1 lists some of them, some of which are important economic and unpleasant insect pests, while others are beneficial weed biological control agents (Parker, 2017). In Jammu & Kashmir, San Jose scale and European red mites have become severe pests of apple trees (Bhalla & Gupta, 1993).

4. Biological Control in Orchards' Pest Management

Plant diversity within agricultural systems and the strength of resources for floral and faunal communities act as food and feasts for natural enemies and honeybees (Boller et al., 2004). Moreover, the control of numerous pests of fruit, viz., aphids, mites, leaf miners, bugs, caterpillars, and beetles, is controlled by the natural enemies (Boller et al., 2004). Natural enemies are more prevalent towards the outskirts of hedgerows than in the centre, and aphid abundance is lowest along orchard edges (Altieri & Schmidt, 1986). Debras et al. (2008) established that natural enemies actively relocate from the hedgerow to the orchard in response to prey availability, even though cultural practices (including pesticide usage) modify their functioning pattern in most orchards. As biological control agents, birds generally avoid parasitised prey and may play an important role in pest management in orchard systems (Jones et al., 2005). The natural environment of orchards aids in the application of various management options that provide suitability for biological control (Brown 2008). Insectivore birds such as the Great Tit *Parus major* (Paridae) have reproduction rates equal to those in pesticide-free forests (Bouvier et al., 2005). Coccinellids are a significant group of natural enemies of numerous horticultural and agricultural crop pests, including scale insects, mealy bugs, aphids, and mites (Dixon, 2000). Natural enemies of mites have been shown to play an important role in apple orchards (Solomon et al., 2000). Coccinellid beetles, predatory mites, *Amblyseius fallacies* (Garman), and *Stethorus aptus* (Kapur) are also recognised as predators of mites (Govindasamy & Khursheed, 2018; Khursheed et al., 2015). Hoverflies, the most important aphid predators, are *Eristalis* sp. and *Syrphus* sp., which are essential for aphid management because they efficiently regulate the prey population (Symank et al., 2008; Khan et al., 2016). Spiders are opportunistic and polyphagous and act as predators of orchard pests in agroecosystems (Birkhofer et al., 2008; Lefebvre et al., 2017; Whitney et al., 2018). Although the role and function of spiders in biological control in orchards have received little attention (Michalko et al., 2019), few natural enemies of insect pests, especially parasitoid wasps, revolutionised the different ecosystems to manage the dreadful insect pests. *Aphelinus mali* (Haldeman) is widely regarded as one of the most significant biocontrol agents of the woolly apple aphid on a global scale. Another two parasitoids, *Aphytis proclia* (De Bach) and *Encarsia perniciosi* (Tower), are effective San Jose scale parasitoids. Parasitoids are the most numerous, diversified, and significant group of natural enemies of apple and pear insect pests, such as *A. mali* to combat the woolly apple aphid, *E. lanigerum* (Cross et al., 1999). The mass production of egg parasitoids, especially *Trichogramma* species, has revolutionised different ecosystems by managing a varied number of insect pests on a variety of crops.

5. Pesticides and Orchards' Pest Management

Pests and illnesses can linger in the orchard all year round due to their host-tree persistence. This encourages an increase in infestation or infection levels from year to year, necessitating ongoing protection—that is, the frequent application of pesticides to keep them under control. Compared to other crops, fruit trees demand significantly higher pesticide dosages and require more intensive care. The major way of application of pesticides in pest management in orchards is spray (Simon et al., 2010). A comprehensive description of pesticides used in pest management is given in Table 1. While some pesticides function as growth regulators or endotoxins, many others affect the neurological system of the insect (cholinesterase inhibition, for example). According to USDA statistics from 2001, 12 per cent of the pesticides used on the crops under study were insecticides (Insecticides 2024, US EPA). Using pesticides safely and effectively is essential for controlling pests and safeguarding the environment. As pesticides are designed to kill pests, which are biological forms, their traits are dangerous for people, animals, and nature. Enhancing the application of pesticides can prevent them from contaminating surfaces and groundwater. Growers have options to lessen the effects of pesticide use at every step, from choosing which pesticide to apply to choosing how much to use. Pesticide residues in fruits changed customers' perceptions of non-contaminated and healthy food, leading to the establishment of zero residue programmes in several nations (Berrie & Cross, 2006). Hence, there is growing and challenging concern in the current scenario about providing consumers with good fruits and maintaining pests below certain economic limits using other green technologies (Hussain et al., 2015). They contribute to the soil's fertile quality, control of pests, and crop production (Zhang et al., 2007). However, it is a time-consuming practice that increases the strength of biodiversity.

6. Orchard Systems and Insects

The system of orchards and their edges are extremely important in terms of contributing plant and animal variety across numerous scales. However, such potentially advantageous conditions must be avoided: Extensive application of the mating disruption strategy to control Lepidopteran insect pests necessitates large surface areas devoid of interplant hedgerows (Witzgall et al., 2008; Hussain et al., 2015). Neumann (1999) showed that the effectiveness of continual surface coverage of homogeneous orchards protected and favoured pheromone diffusion, saving a lot of produce from pest incursions. According to one study (Ricci et al., 2009), the codling moth, *Cydia pomonella*, populations in an orchard are adversely connected with the adjacent areas that are cultivated with apples and administered with chemicals that boost apple development within wide surface areas of the codling moth. Fruit tree security is time-consuming and requires more insecticide than other crops. Current apple production systems in fruit-producing countries rely heavily on pesticides (Eurostat, 2002), which contribute to the emergence of strains that are resistant to certain pests (Sauphanor et al., 2000; Reyes et al., 2008). Climate change increases voltanism and the probability of greater pest assaults (Sauphanor, 2004), as well as the introduction of novel pests (Ahmad et al., 2022). The impact of pest management regimes and pesticides on arthropods is well-studied in orchards (Bogya & Markó, 1999; Pekár, 1999; Bogya et al., 1999; Miliczky et al., 2000; Brown et al., 2003) and ground-living beetles (Pearsall & Walde, 1995; Labrie et al., 2003). However, it is less frequently researched for the entire arthropod group (Sauphanor et al., 1993, 2005; Suckling et al., 1999; Brown & Schmitt, 2001; Debras et al., 2006; Simon et al., 2007b). Pesticides harm hunting spiders (Pekár, 1999), ground-living arthropods (Epstein et al., 2000), and insect-parasitising leaf miners (Prokopy et al., 1996). However, as compared to ecologically friendly approaches, the application of broad-spectrum insecticides has little or very little effect on the total arthropod diversity or richness of the tree canopy (Suckling et al., 1999; Brown & Schmitt, 2001; Simon et al., 2007b). Birds are bioindicators of the environmental impact of cultural practices (Ormerod & Watkinson, 2000). According to an analysis of 143 food webs, they are important markers of global change (Furness & Greenwood, 1993). The research on the impact of pest management measures on birds in apple orchards is still in its early stages. Most research in Northern America (Powell, 1984; Fluetsch & Sparling, 1994; Bishop et al., 2000) and Europe (Bouvier et al., 2005) focuses on the breeding rate of passerine birds, which is lower in carefully managed orchards compared to those that are organic. Pesticides' impact on bird communities has received less attention. Bird variety and abundance were higher in organic German orchards than in IPM (Integrated Pest Management) of orchard pests (Rösler, 2003).

Conclusion

The present information on insect pests poses a greater threat to high-density orchards. Fortunately, many species of predators and parasitoids were discovered in high orchards connected with practically all pests, although pesticides should be avoided. As a result, more research on pesticide selection and use is required to minimise their negative impact on natural enemies. Furthermore, conservation efforts must be made to improve beneficial arthropod populations, such as biological control agents, in high-density agroecosystems. The safety of humans and biodiversity is a growing concern in a sustainable way in the current scenario. Insects are a noxious group that causes tremendous damage to various crops throughout the world. They are the major rivals of humans for fibre and food. Despite their use, approximately 52,000 metric tonnes of technical-grade pesticides are being employed by humans annually, which cause loss and pose a threat to the biodiversity of both soil flora and fauna. There are several ways to apply insecticides more effectively without compromising effective pest management. Certain practices, such as alternating rows or spot spraying, might lower hazards and perhaps save costs. Improved pesticide management and procedures may lessen chemical exposures on farms, which can cause long-term, chronic health issues. A complete reduction of pesticides is crucial for the implementation of sustainable and organic agricultural practices, especially when orchard systems are completely based on pesticides. Due to the increasing population of pests and diseases affecting orchard systems, it is essential to strengthen biological control agents to avoid insect population growth and improve conservation programmes for biological control.

References

Abdessemed, A., Fellak, A., Abdessemed, A. & Awais Khan A. (2022). Status, challenges and opportunities for apple production in Eastern Algeria. *Horticultural Science*, 49(3):147-153 | doi: 10.17221/103/2021. https://www.ishs.org/ishs-article/240_11.

Ahmad, R., Ahmad, T. A., & Hussain, B. (2022). Report of guava fruit fly Bactrocera correcta Bezzi from temperate Kashmir. *Indian Journal of Entomology*, 1-3. https://doi.org/10.55446/ije.2022.533.

Ahuja, D. B. & Chattopadhyay, C. (2015). Pests of fruit trees (Citrus, Banana, Mango, Pomegranate and Sapota) E-Pest surveillance and pest management. Advisory, publishers ICAR-National Research Centre for Integrated Pest Management, New Delhi and State Department of Horticulture, Commissionerate of Agriculture, Pune (Maharashtra): 124.

Altieri M. A. & Schmidt L. L. (1986). The dynamics of colonizing arthropod communities at the interface of abandoned, organic and commercial apple orchards and adjacent woodland habitats, *Agriculture, Ecosystems & Environment*, *16*, 29–43.

Anonymous (2017). Horticultural Statistics at a Glance 2017, Horticulture Statistics Division, Department of Agriculture, Cooperation & Farmers Welfare, Ministry of Agriculture & Farmers Welfare, Government of India, pp. 481.

Bernardes, M. F. F., Pazin, M., Pereira, L. C., & Dorta, D. J. (2015). *Impact of pesticides on environmental and human health*. In: Toxicology studies-cells, drugs and environment. Intech Open London, UK, 195–233. doi: 10.5772/59710.

Berrie, A., & Cross, J. (2006). Development of an integrated pest and disease management system for apples to produce fruit free from pesticide residues - aspects of disease control, Bulletins of IOBC/wprs 29, 129–138.

Bhalla, O. P. & Gupta, P. R. (1993). Insect pest of temperate fruits. Advances in Horticulture - Fruit Crops Malhotra Publishing House, New Delhi, India. 3:1557-1589.

Birkhofer, K., Gavish-Regev, E., Endlweber, K., Lubin, Y. D., Von Berg, K., & Wise, D. H. (2008). Cursorial spiders retard initial aphid population growth at low densities in winter wheat. *Bulletin of Entomological Research*, *98(3)*,249-255.

Boller, E. F., Häni F., & Poehling H. M. (2004). Ecological infrastructures: idea book on functional biodiversity at the farm level, Landwirtschaftliche Beratungszentrale Lindau, Lindau, Suisse.

Bošković, N., Brandstätter–Scherr, K., Sedláček, P., Bílková, Z., Bielská, L., & Hofman, J. (2020). Adsorption of epoxiconazole and tebuconazole in twenty different agricultural soils about their properties. *Chemosphere* 261, 127–637. doi: 10.1016/j.chemosphere.2020.127637.

Bouvier, J. C., Toubon, J. F., Boivin, T., & Sauphanor, B. (2005). Effects of apple orchard management strategies on the great tit (*Parus major*) in Southeastern France, *Environmental Toxicology and Chemistry*, 24, 2846–2852.

Brown, M. W. (2008). Sustainable biocontrol of apple insect pests. Pest Technology 2(2):98-103.

Carriger, J., Rand, G. M., Gardinali, P. R., Perry, W. B., Tompkins, M. S., and Fernandez, A. M. (2006). Pesticides of potential ecological concern in sediment from south Florida Canals: An ecological prioritization for aquatic arthropods. *Soil and Sediment Contamination*, *15*, 21–45. doi: 10.1080/15320380500363095.

Chaudhari, K., Nishant, N., Upadhyay, G., More, R., Singh, N., Vyas, S. P., & Bhattacharya, B. K. (2019). Crop inventory of orchard crops in India using remotely sensed data. The International Archives of the Photogrammetry, Remote Sensing and Spatial Information Sciences, Volume XLII-3/W6, 2019ISPRS-GEOGLAM-ISRS Joint Int. Workshop on "Earth Observations for Agricultural Monitoring," 18–20 February 2019, New Delhi, India. doi: 10.5194/isprs-archives-XLII-3-W6-269-2019.

Clayton-Greene, K. A. (1989). Orchard Design. *Acta Horticulturae*, *240*, 71-82, doi: 10.17660/*ActaHortic*.1989.240.11, https://doi.org/10.17660/ActaHortic.1989.240.11.

Cross, J. V., Solomon, M. G., Babandreier, D., Blommers, L., Easterbrook, M. A., & Jayl, C. N. (1999). Biocontrol of pests of Apples and Pears in Northern and Central Europe: 2. Parasitoids. *Biocontrol Science and Technology*. *9*, 277-314.

Das, A. K., Rao, C. N. Ahuja, D. B. Mane, S. S. Dadmal, S., Bhosle, B. B. Jhanwar, P. R. Surywanshi A. P. and Bhede B. V. (2015). Major insect pests and diseases of citrus and their management. In: *Pests of fruit trees (Citrus, Banana, Mango, Pomegranate and Sapota) E-Pest surveillance and pest management.* (D. B. Ahuja & C. Chattopadhyay Eds.). Advisory, publishers ICAR-National Research Centre for Integrated Pest Management, New Delhi and State Department of Horticulture, Commissionerate of Agriculture, Pune (Maharashtra): 124.

Debras, J. F., Senoussi, R., Rieux, R., Buisson, E., & Dutoit, T. (2008). Spatial distribution of an arthropod community in a pear orchard (southern France), Identification of a hedge effect, *Agriculture, Ecosystems & Environment*, 127,166–176.

Dhaliwal, G. S., Jindal, V., & Dhawan, A. K. (2010). Insect pest problems and crop losses: changing trends. *Indian Journal of Ecology*, *37(1),*1-7.

Dixon, A. F. G. (2000). Insect predator-prey dynamics, ladybird beetles and biological control. Cambridge University Press, London.

FAO (2024). FAO Home. https://www.fao.org.

Gordon, A. (2023). Yield losses due to pests. AGRIVI. https://www.agrivi.com/blog/yield-losses-due-to-pests/.

Govindasamy, M., & Khursheed, S. (2018). A new host and distribution record for the black coccinellid, Stethorus aptus Kapur (Coccinellidae: Coleoptera). *Egyptian Journal of Biological Pest Control 28*,53. https://doi.org/10.1186/s41938-018-0057-9.

Hinsley, A., de Boer, H. J., Fay, M. F., Gale, S. W., Gardiner, L. M., Gunasekara, R. S., Kumar, P., Masters, S., Metusala, D., Roberts, D. L., Veldman, S., Wong, S., & Phelps, J. (2017). A review of the trade in orchids and its implications for conservation. Botanical Journal of the Linnean Society, 186(4), 435–455. https://doi.org/10.1093/botlinnean/box083.

Hussain, B., Abdul Rasheed, W. A. R., & Kandoo, A. A. (2021). First report of an invasive pest, Phyllonorycter populifoliella (Lepidoptera: Gracillariidae) from Ladakh. *Acta agriculturae Slovenica*, *3*(2), 117.

Hussain, B., Ahmad, B., & Bilal, S. (2015). Monitoring and mass trapping of the codling moth, Cydia pomonella, by the use of pheromone baited traps in Kargil, Ladakh, India. *International Journal of Fruit Science*, *15*(1), 1-9.

Insecticides. (2024, February 29). US EPA. https://www.epa.gov/caddis/insecticides.

Jones, G. A., Sieving, K. E., Jacobson, S. K., Avery, M. L., Meagher, R. L. (2005) Parasitized and non-parasitized prey selectivity by an insectivorous bird, *Crop Prot*. 24, 185–189.

Joshi, P. C., Khamashon, L., Kaushal, B. R., and Kumar, K. (2012). New Additions of Coccinellid Beetles (Coleoptera: Coccinellidae) to the already reported Species from Uttarakhand, India. *Nature and Science* 10(6):26- 30.

Khan, A. A., Shah, M. A., and Majid, S. (2016). Functional response of four syrphid predators associated with green Apple aphid (Hemiptera: Aphididae) in Laboratory. *Journal of Economic Entomology*, *109(1),*78-83.

Khursheed, R., Hussain, B., Ahmad, S. B., & Ashraf, M. (2006). Biology and feeding potential of Coccinella septumpunctata on mustard aphid, *Lipaphis erysimi. International Journal of Zoological Research, 2(1)*, 30-33.

Kimothi, M. M., Kalubarme, M. H., Dutta, S., Thapa, R. & Sood, R. K. (1997). Remote sensing of horticultural plantations in Kumarsain tehsil in Shimla district, Himachal Pradesh: *Journal of the Indian Society of Remote Sensing, 25*, pp. 19-26.

Lamichhane, J. R. (2017). Pesticide use and risk reduction in European farming systems with IPM: An introduction to the special issue. *Crop* Protection, *97*, 1–6.

Lefebvre, M., Franck, P., Olivares, J., Ricard, J. M., Mandrin, J. F., & Lavigne, C. (2017). Spider predation on rosy apple aphid in conventional, organic and insecticide-free orchards and its impact on aphid populations. *Biological Control 104,*57-65.

Manosathiyadevan, M., Bhuvaneshwari, V., Latha, R. (2017). Impact of Insects and Pests in Loss of Crop Production: A Review. In: Dhanarajan, A. (eds) Sustainable Agriculture towards Food Security. Springer, Singapore. https://doi.org/10.1007/978-981-10-6647-4_4.

Michalko, R., Pekár, S., & Entling, M. H., (2019). An updated perspective on spiders as generalist predators in biological control. *Oecologia*, 189(1):21-36.

Parker, J. E. (2017). Managing Cruciferous and Solanaceous Flea Beetles in Organic Farming Systems USDA National Institute of Food and Agriculture, Division of Community and Education (USDA NIFA DOCE), William E. Snyder, Washington State University, Department of Entomology https://eorganic.org/node/12461.

Rathee, M. (2018). Emerging insect pests in Indian agriculture. *Indian Journal of Entomology 80(2),* 267-281.

Sharma, J., Suroshe, S., Raghuvanshi, K. S., Borkar, S. G. & Kulkarni, S. R. (2015). Major insect pests and diseases of pomegranate and their management. In: *Pests of fruit trees (Citrus, Banana, Mango, Pomegranate and Sapota) E-Pest surveillance and pest management*. (D. B. Ahuja and C. Chattopadhyay Eds.). Advisory, publishers ICAR-National Research Centre for Integrated Pest Management, New Delhi and State Department of Horticulture, Commissionerate of Agriculture, Pune (Maharashtra): 124.

Shinde, B. D., Makarand, S., Joshi & Ahuja, D. B. (2015). Major insect pests and diseases of sapota and their management. In: *Pests of fruit trees (Citrus, Banana, Mango, Pomegranate and Sapota) E-Pest surveillance and pest management*. (D. B. Ahuja & C. Chattopadhyay Eds.). Advisory, publishers ICAR-National Research Centre for Integrated Pest Management, New Delhi and State Department of Horticulture, Commissionerate of Agriculture, Pune (Maharashtra): 124.

Simon, S., Bouvier, J. C., Debras, J. F., & Sauphanor, B. (2010, March). Biodiversity and pest management in orchard systems. A review. *Agronomy for Sustainable Development*, 30(1), 139–152. https://doi.org/10.1051/agro/2009013.

Solomon M. G., Cross J. V., Fitz Gerald J. D., Campbell C. A. M., Jolly, R. L., Olszak R. W., Niemczyk E., Vogt H. (2000). Biocontrol of pests of apples and pears in northern and central Europe - 3. Predators. *Biocontrol Science and Technology*, *10*, 91–128.

Ssymank, A., Kearns, C. A., Pape, T., & Thompson, F. C. (2008). Pollinating flies (Diptera): A major contribution to plant diversity and agricultural production. *Biodiversity* 9(1, 2): 86-89.

Tudi, M., Daniel Ruan, H., Wang, L., Lyu, J., Sadler, R., Connell, D., Chu, C., & Phung, D. T. (2021). Agriculture development, pesticide application and its impact on the environment. *International Journal of Environmental Research and Public Health,* 18, 1112. https://doi.org/10.3390/ijerph18031112.

Usha, K. & Singh, B. (2013). Potential applications of remote sensing in horticulture-A review. *Scientia Horticulturae*, *153*, pp. 71-83.

Whitney, T. D., Sitvarin, M. I., Roualdes, E. A., Bonner, S. J., & Harwood, J. D. (2018). Selectivity underlies the dissociation between seasonal prey availability and prey consumption in a generalist predator. *Molecular Ecology* 27(7), 1739-1748.

Zhang, W., Ricketts, T. H., Kremen, C., Carney, K., & Swinton, S. M. (2007). Ecosystem services and dis-services to agriculture, *Ecological Economics*, *64*, 253–260.

Chapter 6

Post Harvest Management of Crops Against Stored Grain Pests: New Techniques and Strategies

Pushpa Singh*
and Nishi Keshari
Department of Entomology, Dr. Rajendra Prasad Central Agricultural University,
Pusa, Samastipur, Bihar, India

Abstract

Providing safe and nutritious food has been the top priority, and if food production is self-sufficient, protecting food grains from biotic and abiotic stressors is a significant concern. Post-harvest storage has to ensure the optimum quality of grains and seeds for future use. Among the many biotic stresses, insects, mites, nematodes, and non-insect pests, like rodents and microorganisms, play an important role in causing damage to the quality and quantity of stored products. Nearly a thousand stored grain pests have been reported damaging the grain, most belonging to the coleopteran and lepidopteran orders. The pest bores into the kernel, feeding upon the grain and destroying the germ portion, causing nutritional depletion and reduced market value. It becomes hazardous to human health due to contamination by their excretory product and qualitative loss by chemical changes in protein, carbohydrates, and amino acids affecting the grain's nutritional value. Synthetic pesticides, although reliable and effective, have several drawbacks due to their harmful impact on the environment and humans. This is pushing us to look for other bio-rational, economical, and eco-friendly alternatives, with more thrust given to preventive aspects. For monitoring, traditional conventional methods along with techniques like food-attracting lures, good hygiene, acoustic probes, and critical high or low temperatures can be effectively utilised. New advances have been developed for preventing grain pests, like the use of plant volatile organic compounds (PVOC), ionising radiation, nanotechnology, insect pheromones, IGR (Insect Growth Regulators), etc. PVOC applications impact insects by different methods and act as repellents, antifeedants, and chemo-sterilants, and prevent the use of synthetic pesticides. These techniques provide a new opportunity for stored grain pest management.

Keywords: stored grain pests, PVOC, chemosterilants, nanoparticles, fumigants, IGR (Insect Growth Regulators)

---

* Corresponding Author's Email: pushpa@rpcau.ac.in.

In: Insects as Humans' Frenemies: From Friends to Foes
Editor: Istkhar Rao
ISBN: 979-8-89530-320-7

1. Introduction

In the current environment, ensuring that customers receive food promptly has become a top responsibility. All food commodities must be stored safely before being consumed. The production of food grains has constantly increased to fulfil the increasing demand; however, incorrect storage of the produce has been reported to cause ten to 20 per cent of storage loss (Esther et al., 2014). As per the World Bank report, there is a loss of 12 to 16 million metric tonnes of food grain in India, which impacts badly on our food security. Indian farmers mostly store 50–60 per cent of produce for self-consumption, and the rest is sold to private agencies or the government. The food grains are stored domestically in a traditional manner or by using the storage structures available in the market and government godowns, such as the Food Corporation of India (FCI). In a warm tropical environment, it can be difficult to store agricultural products safely because of the various biotic and abiotic elements that can seriously damage them.

Much of the insect pest infestation takes place in the field during harvesting, carrying over to storage facilities. Many insects and non-insect pests have been reported to cause damage during storage, including more than 600 beetles, 355 kinds of mite species, and 70 moth species (Rajendran & Sriranjani, 2008). By eating the kernels or by causing exuvial accumulation, webbing, and cadavers, insects are said to directly destroy between 5 and 30 per cent of the world's output, rendering it unfit for human consumption. Stored grain insects are categorised as main feeders based on how they feed. They consume the entire grain, and their offspring feed inside it, resulting in undetected infestations that grow in number and are extremely challenging to control. The secondary feeders, called barn bugs, establish themselves on already damaged grains or grains already damaged by primary feeders, surviving on broken kernel debris or high-moisture seeds. Moulds and fungi are also secondary feeders, contaminating the grains with metabolic waste and causing mould development due to high moisture (Magan et al., 2003). Major sources of infestation in storage are old, dilapidated storage structures and old bags, containers, farm machinery, etc. (Perez-Mendoza et al., 2004), which can be easily managed by proper and safe harvesting and storage with low moisture content. Pest infestations vary by crop type and geographical location; for instance, the khapra beetle is a flour beetle in South India and a storage pest of wheat in North India (Ramya et al., 2018). At the farm level, insect infestation can be detected by observing insect movement, clumpy grains, and congregating insects near the walls of storage structures or the bottom surfaces of storage. Novel approaches to assessing stored grain insects include CO_2 determination, chitin analysis from grain samples, grain temperature, headspace analysis of gases, and checking the nutritional quantity of grains, etc. (Banga et al., 2018).

2. Stored Grain Pest Detection

The detection of insects in stored grains is important for effective storage management, as each insect has its unique behaviour, feeding, biology, and ecology. The traditional and basic method is visual inspection, which is done by noticing different stages of insects. Advanced techniques have been developed by different people in stored grain insect detection, as Pearson et al. (2003) observed internal insect infestations in wheat kernels by using electrical conductance.

Sitophilus zeamais larvae inside popcorn kernels were detected with 75 to 81 per cent accuracy by using a modified roller mill with electrical conductivity by Brabec et al. (2017). Hagstrum et al. (1996) used acoustical detection methods to detect infested wheat kernels. Shen et al. (2018) made a trap using a deep learning technique with a vision analysis program, for which initially the database of live *Cryptolestes pusillus*, *Oryzaephilus surinamensis*, *Tribolium confusum*, *Sitophilus oryzae*, and *Rhyzopertha dominica* was used to extract and classify the insects. The use of sensors, biosensor-based detection, and semiochemical detection by gas chromatography are advanced methods of stored grain pest detection.

3. The Role and Importance of Stored Grain Storage Structure

The farmers have been storing their grain in different structures in different parts of the world, both below and above ground. Earlier, different types of mud structures, bamboo baskets, wooden structures, jute, and plastic bags were in common use, which, with time, skill, resources, and requirements, changed into metal bins, automated storage structures, warehouses with elevators, and silos. Airtight storage structures increase the CO_2 concentration and decrease O_2, affecting the metabolic activities of insects and causing their deaths, increasing the focus on hermetic seals and structures (Kumar et al., 2017), like silos, hermetic cocoons, metal drums, hermetic bags, and containers of different capabilities and sizes (Baributsa & Cristine, 2020).

4. Management of Stored Grain Insect Pests

4.1. Ozone (O_3)

The change in the necessary amount of oxygen for inhaling, leading to a higher concentration of carbon dioxide, is dangerous for storage pests. The controlled atmospheric storage causes distress in insects due to prolonged exposure to high CO_2 levels and lower O2 levels (Tutuncu & Emekci, 2019). Ozone (O_3), an effective, highly reactive oxidative agent, is effective in killing both external and internal feeder pests (Isikber & Athanassiou, 2015; Subramanyam et al., 2017). Ozone use has been done to manage the *Tribolium castaneum* and *Rhyzopertha dominica* phosphine-resistant populations (Sousa et al., 2008). A major constraint in the application of ozone as a fumigant is achieving its dosage and continuous purging through the treatment time. As O_3 has proven to be a strong fumigant, the selection of produce must be carefully made. There is also a need for developing and designing compatible storage structures and many pilot studies for large-scale applications.

4.2. Inert Dust

Some inert dust, such as (a) clays, sands, and ashes; (b) minerals; (c) dust containing synthetic silica (silicon dioxide—SiO_2); and (d) dust containing diatomaceous earth (DE), are effective against stored grain insects. These dusts work through different mechanisms, such as obstructing the digestive tract and spiracles, absorbing moisture and lipids from the cuticula, and abrasion of the cuticle (Banks & Fields, 1994). Numerous studies have shown that diatomaceous earth (DE) containing natural silica is an excellent alternative to chemical pesticides (Arthurs et al., 2004). The study conducted by Zhang et al. (2014) examined the use of food-grade inert dust to control five major insect species that feed on stored grains. The insects included Rhizopertha dominica, *Sitophilus zeamais*, *Oryzaephilus surinamensis*, *Tribolium castaneum*, and *Cryptolestes ferrugineus*. The results showed that the inert powder was uniformly dispersed throughout the grain and killed all five species within three days. Inert dust has to be widely used by incorporating it into other management strategies.

4.3. Heat or Cold Temperature and Irradiation

Insect metabolism is sensitive to changes in temperature, and insects need a certain temperature range to develop and grow normally. Pest control in stored grain has historically made use of very high temperatures (Bell, 2014). Grain temperature may be controlled effectively in storage by either raising it by artificial heating (e.g., by passing hot air) or reducing it (e.g., by using dielectric heating). Radiofrequency (RF) at 13,27 or 40 MHz, dielectric heating at 2450 MHz (microwave heating) or 915 MHz, and the subsequent sterility of stored grain insects have all been documented (Wang et al., 2001; Paul et al., 2020). Using 10- and 40-MHz RF, the dielectric effects of adult *Sytophylus oryzae* and *Tribolium confusum* were investigated (Nelson & Whitney, 1960). According to Singh et al. (2012), the insect species *Callosobruchus chinensis* was rendered extinct when exposed to microwave radiation (2,450 MHz) for 100 seconds at 700 W. These non-thermal heating methods have recently shown promise for improved grain quality and environmental safety in stored-grain pest control (Paul et al., 2020). Cell membrane shape and cell division are both impacted by microwave heating (Lu et al., 2011). Microwave radiation also produces cellular alterations in adult *Tribolium castaneum* cells, including fat body cell aggregation, enlarged nucleus cavity appearances, mitochondrial dysfunction, and the loss of the Golgi apparatus. When subjected to microwave radiation, the target insects perished due to these thermal and non-thermal impacts.

Some ionising radiations, such as gamma rays, x-rays, electric beams, and others, have been used to protect stored grain from insects. These radiations work at the molecular level, disrupting the DNA structure and causing sterilisation, mutations, and other problems. Although they are efficient and simple to use, they may be expensive and pose a danger to people (Hasan & Khan, 1998). Researchers studied the effects of gamma radiation on *C. chinensis* adults at doses ranging from 0 to 600 Gy, finding that the highest treatment of 300 Gy lowered adult fertility, hatchability, and female fecundity (Chiluwal et al., 2019). According to Hassan et al. (2019), the standard fatal dosage of fast neutron irradiation (FNI) was determined to be 256 Gy, which effectively prevented the development of all embryonic stages. As a successful alternative to currently utilised techniques, there is a need for an in-depth understanding of irradiation through the use of gamma rays and x-rays, as well as their integration with other suitable methods and safer handling.

4.4. Biological Methods

4.4.1. Semiochemicals

Nearly forty species of stored grain insects have semiochemicals that may be used as lures for mass trapping, mating disruption formulations, deterrents to feeding and oviposition, and attractants (Phillips & Throne, 2010). These signals aid insects in their search for food, mates, and egg-laying sites; they are chemical messages that elicit behavioural or physiological reactions in the receiving creature. Stored grain insects are attracted to synthetic lures in a variety of traps (Sambaraju & Phillips, 2008). These traps mostly target particular species; however, there have been reports of cross-attracting insects (Cox, 2004). The study conducted by Losey et al. (2019) examined *Cryptolestes ferrugineus* using several synthetic semiochemicals in both laboratory and field settings. The results showed that aggregation pheromones, whether applied alone or in combination with cucujolide I and cucujolide II, enhanced appeal in adults. According to Sambaraju and Phillips (2008), pheromone traps attracted a considerable majority of *Plodia interpunctella* adults (71%). These adults tended to cluster in dimly lit areas. The sex lure used in the experiment was (Z, E)-9,12-tetradecadienyl acetate (TDA/ZETA) in conjunction with ultraviolet (UV), green (green), and white light. As a component of integrated stored grain pest control, semiochemicals change the biology and behaviour of both sexes of an insect. As our understanding of its chemistry, biochemistry, physiology, and genetics grows, it will likely become one of the most effective long-term strategies for controlling insects that feed on stored grain. Having things readily available on the market is also important.

4.4.2. Plant Volatile Organic Compounds (PVOC)

As a time-honoured method of controlling stored grain pests, these organic, volatile bioactive compounds provide a pungent scent. Alkanes, alcohols, aldehydes, terpenoids, particularly monoterpenoids, fumigants, growth inhibitors, suppressors of reproductive behaviour, and loss of fecundity and fertility are all ways in which phytochemicals, which include extracts from numerous plants, influence the insect population (Rajashekar et al., 2012). Insects' physiology and behaviour are permanently altered by PVOC, leading to altered phases of growth in larvae and pupae as well as adult weight loss. Different plant extracts decreased the percentage of pupation after the treatment of larvae. To prevent the development of eggs and immature stages inside grain kernels, botanicals produced from plants were also used.

It has been shown that phytochemicals may affect the life cycle of insects, including their growth, development, and transformation. Insects' physiology and behaviour are permanently altered by these, leading to altered times of weight loss throughout development (in larvae, pupae, and adults) and lengthened periods of inactivity (in pupae and adults). Different plant extracts decreased the percentage of pupation after the treatment of larvae. Another discovery that prevented eggs and immature stages from growing within grain kernels was the use of botanicals produced from plants. Roy et al. (2014) and Tatun et al. (2014) found that aqueous extracts of common cocklebur (*Xanthium strumarium* L.) (Asteraceae) leaves had several insecticidal effects, such as repellency, inhibition of fecundity and adult emergence, toxic effect, and protection of grains against *Callosobruchus chinensis*. PVOC has a variety of potential uses in the fight against stored grain pests, including growth inhibitors, oviposition deterrents, antifeedants, repellents, and chemosterilants. "Oviposition deterrents" are substances that hinder insects from laying eggs. As an effective initial line of defence against pest insects, oviposition deterrents provide great promise for avoiding infestations. Experts agree that polyvinyl chloride (PVC) may inhibit egg laying and promote the spread of many insect pest groups that feed on stored grain. Cineole, a volatile component of essential oils derived from many Lamiaceae plants, was shown to influence the oviposition rate of many insects. One frequent method for managing stored grain pests is to use garlic oil as an oviposition deterrent (Sousa et al., 2015; Subramanyam et al., 2012). Insects' ability to feed is disrupted by antifeedants, often called "feeding deterrents," which make the treated food taste bad. One kind of pesticide that tries to get rid of bugs without messing with the ecosystem too much is an antifeedant. Antifeedants work by exposing the target insect to its natural enemies rather than killing them. Perhaps the unintended repercussions will be taken into consideration. The potential of azadirachtin and neem seed extracts as natural insecticides has been uncovered due to their effectiveness as feeding deterrents (Casida, 2012). Chemicals called repellents make plants or stored grain products taste bad or unattractive to pest insects, thereby protecting them from harm. An excellent repellent action against *Sitophilus zeamais* was shown in the arturmerone that was isolated from the rhizomes of *Curcuma longa* (Zingiberaceae). Chemicals known as chemosterilants are effective in preventing insect reproduction. It is a crucial part of IPM strategies, and it has proven to be so. According to Tavares et al. (2013), two compounds found in the *Acorus calamus*, asarone from the root system and 1, 3, and 7-trimethylxanthine from the seed extract, had chemosterilant effects on insects. Several plant volatiles may be ovicidal, meaning they may harm eggs. This is particularly true of bug and mite eggs. All phases of an insect's life cycle may be managed using this plant characteristic, which also helps to avoid harm from earlier stages. L-menthol, a component derived from the essential oils of several plants, has been shown to inhibit the development of *Tribolium castaneum* eggs (Hong et al., 2018; Mondal et al., 2009). These plants include *Mentha arvensis* L. (Lamiaceae), *Elletaria cardamomum* L. (Zingiberaceae), *Cinnamomum zeylanicum* Blume (Lauraceae), *Sygium aromaticum* L. (Myrtaceae), *Eucalyptus* spp., and *Azadirectica indica*, A. juss (Meliaceae).

4.4.3. Natural Enemies

The use of parasitoids and predators for grain insect management has been explored since long ago. In the cacao warehouses of Bahia, *Habrobracon hebetor* was mass-released for the management of infestations of *Cadra cautella* (Flinn & Scholler, 2012). Knowledge of the behaviour and biology of entomophagous insects attacking stored-grain pests is the prerequisite for biocontrol. Simultaneously, their adaptation to varied conditions in association with the hosts is one of the most promising areas of research. Information regarding the biology and nature of a few parasitoids, entomopathogens, and predators is known to attack stored grain pests. The major parasitoids are pteromalids and braconids (70%), followed by bethylids, ichneumonids, and trichogrammatids (24%), along with chalcidids and eulopids (3%), which are prerequisites for biocontrol (Cassi, 2017). More research is needed in this field, as biocontrol is one of the most sustainable components, along with studies on its biosafety and security.

4.4.4. Microbial-Based Pesticides

A range of bioagents like fungus, bacteria, viruses, and protozoa have been used against stored grain insects in laboratory conditions, giving very promising results, but since they don't adapt to different climates and are very host-specific, they are not in common use (Batta & Kavallieratos, 2018; Kumar et al., 2019). Microbial like *Beauveria bassiana*, *Metarhizium anisopliae* (entomopathogenic fungi), and *Bacillus thuringiensis* (Bt) have been tried against major stored grain pests (Batta, 2016). *Bacillus thurengiensis* var. *kurstaki* (Btk) has been reported to be effective against many stored grain insects like *Plodia interpunctella* and other moths (Lord et al., 2007), but due to the high dose, area-specific strain separation of carriers after treatment, and other problems, its commercial use is very limited (Arthurs & Dara, 2019). These pathogens are lower in efficacy and slower to act as compared to conventional insecticides, and more research is needed for testing field efficacy, isolation of local strains, and finding combinations of multiple strains against the host range of stored grain insects.

4.4.5. Entomopathogenic Nematodes (EPNs)

The use of nematodes in plant protection is a new approach, and *Steinernema* and *Heterhorhabditis* genera are also found effective against stored grain insects (Rumbos & Athanassiou, 2017). The three stains UK 76, USA/SC, and Hawaii of *Steinernema feltiae* were tested against *Tribolium confusum* and *Ephestia kuehniella* on wheat in laboratory conditions, and after an exposure of 14 days, 100 percent larval mortality was observed with the Hawaii stain in *Tribolium confusum*, while 69% mortality was seen after 14 days against *Ephestia kuehniella* larvae by the USA/SC strain (Athanassiou et al., 2008). In another experiment, the efficacy of EPN was tested in laboratory conditions against *Ephestia kuehniella*, *Rhyzopertha dominica*, *Sitophilus oryzae*, and *Tribolium confusum* of three nematode species: *Heterorhabditis bacteriophora*, *Steinernema carpocapsae*, and *Steinernema feltiae*. The efficacy was found effective against *Ephestia kuehniella* larvae by *Steinernema. feltiae*; the other two stains caused less than 23% mortality. (Athanassiou et al., 2010). All these trials have been done in laboratory conditions; bioassays and studies are required to study their field efficacy. A single strain of nematode species cannot be effective against a wide insect base of stored grain as they are species-specific, and a mixture of formulations can be effective. The need is the isolation of local virulent strains, efficacy evaluation against laboratory hosts, and formulation technologies for its success (Rumbos & Athanassiou, 2017).

4.5. Chemical Management

Fumigation of chemical insecticides has been followed as a wide approach for controlling any infestation detected, in addition to spraying for the prevention of infestations that prove hazardous to human health and the environment in the long run. Dichlorvos has been the most commonly used insecticide for grain disinfestation and has been proven to be very efficient as a surface spray. Resistance against *Rhyzopertha dominica* has been reported (Zettler & Cuperus, 1990). The rational insecticide spinosyn group's efficacy has also been widely studied (Vassilakos et al., 2015). In storage, fumigation methodology has been commonly used, as the fumigants toxic to insects with higher vapour pressure can penetrate through the stored commodity but require airtight conditions. Other suitable conditions for its effectiveness are temperature and relative humidity in storage, moisture content of the container, and air tightness (Zettler & Arthur, 2000; Daglish et al., 2018). Many of the registered fumigants have slowly been phased out, and phosphine and methyl bromide, due to their toxicity and ozone depletion, are also being slowly removed from use (Fields & White, 2002). Resistance to phosphine has also been reported, and some recent advancements and studies have been conducted for the use of alternative fumigants and chemicals. Although the focus is shifting towards a non-chemical approach to stored grain insect management, Fumigants like carbonyl sulphide, chlorine dioxide, ethanedinitrile, ethyl formate, methyl salicylate, hydrogen cyanide, and methyl benzoate propylene oxide (PPO) are mostly used in laboratories in search of novel, safer fumigants for storage grain insect management.

5. Novel Approach in Stored Grain Insect Pest Management

5.1. Nanotechnology

The properties of nanotechnology, such as changes in electrical conductivity, surface chemistry, and reactivity, as well as the fact that many metallic oxides, such as zinc, silver, aluminium, and silica, are formulated as nanoparticles (NPs), are leading to an upsurge in experiments involving the use of nanotechnology to protect stored grain (Zayed, 2018; Stadler et al., 2010). Because of their photochemical, antimicrobial, UV-filtering, and stronger catalytic and antifungal activities, zinc oxide nanoparticles have found widespread use (Meruvu et al., 2011). According to Malaikozhundan et al. (2017), Bt-ZnO NPs were created and tested against *Callosobruchus maculatus*. The results showed that the development time of the larvae, pupae, and the whole organism was extended to 25 μg/mL and that the fecundity and hatchability of the eggs were reduced to 10.71 μg/mL. El-Saadony et al. (2020) also found that nanoparticles were effective against *Tribolium. castaneum*, *Sitophylus oryzae*, *Rhyzopertha dominica*, and *Callosobruchus maculatus* in laboratory settings in terms of adult insect mortality, fecundity, hatchability, and biology. Blackgram infected with *Callosobruchus maculatus* showed decreased fecundity and an extended developmental period at dosages of 5, 10, 20, and 40 ppm NiO NPs (nickel-oxide nanoparticles). To better understand how to employ nanoparticles for storage grain preservation and to create relevant technologies, extensive field and laboratory research is required.

5.2. Molecular Technology

Novel molecular techniques like RNA interference (RNAi) and clustered regularly interspaced short palindromic repeats (CRISPR) are used to disrupt the normal gene expression of an organism (Perkin et al., 2016). *Tribolium castaneum* has been preferred for testing these techniques as they can be easily multiplied and, in all its developmental stages, show a strong RNAi response by inhibiting further gene expression as CRISPR deactivates Cas9 (the DNA binder) and RNAi acting at dsRNA (Qi et al., 2014). These futuristic techniques can only be used after identifying insect-specific genes infesting the stored grain and extensive research.

Conclusion

Providing safe and nutritious food is the most important agenda item today. We, as a nation, have attained food security, and the safe storage of agricultural produce is also of utmost importance; plant protection scientists are always in search of safe methods for achieving this goal. From harvesting to storage facilities, the grain is infested by a variety of insects. Farmers have traditionally used storage units to secure their produce. With time, chemicals have been used in general to prevent and control the stored grain insects, with various demerits to human health as well as environmental impact. The focus has shifted to a more scientifically developed storage structure for ozone gas, PVOC, semiochemicals, nanotechnology, etc. It is important to combine all the methods in an integrated way to use technology for the safe storage of grains and the management of insect pests.

References

Arthurs, S. & Dara, S. K. (2019). Microbial biopesticides for invertebrate pests and their markets in the United States. *Journal of Invertebrate Pathology*, 165, 13-21. doi: 10.1016/j.jip.2018.01.008.

Arthurs, S., Heinz, K. M. & Prasifka, J. R. (2004). An analysis of using entomopathogenic nematodes against above-ground pests. *Bulletin of Entomological Research*, 94, 297-306. doi: 10.1079/BER2003309.

Athanassiou, C. G., Kavallieratos, N. G., Menti, H. & Karanastasi, E. (2010). Mortality of four stored product pests in stored wheat when exposed to doses of three entomopathogenic nematodes. *Journal of Economic Entomology*, 103, 977-984. doi: 10.1603/EC09202.

Athanassiou, C. G., Palyvos, N. E. & Kakouli-Duarte, T. (2008). Insecticidal effect of *Steinernema feltiae* (Filipjev) (Nematoda: Steinernematidae) against *Tribolium confusum* du Val (Coleoptera: Tenebrionidae) and *Ephestia kuehniella* (Zeller) (Lepidoptera: Pyralidae) in stored wheat. *Journal of Stored Production Research*, 44, 52-57. doi: 10.1016/j.jspr.2007.04.002.

Banga, K. S., Kotwaliwale, N., Mohapatra, D. & Giri, S. K. (2018). Techniques for insect detection in stored food grains: an overview. *Food Control*, 94, 167-176. doi: 10.1016/j.foodcont.2018. 07.008.

Banks, J. & Fields, P. (1994). Physical methods for insect control in stored-grain ecosystems in *Stored-Grain Ecosystems* (eds. D. S. Jayas, N. D. G. White, and W. E. Muir) (New York, NY: Marcel Dekker, Inc.), 353-410.

Baributsa, D. & Cristine, M. (2020). Developments in the use of hermetic bags for grain storage in *Advances in Postharvest Management of Cereals and Grains* (Phillipines: Burleigh Dodds Science Publishing), 171-198. doi: 10.19103/AS.2020.0072.06.

Batta, Y. A. (2016). Recent advances in formulation and application of entomopathogenic fungi for biocontrol of stored-grain insects. *Biocontrol Science Technology*, 26, 1171-1183. doi: 10.1080/09583157. 2016.1201458.

Batta, Y. A. & Kavallieratos, N. G. (2018). The use of entomopathogenic fungi for the control of stored-grain insects. *International Journal of Pest Management*, 64, 77-87. doi: 10.1080/09670874.2017.1329565.

Bell, C. H. (2014). Pest control of stored food products: insects and mites in *Hygiene in Food Processing: Principles and Practice, 2nd Edn*., eds. H. L. Lelieveld, J. Holah, and D. Napper (York: Woodhead Publishing Series in Food Science, Technology and Nutrition), 494-538. doi: 10.1533/ 9780857098634.3.494.

Brabec, D., Dowell, F., Campbell, J. & West, M. (2017). Detection of internally infested popcorn using electrically conductive roller mills. *Journal of Stored Production Research*, 70, 37-43. doi: 10.1016/j.jspr.2016.11.002.

Casida, J. (2012). Pyrethrum: The Natural Insecticide (Ed.) Elsevier.

Cassi, M. S. (2017). *Approaches for the Biological Control of Stored Product Pests*. Barcelona: Universitat Autònoma de Barcel.

Chiluwal, K., Kim, J., Bae, S., Do, R. oh, G. H., Park, H. J. & Park, C. G. (2019). Effect of gamma irradiation on fecundity, sterility and female sex pheromone production of *Callosobruchus chinensis* (Coleoptera: Bruchidae). *Journal of Economic* Entomology, 112, 156-163. doi: 10.1093/jee/toy317.

Cox, P. D. (2004). Potential for using semiochemicals to protect stored products from insect infestation. *Journal of Stored Production Research*, 40, 1-25. doi: 10.1016/S0022-474X(02)00078-4.

Daglish, G. J., Nayak, M. K., Arthur, F. H. & Athanassiou, C. G. (2018). Insect pest management in stored grain in *Recent Advances in Stored Product Protection*, eds. C. G. Athanassiou and F. H. Arthur (Brisbane, QLD: Springer-Verlag GmbH Germany), 45-63. doi: 10.1007/978-3-662-56125-6_3.

El-Saadony, M. T., Abd El-Hack, M. E., Taha, A. E., Fouda, M. M. G., Ajarem, J. S. & Maodaa, S. N. (2020). Ecofriendly synthesis and insecticidal application of copper nanoparticles against the storage pest *Tribolium castaneum. Nanomaterials* 10, 587. doi: 10.3390/nano10030587.

Esther, M., Sharon, M., Abirami, C. V. K. & Alagusundaram, K. (2014). Grain storage management in India. *Journal of Postharvest Technology*, 2, 12-14. http://jpht.in/MenuscriptFile/03b17047- fa10- 48ed-b0cb-95fb9222798f.pdf.

Flinn, P., & Scholler, M. (2012). "Biological control: Insect pathogens, parasitoids, and predators," in *Stored Product Protection*, eds. D. W. Hagstrum, T. W. Phillips, and G. Cuperus (Braunschweig: Kansas State University Publications), 203–212.

Hagstrum, D. W., Flinn, P. W. & Shuman, D. (1996). Automated monitoring using acoustical sensors for insects in farm-stored wheat. *Journal of Economic Entomology*, 89, 211-217. doi: 10.1093/jee/89.1.211.

Hasan, M. & Khan, A. R. (1998). Control of stored-product pests by irradiation. *Integrated Pest Management Review*, 3, 15-29. doi: 10.1023/A:1009621606024.

Hassan, R. S., Mahmoud, E. A., Sileem, T. M. & Sayed, W. A. A. (2019). Evaluation of fast neutron irradiation as a new control method against the Indian meal moth, *Plodia interpunctella* (Hübner). *Journal of Radiation Research and Applied Science*, 12, 24-30. doi: 10.1080/16878507.2019.1594133.

Hong, T. K., Perumalsamy, H., Jang, K. H., Na, E. S. & Ahn, Y. J. (2018). Ovicidal and larvicidal activity and possible mode of action of phenylpropanoids and ketone identified in *Syzygium aromaticum* bud against *Bradysia procera*. *Pesticide Biochemistry and Physiology*, 145:29-38.

Isikber, A. A. & Athanassiou, C. G. (2015). The use of ozone gas for the control of insects and micro-organisms in stored products. *Journal of Stored Production Research*, 64, 139-145. doi: 10.1016/j.jspr.2014.06.006.

Kumar, K. K., Sridhar, J., Murali-Baskaran, R. K., Senthil-Nathan, S., Kaushal, P. & Dara, S. K. (2019). Microbial biopesticides for insect pest management in India: current status and prospects. *Journal of Invertebrate Pathology*, 165, 74-81. doi: 10.1016/j.jip.2018.1 0.008.

Kumar, S., Mohapatra, D., Kotwaliwale, N. & Singh, K. K. (2017). Vacuum hermetic fumigation: a review. *Journal of Stored Products Research*, 71, 47-56. doi: 10.1016/j.jspr.2017.01.002.

Lord, J. C., Campbell, J. F., Sedlacek, J. D. & Vail, P. V. (2007). Application and evaluation of entomopathogens for managing Lepidoptera in stored products in *Field Manual of Techniques in Invertebrate Pathology*, eds. L. A. Lacey and H. K. Kaya (Frankfort: Springer), 677-693. doi: 10.1007/978-1-4020-5933-9_33.

Losey, S. M., Daglish, G. J. & Phillips, T. W. (2019). Orientation of rusty grain beetles, *Cryptolestes ferrugineus* (Coleoptera: Laemophloeidae) to semiochemicals in field and laboratory experiments. *Journal of Stored Products Research*, 84, 101513. doi: 10.1016/j.jspr.2019.101513.

Lu, P. J., Lin, H. C., Yu, W. T. & Chern, J. M. (2011). Chemical regeneration of activated carbon used for dye adsorption. *Journal of Taiwan Institute of Chemical Engineering*, 42, 305-311. doi: 10.1016/j.jtice.2010.06.001.

Magan, N., Hope, R., Cairns, V. & Aldred, D. (2003). Post-harvest fungal ecology: impact of fungal growth and mycotoxin accumulation in stored grain. *European Journal of Plant Pathology*, 109, 723-730. doi: 10.1023/A:1026082425177.

Malaikozhundan, B., Vaseeharan, B., Vijayakumar, S. & Thangaraj, M. P. (2017). *Bacillus thuringiensis* coated zinc oxide nanoparticle and its biopesticidal effects on the pulse beetle, *Callosobruchus maculatus*. *Journal of Photochemistry and Photobiology B-Biology*, 174, 306-314. doi: 10.1016/ j.jphotobiol.2017.08.014.

Meruvu, H., Vangalapati, M., Chaitanya Chippada, S. & Rao Bammidi, S. (2011). Synthesis and characterization of zinc oxide nanoparticles and its antimicrobial activity against *Bacillus subtilis* and

Escherichia coli. *Rasayan Journal of Chemistry*, 4, 217-222. Available online at: http://rasayanjournal.co.in/vol-4/issue-1/33.

Mondal, M. & Khalequzzaman M. (2009). Ovicidal activity of essential oils against red flour beetle, *Tribolium castaneum* (Coleoptera: Tenebrionidae). *Journal of Bioscience*, 17: 57-62.

Nelson, S. O. & Whitney, W. K. (1960). Radio-frequency electric fields for stored-grain insect control. *Transactions of American Society of Agricultural and Biological Engineers*, 3, 133-137. doi: 10.13031/2013.41143.

Paul, A., Radhakrishnan, M., Anandakumar, S., Shanmugasundaram, S. & Anandharamakrishnan, C. (2020). Disinfestation techniques for major cereals: a status report. *Comprehensive Reviews in Food Science and Food Safety*, 19, 1125-1155. doi: 10.1111/1541-4337.12555.

Pearson, T. C., Brabec, D. L. & Schwartz, C. R. (2003). Automated detection of internal insect infestations in whole wheat kernels using a perten SKCS 4100. *Applied Engineering in Agriculture*, 19, 727-733. doi: 10.13031/2013.15654.

Perez-Mendoza, J., Flinn, P. W., Campbell, J. F., Hagstrum, D. W. & Throne, J. E. (2004). Detection of stored-grain insect infestation in wheat transported in railroad hopper-cars. *Journal of Economic Entomology*, *97*, 1474-1483. doi: 10.1093/jee/97.4.1474.

Perkin, L. C., Adrianos, S. L. & Oppert, B. (2016). Gene disruption technologies have the potential to transform stored product insect pest control. *Insects* 7, 46. doi: 10.3390/insects7030046.

Phillips, T. W. & Throne, J. E. (2010). Biorational approaches to managing stored-product insects. *Annual Review of Entomology*, 55, 375-397. doi: 10.1146/annurev.ento.54. 110807.090451.

Qi, L., Xu, M., Fu, Z., Mira, T. & Zhang, X. (2014). C2SLDS: A WSN-based perishable food shelf-life prediction and LSFO strategy decision support system in cold chain logistics. *Food Control*, 38, 19-29. doi: 10.1016/j.foodcont.2013.09.023.

Rajashekar, Y., Bakthavatsalam, N. & Shivanandappa, T. (2012). Botanicals as grain protectants and Psyche. *Journal of Entomology*, 640-648.

Rajendran, S. & Sriranjini, V. (2008). Plant products as fumigants for stored-product insect control. *Journal of Stored Product Research*, 44: 126-135.

Ramya, R., Srivastava, C. & Subramanian, S. (2018). Monitoring of phosphine resistance in Indian populations of *Tribolium castaneum* (Herbst) from stored wheat. *Indian Journal of Entomology*, 80, 19–23. doi: 10.5958/0974-8172.2018.00005.6.

Roy, B., Amin, M. R., Jalal, S., Kwon, Y. J. & Suh, S. J. (2014). Evaluation of common cocklebur *Xanthium strumarium* leaf extract as post-harvest grain protectant of black gram against pulse beetle, *Callosobruchus chinensis* (Coleoptera: Bruchidae) and isolation of crude compound. *Entomology of Research*, 44: 254-261.

Rumbos, C. I. & Athanassiou, C. G. (2017). Use of entomopathogenic fungi for the control of stored-product insects: can fungi protect durable commodities? *Journal of Pest Science*, 90, 839-854. doi: 10.1007/s10340-017-0849-9.

Sambaraju, K. R. & Phillips, T. W. (2008). Responses of adult *Plodia interpunctella* (Hübner) (Lepidoptera: Pyralidae) to light and combinations of attractants and light. *Journal of Insect Behaviour*, 21, 422-439. doi: 10.1007/s10905-008-9140-5.

Shen, Y., Zhou, H., Li, J., Jian, F. & Jayas, D. S. (2018). Detection of stored-grain insects using deep learning. *Computers and Electronics in Agriculture*, 145, 319-325. doi: 10.1016/j.compag.2017.11.039.

Singh, R., Singh, K. K. & Kotwaliwale, N. (2012). Study on disinfestation of pulses using microwave technique. *Journal of Food Science and Technology*, 49, 505-509. doi: 10.1007/s13197-011-0296-1.

Sousa, A. H., Faroni, L. R. D. A., Guedes, R. N. C., Tótola, M. R. & Urruchi, W. I. (2008). Ozone as a management alternative against phosphine-resistant insect pests of stored products. *Journal of Stored Products Research*, 44, 379-385. doi: 10.1016/j.jspr.2008. 06.003.

Sousa, R. M. O., Rosa, J. S., Oliveira, L., Cunha, A. M. & Fernandes, F. (2015). Activities of *Apiaceae* essential oils and volatile compounds on hatchability, development, reproduction and nutrition of *Pseudaletia unipuncta* (*Lepidoptera: Noctuidae*). *Indian Crop Production*, 63: 226-237.

Stadler, T., Buteler, M. & Weaver, D. K. (2010). Novel use of nanostructured alumina as an insecticide. *Pest Management Science*, 66, 577-579. doi: 10.1002/ps.1915.

Subramanyam, B., Hartzer, M. & Boina, D. R. (2012). Performance of pre-commercial release formulations of spinosad against five stored-product insect species on four stored commodities, *Journal of Pest Science*, 85: 331-339.

Subramanyam, B., Xinyi, E., Savoldelli, S. & Sehgal, B. (2017). Efficacy of ozone against *Rhyzopertha dominica* adults in wheat. *Journal of Stored Products Research*, 70, 53-59. doi: 10.1016/j.jspr.2016.12.002.

Tatun, N., Vajarasathira, B., Tungjitwityakul, J. & Sakurai, S. (2014). Inhibitory effects of plant extracts on growth, development and [alpha]-amylase activity in the red flour beetle, *Tribolium castaneum* (Coleoptera: Tenebrionidae). *European Journal of Entomology*, 111: 181-186.

Tavares, S. W., Sousa Freitas, S., Grazziotti, G. H., Parente, L. M. L., Li~ao, L. M. & Zanuncio, J. C. (2013). Ar-turmerone from *Curcuma longa* (*Zingiberaceae*) rhizomes and effects on *Sitophilus zeamais* (Coleoptera: Curculionidae) and *Spodoptera frugiperda* (Lepidoptera: Noctuidae). *Indian Crop Production*, 46: 158-164.

Tutuncu, S. & Emekci, M. (2019). Comparative efficacy of modified atmospheres enriched with carbon dioxide against *Cadra (=Ephestia) cautella. Journal of Science Food and Agriculture*, *99*, 5962-5968. doi: 10.1002/jsfa.9871.

Vassilakos, T. N., Athanassiou, C. G. & Tsiropoulos, N. G. (2015). Persistence and efficacy of spinetoram against three major stored grain beetles on wheat. *Crop Protection*, 69, 44-51. doi: 10.1016/j.cropro.2014.08.010.

Wang, S., Ikediala, J. N., Tang, J., Hansen, J. D., Mitcham, E. and Mao, R. (2001). Radio frequency treatments to control codling moth in in-shell walnuts. *Postharvest Biological Technology*, *22*, 29-38. doi: 10.1016/S0925-5214(00)00187-3.

Zayed, G. M. M. (2018). Nano-particles: a recent approach for controlling stored grain insect pests. *Academia Journal of Agricultural Research*, 6, 88-94. doi: 10.15413/ajar.2017. IECCNA.14.

Zettler, J. L. & Arthur, F. H. (2000). Chemical control of stored product insects with fumigants and residual treatments. *Crop Protection*, 19, 577–582. doi: 10.1016/S0261 -2194(00)00075-2.

Zettler, J. L. & Cuperus, G. W. (1990). Pesticide resistance in *Tribolium castaneum* (Coleoptera: Tenebrionidae) and *Rhyzopertha dominica* (Coleoptera: Bostrichidae) in wheat. *Journal of Economic Entomology*, *83*, 1677-1681. doi: 10.1093/jee/83.5.1677.

Zhang, T., Cao, Y., Li, Y. Y., Gao, Y. S. & Feng, J. H. (2014). Food-grade inert dust as structural treatment against insect pests. In: *11th International Working Conference on Stored Product Protection*, 883–884. doi: 10.14455/DOA.res.2014.155.

Chapter 7

Stored Grain Pests: Present Scenario and Management Practices with Reference to *Callosobruchus maculatus*

Rajesh Kumar[1]
and Younis Ahmad Hajam[2,*]
[1]Department of Biosciences, Himachal Pradesh University, Shimla, H.P., India
[2]Department of Life Sciences and Allied Health Sciences, Sant Baba Bhag Singh University, Punjab, India

Abstract

Agriculture is the main source of income, especially in rural areas around the globe. In India, approximately 70% of the population depends on farming. With technological advancements, India has swiftly increased its food production during the past few decades, but crop losses during both pre- and post-harvesting are pretty stable at 10%. Proper grain storage arrangements are not available, especially in developing countries, and cause nearly 6% of grain loss. Various bruchid pests start damaging the grain at the maturing stage of crops and remain in warehouses. Storage grain insects have been known as severe pests of dry and stored, long-lasting agricultural goods, various food products, and non-food commodities all around the world. They cause tremendous post-harvest losses, which are approximately 9% in developed countries and 20% in developing countries. These insect pests start damaging the grain at the maturing stage of crops and continue into warehouses. They could contaminate the cereal grains through their occurrence, excretion, and also their dead bodies. They deteriorate the stored products and also cause enormous nutritional depletion. Because of this, consumer acceptability, as well as the market value of grains, are reduced, and the spoilage of grains through their waste products and resulting dead bodies could also prove unhealthy to mankind. Chemical insecticides are used to control stored grain pests in every part of the globe, and they have been proven quite effective against crop pests. However, with time, the ill effects of chemical insecticides have been observed among animals, including humans, as well as in the food chain. A large number of chronic diseases related to the nervous, metabolic, reproductive, and respiratory systems have been reported by various researchers around the globe. The time has come to realise the impact of using chemical insecticides and adopting biological integrative approaches for controlling crop pests for a safe environment and better health. Several management practices have been adopted to control the infestation of stored grain pests since ancient times. These are supposed to be refined as per current needs so that they can be used to combat pest infestations in a changing world. In the present chapter, the biology, damage/infestation, and management strategies of stored grain pests shall be summarised

* Corresponding Author's Email: younismajeed64@gmail.com.

In: Insects as Humans' Frenemies: From Friends to Foes
Editor: Istkhar Rao
ISBN: 979-8-89530-320-7

with special reference to *Callosobruchus* spp. The latest advancements in the field of bruchids and future perspectives will also be discussed.

Keywords: pest, bruchids, *Callosobruchus*, pulses, management

1. Introduction

Stored grains are principally used by billions of people globally to fulfil nutritional requirements. Among stored grains, pulses constitute a major portion of our balanced diet as well as contribute as a source of energy from a nutritional point of view. Therefore, the ingestion of pulses has increased during the past few years, and there has also been a decline in the production of pulses compared to their demand by the globally increasing population. Taxonomically, pulses are included in the Fabaceae family and collectively include all types of proteinaceous grains that are used by the common man in the form of dry grains. World Pulses Day is celebrated on February 10 every year since 2019 by the UN General Assembly to acknowledge the significance of pulses as a global food. Pulses and legumes are two interchangeable terms; however, all leguminous grains cannot be considered pulses. Approximately 85.40 ha of the total world area has been covered by pulses, with an average production of 87.40 MT at 1023 kg/hectare. India is the biggest producer of pulses, with an average of 34 per cent of the total cultivating area and 26 per cent of total production all over the world. Most of the pulse cultivation area is rainfed, and farmers are dependent on weather conditions, which is a major drawback as it directly affects overall pulse production. In cases of drought conditions or less rainfall, production drops drastically. Another main problem with stored grains, especially pulses, is the attack of pests at almost every stage of production. Pests are used to cause damage both in fields and in stored conditions (Sharma, 1989). The pulse seeds are extremely prone to pest infestation; they reproduce very quickly under storage conditions due to sufficient hiding and feeding material. Ultimately, insect pests cause huge losses to grain, and the numbers of pests rise drastically under stored conditions in comparison to open fields. More than 50% of grain loss takes place in stores due to pest attacks or infections in India (Allotey & Oyewo, 2004).

Stored grain pests are the main destroyers of pulses (Sherma, 1989; Allotey & Oyewo, 2004; Phillips & Throne, 2009). More than 600 insect species of the Coleoptera order have been reported to cause infestation under storage conditions (Rajendran & Sriranjini, 2008). These insect pests are usually shifted to stores from crop fields in the form of eggs or larvae and acclimatise very well to suitable conditions in both habitat and food (Hagstrum & Phillips, 2017). As the population of India increases, demand for food is also high, and food production has to increase subsequently. The government of India is making constant efforts to promote agriculture and crop production activities through various schemes. Farmers are also adopting new methods of cropping and harvesting by making use of the latest technology. Still, stored-gain pests, especially bruchids, are the main destructors of harvested grains and affect overall productivity. More than 200 insect species have been reported to cause infestations in stored grains. The majority of these insect pests are bruchids (117 species) that have a high range and level of infestation in storage conditions (Arora, 1977; Singh & Ahlawat, 2005). Among bruchids, *Callosobruchus* infests stored grain in both fields and stores. It is a cosmopolitan insect found all over the world. In India, *Callosobruchus maculatus*, *Callosobruchus analis*, and *Callosobruchus chinensis* are the dominant species that immensely contribute to damaging stored grains (Dias, 1986).

Insect pests pass the initial stages of their lives inside grains by feeding upon them and hurting their germination potential and nutritive quality as well. *Callosobruchus* is responsible for major pulse loss by consuming protein parts, making it unusable for common use (Tsedeke, 1985). The adult pulse beetles are small in size and vary approximately 2–3 mm in length, with a typical round outer appearance in comparison to other members of the family Bruchidae. These are reddish brown (Fox, 1993). Infestation of pulses by *Callosobruchus* spp. in fields is quite common (Raina, 1970; Mohan & Subbarao, 2000). This insect pest is the most potential pest of pulses, also known as bean beetles (Onyido et al., 2011), that belong to the family Chrysomelidae and the subfamily Bruchinae (Kergoat et al., 2008). Pests are used to damage the grains in stored facilities irrespective of region and climate; however, the rate of infestation is high in developing countries due to poor storage conditions (Ekeh et al., 2013). Infection by bruchids (*Callosobruchus* spp.) begins in the open fields and continues to the post-harvest period (Messina, 1991). The rate of infestation is very slow during the initial phase but slowly increases, and an ample population of insect pests develops to damage the grains during the storage phase (Khavilkar & Dalvi, 1984).

All over the world, agriculture is the principal source of revenue for the common people. In India, about 70% of the population depends on agriculture. With the use of technological advancements, the production of grains, including pulses, has increased manyfold. But still, the losses to grains both in fields and stores are more than 10%, which is a very big number. The loss of grains is more due to storage conditions than to fields because of poor storage and management facilities (Prakash et al., 2016). Hence, grain storage facilities play a significant role in preventing losses to stored products due to one or another enemy. The stored grains are infested by insect pests during various developmental stages and cause huge losses to grains, which are otherwise of immense use to common people. In developing countries, meagre or poor facilities are available for the packing and storage of grains, both domestically and commercially. With the changing times, some efforts have been made to manufacture special gunny bags for packing and some large metal containers for the storage of grains for longer periods; however, losses are still stable because of various factors. Insect pests are the main destroyers of pulses, accounting for more than 10–20 per cent of storage losses (Phillips & Throne, 2009).

Reports have shown that various bruchid pests start damaging the grain at the maturing phase of crops and remain until it reaches warehouses. However, the principal causes of infestations are inappropriate storage facilities like old containers and, bags, and storage rooms. These are also the reasons for cross-infestation (Perez-Mendoza et al., 2004). A vast population in the world depends on an agro-based economy for their livelihood, in which a very significant portion of the gross domestic product comes from agriculture. However, in most underdeveloped countries, as a result of subsistence farming, low education levels, and low technology inputs in agriculture, agricultural activities are not advanced, which leads to suboptimal production and shoddy storage of these products. It has been estimated that the losses caused by various types of enemies, including insect pests, may be reduced by taking various measures on a domestic and commercial scale. Effective threshing, processing, packing, and storing of grains may be adopted to minimise attack and loss by pests during post-harvest time.

Some pest-resistant, better-yielding seed varieties were introduced in India during the 1960s under the aegis of the Green Revolution. This has led India to become a major producer of grains, reaching about 230 million metric tonnes per year. Notwithstanding excess and surplus grain production, we are still unable to cater to the needs of the common people, and poverty is still at its peak in various parts of the country. Thousands of tonnes of grains get wasted and spoilt, leading to huge financial losses every year. Storage grain insects have been known as severe pests of dry and stored long-lasting agricultural goods, various food products, and non-food commodities all around the world. These cause tremendous post-harvest losses, which are approximately 9–20% in different countries (Pimentel, 1991). These insects could contaminate the cereal grains through their occurrence, excretion, and dead bodies.

All over the world, yearly damage of approximately 8–10% is found in food commodities under stored conditions. Most of the grains are lost due to these pests and improper storage. So, there should be reliable biopesticides and properly stored conditions to control the damage to storage grains in various developing countries. Sometimes, insects, pathogens, and mites release toxins (myotoxins and entomotoxins) in stored pulses and spoil them. Jacobson (1982), and Ahmed and Grainage (1986) reported that approximately 20,000 species of insect pests are present, which destroy approximately one-third of the world's grain production. Several pests are mostly related to the orders Coleoptera and Lepidoptera, which cause 60–100% losses, respectively (Atwal & Dhaliwal, 2008). These pests damage the pulse grain by creating a hole in the seed and entirely feeding on the inner germ part. It deteriorates the stored products, and enormous losses are also due to nutritional depletion. Because of this, the market value was also reduced. Even the spoilage of grains through their waste products and the remains of dead bodies could be unhealthy for mankind. Therefore, quantitative as well as qualitative food grain loss occurs. Quantitatively, seed weight loss, damage, and holes in the grains were found, which decreased the grain mass. Qualitatively, the biochemical alterations were observed in various biochemical as well as nutritive parameters, thus degrading the nutritional properties of grains.

In this modern time, pre-and post-harvest management techniques have achieved adequate importance. This is because of the high demand for the import and export of grains, seeds, and other agricultural products. Therefore, insect pests have crossed their geographical boundaries and are now responsible for the spoilage of food products and economic losses. The current scenario of higher demand for food grains and insect pest resistance against pesticides are the main reasons to explore alternative measures of pest management. Also, due to health awareness, the demand for organic food, especially chemical-free food, has been increasing all over the world (Paul et al., 2020).

2. The Origin of Stored Grain Insect Pests

The insect pests remain on the crop farm, but some of them start entering the stored grains. This provides hidden infestation to grain and sidewise protection of pests from other biotic agents. According to their adaptation to the pest body, some of them feed on dried and decaying plant and animal materials, while others enter the roots, shoots, and tubers of plants. In an earlier time, when traditional methods were used to store the seeds, it provided an easy and hidden way of life for the insects. Thus, accidentally, these insect pests become stored grain pests. These conditions in storehouses for the easy development and reproduction of insect pests have made it useless to fly long distances searching for food. This has resulted in an easy life, and hence several stored grain pests have entirely lost the power to fly. Shankar and Abrol (2012) provide evidence that indicates that the animals that trouble grain stores today were more common in prehistoric times.

Several pulses have been attacked by various pests that infest grains to complete their physiological necessities, resulting in nutritional and financial losses. Tropical and subtropical weather conditions in our country are pretty suitable for the survival of these insect pests and their unremitting abundance throughout the year. Temperature changes has caused a major impact on the growth and development of beetles. The duration of eggs, larvae, pupae, and adults is reported to decrease with an increase in temperature (Omar & Mahmoud, 2020). Even some insect pests start damaging the crops during the ripening stage, which also continues during storage. Pruthi and Singh (1950) reported that significant infestation sources are storage bags, old containers, and, most crucially, cross-infestation. The import, export, and distribution of stored products have also paved the way for transboundary pest transmission. Around the world, nearly 1000 species of insects are engaged with stored grains and their byproducts. These can damage grains and pollute the remaining products with an undesirable odour and flavour.

2.1. Classification

Based on feeding habits, the stored grain insect pests could be divided into external and internal feeders.

External feeders: External feeders (or "bran bugs") consume grain powders such as the floor, damaged seeds, and a waste of grain. These pests did not enter the seed. These insect pests did not enter into seeds; hence, their developing stages can be seen (Cotton, 1960).

Internal feeders: These insect pests commonly lay eggs either inside or outside on the outer surface of grains. They spend most of their active larval phase over there and keep on feeding from inside, thereby contributing to potential germination losses that are not detected otherwise, for example, *Sitophyllus* and *Callosobruchus*, etc. (Cotton, 1960).

3. Bruchids

Bruchids are serious pests of stored grains, especially pulses. Genus *Callosobruchus* is distributed worldwide and is ranked as a potential stored grain pest. It causes both qualitative and quantitative damage to seed grains by infecting seeds, their weight, and their germination potential. Females lay eggs on the outer seed cover that enters inside after hatching into the first instar larvae. These larvae have the potential to pierce seed coverings, and the larvae bore the seed vertically for a short distance (Augustine & Balikai, 2019). There are approximately 117 species of bruchids, which belong to the 11 genera and have also been reported to infest various pulse grains (Arora, 1977). All around the world, the genus *Callosobruchus* harms grain legumes during pre- and post-harvest storage. In India, *Callosobruchus maculatus*, *Callosobruchus chinensis*, and *Callosobruchus analis* are the dominant pest species of the genera (Dias, 1986). Bruchids, the seed beetles, are the major pests and internal feeders of stored grain legumes. They are well-known to reproduce on all continents except Antarctica (Southgate, 1979). The most damaging and economically important bruchid species are members of the genera *Callosobruchus*, *Acanthoscelides*, and *Zabrotes* and are severe pests of stored grains (Credland, 1994). Infestation by Callosobruchus in open fields is also a common observation (Raina, 1970; Mohan & Subbarao, 2000).

3.1. *Callosobruchus Maculatus*

Callosobruchus maculatus is generally known as the cowpea beetle (Onyido et al., 2011). This is the most significant pest of stored grain legumes and is also scattered worldwide. The beetle originated in West Africa and accidentally crossed geographical boundaries due to transboundary trade.

3.1.1. Nomenclature and Distribution

Callosobruchus was named by Pic (1902) as a new subgenus of the *Bruchus*. Later, Bridwell (1929) proposed its generic status and named it *Bruchusscutellaris*. Fabricius (1775) recognised this species maculatus from America and put it under the genus *Bruchus*. Although it had six generic synonyms, Hoffman (1945) finally altered the name to *Callosobruchus maculatus*. It is commonly known as Dhora locally. The insect is kept in the order of Coleoptera. This insect is ubiquitous and is distributed throughout tropical and subtropical regions (Dobie et al., 1984).

3.1.2. Morphology

Adult bruchids can be simply differentiated by the shape, size, and length of their abdominal part. Adults are dark brown or black. The head includes black, elongated, frons, carinate, and serrated antennae. The antennae of males are longer than the antennae of females. Elytra is dull-coloured. Males have a relatively shorter abdomen, while the dorsal region of the terminal section is stridently curved downwards as well as inwards. On the contrary, the female bruchids have a fairly longer abdominal region, while the dorsal side of the posterior region moves a little bit downward. The adult female elytra has a white pubescence, forming two different C-shaped regions (Bandaara & Saxena, 1995).

3.1.3. Developmental Biology of C. Maculatus

The adult female beetles usually live for more than 12 days. During this survival period, the adult females oviposit several eggs (*Callosobruchus maculatus* (115), *Callosobruchus chinensis* (70), and the number of eggs laid is reduced if the seeds that are already infested are provided to the beetle (Parr et al., 1998). Few adult females may possess the ability and build up the senses to differentiate their oviposition markers and flout the oviposition markers given by other females (Wijeratne & Smith, 1998). *Callosobruchus maculatus* females lay eggs on the exterior surroundings of pulse grains; after completing the incubation period, most of the eggs hatch into larvae. The adult female deposited the egg with glue, which helps in the firm attachment of the egg to the surface of the seed. Many times, the female prefers the smooth seeds for ovipositing the eggs and avoids the other rough variety seeds (Parr et al., 1996). The eggs of *Callosobruchus maculatus* were oval with flat bases. When pests emerge as adults and come outside the grain, they become completely mature within a day. The adult survived for approximately two weeks. During declining and low temperatures, the range of developmental egg larval and pupal stage durations and vice versa increases (Raina, 1970). The most important point is that this beetle can tolerate a range of humidity and temperature, which makes it adjustable in different climates worldwide. The range of 25% to 80% humidity was appropriate for each developing life stage. Most egg hatching and endurance occur between 44% and 63% humidity. Even adults can live longer at 81% to 90% humidity (Xu, 1999; Fox & Reed, 2011).

3.1.4. Oviposition Period

Callosobruchus maculatus survives well in pulses (Figures 1 and 2). Zalavadia (1971) observed 5.10 hours, 4.5 days, and 0.66 days of pre-oviposition, oviposition, and post-oviposition periods in *Collosobruchus. maculatus*. While these values in the same ovipositional phases were observed to be 7.40 ± 0.99 hours, 7.88 ± 1.20 days, and 1.56 ± 0.58 days by Vyas (2004). According to Moreno et al. (2000), the adult female also shows an oviposition period with an average duration of 10.2 days. Patil (2007) studied the oviposition period for *Callosobruchus maculatus* and found 7.33 ± 0.57 days between a range of 5 and 10 days. Sharma et al. (2007) studied the life cycle of *Callosobruchus maculatus* on the grains of the soybean variety JS-335 and recorded an oviposition period in the range of 4–7 days with an average of 5 days. Kazemi et al. (2009) studied the ovipositional periods for *Callosobruchus maculatus* on different pulses, but the pre-ovipositional period was not observed in this bruchid pest. Ranaware (2009) studied the life table for *Callosobruchus serratus*, and the pre-oviposition and oviposition periods were 1 and 14 days, respectively. The oviposition in *Callosobruchus chinensis* was also studied by Chakraborty et al. (2015) and found 6.36 hours, 4.21 days, and 4.15 days, respectively, in F1 generations, while these parameters were varied for F2 generations. Savita and Anandhi (2010) studied the pre-oviposition, oviposition, and post-oviposition periods for *Callosobruchus chinensis* and observed 0.4, 8.0, and 2.2 days, respectively. The oviposition period and post-oviposition period were found in the range of 6.6 to 7.9 and 1.09 to 2.6 days, respectively, on different studied pulses (Sharma et al., 2016; Kavitha et al., 2018). Sharma et al. (2018) studied the biology of the pulse beetle (*Callosobruchus maculatus*) in laboratory conditions. The average oviposition and post-oviposition periods were recorded at 8.2±0.5 and 2.8±0.5 days, respectively. Similar results were recorded by Hosamani et al. (2018), Augustine and Balikai (2019), and Mehta and Negi (2020).

Figure 1. Pest infestation on cowpeas.

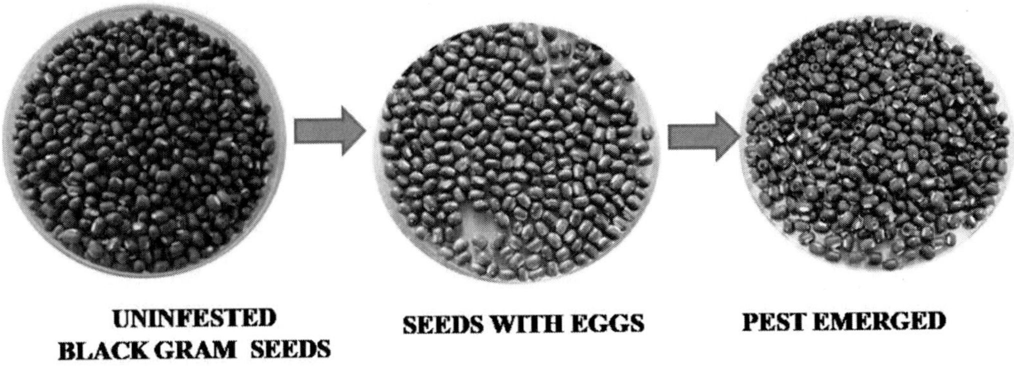

Figure 2. Pest infestation on black grams.

3.1.5. Incubation Period of C. Maculatus

The average incubation period of *Callosobruchus maculatus* was recorded at 3.5 days in the moog bean (Raina, 1970). During the study of the life of *Callosobruchus* on other various pulses, the average incubation period was 10.65, 11.39, 11.12, 10.62, and 10.82 days in mung, moth, urad, tur, and soybean, respectively (Zalavadia, 1971). Singh and Kumari (2000) found an incubation period within the range of 4-5 days. Vyas (2004) observed that the average incubation period was 3.98 ± 0.48 days. Patel et al. (2005) reported that incubation periods varied significantly among the different host grain pulses. The lowest incubation period (4.10 days) was recorded on lentils among all pulses. The mean incubation period was observed to be 5.03 ± 0.11 days by Patil (2007). The mean incubation period was the shortest on cowpea seeds (4.40 ± 0.54) and the longest on soybean seeds (7.20 ± 0.44). Varma and Anandhi (2010 studied the biology of *Callosobruchus chinensis* and found the average incubation period was 4.0 days (Bhargava et al., 2008).

The average incubation period of *Callosobruchus chinensis* varies between 4.16 and 4.27 days (Borude et al., 2012). At 30° C, the average incubation period was 6.27–7.08 days, while the incubation period decreased and increased with the temperature. Sharma et al. (2016) studied the life table of *Callosobruchus maculatus* in different pulses (black gram, green gram, chickpea, cowpea, and pea) and found an incubation period of 5-7 days. Sharma et al. (2017) studied the biology of pulse beetles (*Callosobruchus maculates*), and the average incubation period was found to be 4.2±0.2 days with hatching at 98.2±0.3 per cent. Jaiswal et al. (2018) studied the biology of *Callosobruchus chinensis* on chickpeas and reported the incubation period to be 4.15 ± 0.87 days. A similar study was done by Kumar and Kumar (2018), Augustine and Balikai (2019), and Mehta and Negi (2020).

3.1.6. Larval Pupal Period

Raina (1970) studied the combined larval and pupal periods for *Callosobruchus maculatus* and found 18.8 days. The duration of the larval stage was 11.0 days when larvae were reared on mung, moth, and urad, whereas it was 12.41 days when larvae were reared on cowpea. The pupal development duration was 8.35 ± 3.01, 7.16 ± 2.10, and 8.98 ± 2.08 days in mung, moth, and urad, respectively, according to Zalavadia (1971). Singh and Kumari (2000) found the larval and pupal periods to last approximately 20 days. Joshi and Ghorpade (2001) constructed life tables for the pod borer, *Callosobruchus serratus*, reared on groundnut kernels in the laboratory.

Vyas (2004) studied the life cycle of Callosobruchus chinensis on cowpeas and found the average duration of the larval stage was 16.11 ± 0.88 days. Varma and Anandhi (2010) studied the larval and pupal periods of *Callosobruchus chinensis* and found a total of 16.4 days. The highest larval-pupal duration (29.33 days) was observed in soybean grains, whereas the lowest period was observed in gram grains (19.37 days) (Borude et al., 2012). The developmental biology of *Callosobruchus maculatus* in laboratory conditions was carried out on five different pulses. Larval-pupal periods were in the range of 20.1 to 37.2 and 31.6 to 39.1 days, respectively, on various pulses (Sharma et al., 2016; Kavitha et al., 2018). Sharma et al. (2018) also assessed the biology of *Callosobruchus maculatus* in laboratory conditions and found an average larval-pupal period of approximately 21.3±0.3 days. Similar to Sharma et al. (2017), Hosamani et al. (2018), Augustine and Balikai (2019), and Mehta and Negi (2020).

3.1.7. Developmental Durations

Srivastav and Pant (1989) observed that the duration of emergence of *C. maculatus* from seed grain was the shortest on cowpeas (28.81 days) and the longest on soybean (50.25 days). Wijenayake and Karunaratne (1999) studied the comparative development response of *Callosobruchus chinensis* for six different pulses. They found that development was comparatively faster on mung beans in comparison to black and white varieties of cowpea and chickpea. Patel (2002) recorded the developmental period of the pulse beetle higher on pea than green gram values, 23.49 days and 17.19 days, respectively. Vyas (2004) observed the life cycle of *Callosobruchus chinensis* on cowpea and the total developmental period was 20.09 ± 0.91 days.

Sharma et al.'s (2007) laboratory assessed the biology of *Callosobruchus maculatus* with a total developmental period of 40–50 days (average 43.9 days). Bhargava et al. (2008) studied the biology of *Collosobruchus chinensis* on different pulses and found the total developmental period to be shortest on cowpea (24.60 ± 1.81 days) and longest on soybean (44.80 ± 3.14 days). Malaikozhundan and Raj (2012) observed the life cycle of the pulse beetle on several pulse varieties and observed the developmental durations of 28, 27, 33, 35, and 20 days on the green gram (Km-2, VBN2), black gram (VBN3, VBN4), and cowpea (CoCp-7, Co-6), respectively. Adhikary and Anandamay (2012) found that the developmental period of the cowpea beetle (*Callosobruchus maculatus*) was the fastest on Bio-L-212 Ratan at 30 °C, with greater damage due to the recurrence of numerous generations yearly. Borikar and Pawar (1996) suggested that the rate of reproduction of *Callosobruchus chinensis* was increased by 0.080 (rm) and 1.08 (λ) per day. The beetle population multiplied 11.34 times between two succeeding generations, and a generation completed in approximately 30.51 days. Chakraborty et al. (2015) studied the life cycle of the pulse beetle *Callosobruchus chinensis* on the stored seeds of cowpeas. The total development time was found to be 33 days (July–August) and 37.3 days (April–May). Kumar and Kumar (2018) studied the biology of *Callosobruchus chinensis* in Gram. The total developmental period was in the range of 31.6 to 39.1 days on various pulses (Sharma et al., 2016; Kavitha et al., 2018). Sharma et al. (2018) observed an average total life period of 33.3 ±2.4 days. Similar to Hosamani et al. (2018), Augustine and Balikai (2019), Mehta and Negi (2020), and Kalpna et al. (2022).

3.1.8. Longevity

Raina (1970) studied the life span of *Callosobruchus* spp. on mung beans and recorded those adult males survived for 6 to 11 days, while adult females lived approximately 5 to 10 days.

Zalavadia (1971) observed that the average adult longevity was 11.75, 8.66, 8.0, and 7.83 days for males, while for females, it was recorded at 10.33, 8.0, 8.0, and 6.0 days when the adults were reared on mung, moth, gram, and pea, respectively. Singh and Kumari (2000) studied longevity and reported statistical similarity in the life span of adult male and female pulse beetles. Moreno et al. (2000) also recorded the period of the different states of the cowpea beetle, Callosobruchus maculatus, and recorded those adult females lived for 11.45 days and adult males for approximately 10.9 days. Vyas (2004) observed the developmental biology of *Callosobruchus chinensis* on cowpeas and found the life span of males and females was 9.76 ± 1.17 and 9.44 ± 1.16 days, respectively.

Kazemi et al. (2009) recorded the developmental biology of *Callosobruchus maculatus* and found that the adult life span was the longest on green gram pulse seeds (8.29 ± 0.38 days) while the shortest on cowpea pulse seeds (5.87 ± 0.08 days). Varma and Anandhi (2010 studied the adult longevity of male and female *Callosobruchus chinensis*, which was found to be 11.0 and 9.6 days, respectively. Jaiswal et al. (2018) reported the developmental period of *Callosobruchus chinensis* on chickpea seeds as 9.30 ± 1.08 days, whereas for adult females, it was 10.15 ± 0.98 days. Kumar and Kumar (2018) also recorded the biology of *Callosobruchus chinensis* in grams. The mean adult longevity of males and females was 11.0 and 9.6 days, respectively. The highest adult *Callosobruchus maculatus* survival period was about ten days (Barde et al., 2014; Devi & Devi, 2014; Radha & Susheela, 2014). Sharma et al. (2017), Augustine and Balikai (2019), and Mehta and Negi (2020) also studied the biology of the pulse beetle (*Callosobruchus maculatus*) in laboratory conditions. The average adult life span was recorded at 12.0±2.1 days.

3.1.9. Ovipositional Preference

Dina (1971) observed that *Callosobruchus maculatus* prefers cowpea, pea, black gram, and green gram over Bengal gram and lentil for egg laying. Zalavadia (1971) studied the oviposition preference of *Callosobruchus maculatus* and found that the moth pulse was preferred, with the highest eggs recorded (87.00 eggs) in comparison to other studied pulses such as urad (71.57 eggs), soybean (61.00 eggs), mung (56.85 eggs), tur (49.00 eggs), and gram (35.20 eggs) at a temperature of 28.8°C and relative humidity of 63.6%. Sulehrie et al. (1998) observed the ovipositional response of the Pakistani biotype of *Callosobruchus maculatus* for two different Pakistani varieties, each of *V. radiata* (NM-92 and NCM-209) and *V. mungo* (NARCl and NARC-3), and reported maximum egg laying on NM-92 and minimum on NARC. Raghwani (1998) studied the life cycle of *Callosobruchus maculatus* with different varieties of pigeon peas and observed that *Callosobruchus maculatus* preferred pigeon pea variety ICPL-87119 and laid the highest eggs (220.85 eggs), which were further followed by ICPL-87 (213.89 eggs), GAUT 88-13 (202.55 eggs), and GAUT 85-18 (199.54 eggs), while the lowest number of eggs was recorded in variety BDN 2 (123.85 eggs). Kaur and Ramzan (2001) studied the correlation between seed size and oviposition and found that seed size was positively correlated with egg laying by *Callosobruchus maculatus*.

Lal and Raj (2012) found that small, seeded varieties were preferred for egg laying as compared to large seeds. They also reported that a higher number of eggs (68.23) were laid by the adult female of *Callosobruchus maculatus* on the seeds of the pigeon pea variety PDA-9, while it was lesser on the large seed size variety AMAR (33.78). The adults of *Callosobruchus maculatus* have a great preference for clean and healthy seeds for egg-laying. The order of oviposition preference was cowpea > green gram > black gram > horse gram > green peas

(Barde et al., 2014; Devi & Devi, 2014; Radha & Susheela, 2014). Bharathi et al. (2017) gave a comparative account of the susceptibility of various legume seeds infested by *Callosobruchus maculatus*. Similarly, Sindhura & Godhani (2020) also evaluate the effect on the developmental stages, adult longevity, and fecundity. Results revealed that the order of suitability was observed to be highest for cowpeas, which was further followed by green gram and chickpeas.

4. Management Strategies to Control *Callosobruchus maculatus*

Currently, bruchids have been recognised as the most potential insect pests of stored grains, especially pulses. Lots of strategies have been in practice for many decades. Also, many new management strategies have been developed keeping in mind insect resistance against traditional practices (Figure 3).

4.1. Traditional Pest Management Approaches

Traditional management of insect pests includes alteration and manipulation of the type of environment in stores. By doing so, reproduction and the overall population of insect pests may be controlled to a certain extent. The use of tightly sealed containers, painting walls of storage areas, fumigating in and around stores, separating infected grains, etc. are a few practices that are part of traditional pest management approaches followed, especially in developing countries.

4.1.1. Storage in Sealed Containers

Insect pests may easily enter normal gunny bags or containers; therefore, lined cotton or jute bags are used for packing grains. Internally lined packing bags/ containers may cause choking of the respiratory system in insect pests due to oxygen deficiency. It may cause a decline in the pest population to some extent. One of the main precautions is the safety of such bags from rodents that may eat upon and rupture the bags, giving air passage to insect pests inside. Bags must be filled with grains and tightly sealed so that there is no space for insect pests to enter. It also leads to the asphyxiation of developing larvae, which would otherwise be affected by such practices (Mohapatra et al., 2014).

4.1.2. Sand/Ash

The mixing of inert components in grains is a traditional practice that may act as a barrier between males and females and hinder the process of copulation. It also acts as an asphyxiating agent for various developing stages of insect pests, causing cessation of their development (Tabu et al., 2012; Karimzadeh et al., 2020). Various materials, like paddy, husk, and ash, may be used against *C. maculatus*. (Ashamo et al., 2021).

4.1.3. Intercropping

It is the oldest management practice that has been in use since time immemorial. A mixture of crops that have similar climatic requirements can be sown at the same time to confuse the insect pest. Maize has been considered the best crop to be intercropped with pulses to reduce infection by bruchids (Kabeh & Lale, 2008).

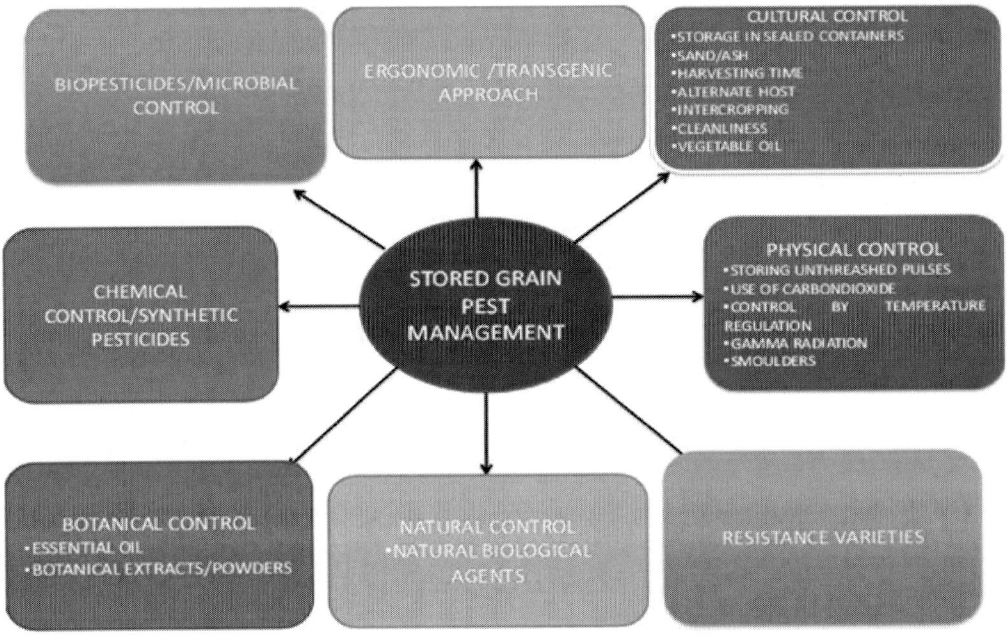

Figure 3. Common management practices for stored grain pests.

4.1.4. Cleaning before Storage

Cleaning the grains and disinfecting them with some organic disinfectants is also a common practice. Clean or disinfected grains are less susceptible to the attack of insect pests. Also, storage areas/ sheds/ stores should be properly washed, cleaned, and disinfected before shifting the grains from the fields. It has been reported that pat drying the grains also prevents the attack of bruchids (Mohapatra et al., 2014).

4.1.5. Smoke Treatment

Smoke therapy is another very traditional method of conserving stored grains. Many household items, like dung cakes and herbal plant leaves, have been in use for a long time. Farmers used to burn dung cakes in and around storage facilities to asphyxiate respiration and kill various developing stages of insect pests. This also does not have any negative effect on seed germination potential or quality. It is considered a very good alternative to chemical pesticides and fumigants (Yadav & Tiwari, 2018). The smoke from cow dung cakes is effective in controlling bruchids, especially in pulses (Kishor & Tiwari, 2021).

4.1.6. Vegetable Oil Coatings

Oil from vegetables or various medicinal plants is one of the best practices that hinder the process of oviposition. Insect pests usually lay eggs on the outer coating of grains. If the surface is coated with oil, the females may not get a proper place for egg laying, and the population of pests will automatically decline. Due to their edible nature, oil coatings do not have any side effects on nutritional or germination aspects. Neem, castor, and cinnamon oils are used in many parts of the world to safeguard stored grains from bruchids (Boeke et al., 2004; Wanderley et al., 2019). All these traditional practices of managing insect pests in stored areas are now less in use. Small-scale farmers in developing countries are still practising these methods to protect their grains from the attack of various pests, but they have certain drawbacks (Gupta et al., 2016). However, modern approaches are now much more frequently and widely used by seed growers or warehouse owners, as they are considered to be quick, reliable, and cost-effective in the present scenario.

4.2. Modern Strategies

4.2.1. Physical Control

Insect pest management through physical means like altering temperature, humidity, pressure, etc. is one of the common practices that are adopted by farmers or warehouse owners (Sahadia & Aziz, 2011). All living beings, including insect pests, require a specific growing environment, especially physical conditions. In the absence of these conditions, eggs may not hatch into larvae, or larvae may not grow into full-fledged adults. Eggs and larvae of *Callosobruchus* spp. can be destroyed by treating them at more than 60–65 °C (Sahadia & Aziz, 2011). Upadhyay and Ahmad (2011) reported a decline in the population of bruchids at humidity levels lower than 10%. Solar heating has also been reported as the best measure of physical control for insect pests (Moumouni et al., 2014). Also, such solar-treated grains were reported to have a higher germination potential than untreated seeds (Maina & Lale, 2004).

4.2.2. Radiation Treatments

Treating grains with various radiation is mostly adopted in large-scale storage areas. Females get sterilised with gamma radiation treatments, and oviposition is automatically reduced, thereby controlling the population to a greater extent. Adult mortality also takes place due to their exposure to harmful radiation; however, this approach has certain limitations like high cost, radiation frequency, and handling. Sometimes, germination also gets affected, due to which this approach to managing insect pests has gained much importance (Mishra et al., 2018). Valizadegan et al. (2009) reported the radiation technique as effective in managing insect pests.

4.2.3. Use of Resistance Varieties

The use of pest-resistance varieties is the first and primary approach that may help in the complete eradication of pests. Domestic and wild varieties may be crossed over to have new varieties that are pest-resistant due to the presence of mixed traits (Badii et al., 2013). Various characteristics, like the outer seed coat, hardness, etc., may act as barriers to pest growth. The presence of a hard seed coat does not allow the oviposition of eggs by females and inhibits the pest population. It may be the easiest, most cost-effective, and most valid approach to controlling insect pests on grains, and nothing much is involved in this. No special training is required, as it is simply the sowing of resistant seeds.

4.2.4. Biological Control

In the modern world, the use of living organisms in the area of pest control is a very common practice (Altieri et al., 2005). It is a popular technique to control the population of stored grain pests by controlling their population at various stages of development. However, control is not 100 per cent (Gahukar, 1994). *Dinaramus* spp. has been used against *Callosobruchus maculatus* (Soundararajan et al., 2012), and *Apanteles flavipesis* is effective in controlling the population of *Bruchus Chinensis* (Eliopoulos, 2006). The practice is good but not popular among low-income farmers due to a lack of awareness about which organisms can be deployed to control insect pests.

4.2.5. Chemical Pesticides

The use of chemicals is the most common practice for controlling insect pests (Dent, 1991; Megerssa, 2010). There are many ways and means, i.e., the use of fumigants, mist, spray, etc., to use chemical insecticides in storage facilities. It has been seen that many bruchid pests have made themselves resistant to chemical pesticides (Talukder, 2009). Chemical pesticides are quick in action; however, they do have some health problems if used in incorrect concentrations. Villagers and farmers usually do not have any idea about the correct doses of pesticides that may prove fatal to consumers. Also, beneficial insects are killed along with harmful ones (Ramzan, 1994). Therefore, in the present world, scientists and researchers are suggesting either no use or minimum use of chemical pesticides for pest control.

4.2.6. Biological Pesticides

The use of microorganisms is the most environmentally friendly and simple technique for pest control. Various types of fungal treatments are being used to control the population of pests. The use of *Trichoderma harzianum* against *Callosobruchus maculatus* and *Callosobruchus chinensis* has been evaluated by Abdelgaleil et al. (2021). However, there is less microbial research, less space and infrastructure, and meagre extension activities in developing countries (Gahukar, 2014).

4.2.7. Transgenic Techniques

These are the modern and emerging techniques that use recombinant DNA technology. It includes gene alteration, modification, DNA marker-assisted breeding, etc. (Collard & Mackill, 2008). Various inhibitors like arcelins, phytohemagglutinin, and β-amylase have the most potential for controlling bruchids (Peumans & Van Damme, 1995; Lioi et al., 2003).

4.2.8. Plant Products

Grain and legumes suffer great losses due to pest attacks around the globe. During the last two decades, researchers have realised the significance of plant products, and more efforts have been made to formulate biopesticides using plant parts. Plant-based pesticides are considered safer and eco-friendly in comparison to chemical and other means to control pest infestations (Table 1). Various plants that are used as whole or in part are *Ocimum* spp., *Azadirachta indica*, *Dennittia tripetala*, *Curcuma longa*, *Anacardium occidentale*, *Alpinia galangal*, *Clusia palmicida*, *Curcuma zedoaria*. The best thing about plant-based pesticides is that their preparation does not require any technicalities. Also, they don't interfere in the food chain, making them highly eco- and health-friendly. Lots of plants have been scrutinised for their insecticidal properties, especially against bruchids (Kalpna et al., 2022).

Table 1. The details of various plants effective against *C. maculatus* in stored pulse (Black gram)

S. No.	The scientific name of the plant	English name	Common name/Local name	Family	Parts used
1	*Terminalia bellirica*	Bahera	Bahera	Combretaceae	Leaves,
2	*Terminalia bellirica*	Bahera	Bahera	Combretaceae	Fruits
3	*Terminalia chebula*	Harad	Harad	Combretaceae	Leaves,
4	*Terminalia chebula*	Harad	Harad	Combretaceae	Fruits
5	*Grewia optiva*	Bhimal	Biul	Tiliaceae	Leaves
6	*Morus alba*	Mulberry	Toot	Moraceae	Leaves
7	*Murraya koenigii*	Curry Leaves	Kari Pata	Rutaceae	Leaves
8	*Ficus carica*	Common fig	Khasra	Moraceae	Leaves,
9	*Ficus carica*	Common fig	Khasra	Moraceae	Fruits
10	*Mentha piperita*	Mint	Pudina	Lamiaceae	Leaves
11	*Azadirachta indica*	Neem	Neem	Meliaceae	Leaves
12	*Ricinus communis*	Castor bean	Arand	Euphorbiaceae	Leaves
13	*Phyllanthus emblica*	Indian gooseberry	Amla	Phyllanthaceae	Fruits
14	*Syzygium cumini*	Jamun	Jamun	Mirtacae	Leaves
15	*Citrus limon*	Lemon	Nimbu	Rutaceae	Leaves
16	*Mangifera indica*	Mango	Aam	Anacardiaceae	Leaves
17	*Anacyclus pyrethrum*	Akarkra	Karkra	Asteraceae	Leaves,
18	*Anacyclus pyrethrum*	Akarkra	Karkra	Asteraceae	Flower
19	*Aegle marmelos*	Stone apple	Bel	Rutaceae	Leaves
20	*Leucaena leucocephala*	White lead tree	Safed Babool	Fabaceae	Leaves
21	*Carissa carandas*	Karunda	Karunda	Apocynaceae	Leaves
22	*Cassia fistula*	Indian Laburnum	Amaltas	Fabaceae	Leaves
23	*Ocimum sanctum*	Basil	Tulsi	Lamiaceae	Leaves
24	*Cannabis sativa*	Hemp	Bhang	Cannabaceae	Leaves
25	*Eucalyptus terticornis*	forest red gum	Safeda	Mirtacae	Leaves
26	*Ricinus communis*	Castor bean	Arand	Euphorbiaceae	Seeds
27	*Leucaena leucocephala*	White lead tree	Safed Babool	Fabaceae	Seeds
28	*Citrus limon*	Lemon	Nimbu	Rutaceae	Seeds

The use of the powder formulation may provide a platform for analysing insecticidal properties among plant species. Various medicinal and insecticidal properties are found in the plants (Bahera, Harad, Karipatta, Mint, Safed babool, mulberry, biul, arand, common fig leaves, etc.) that have been investigated against the attack of the bruchid, *Callosobruchus maculates* (Kalpna et al., 2022). Therefore, available botanicals might be used as protectants against the attack of the bruchid, *Callosobruchus maculatus*, on pulse seeds. Plant parts may be used as pest control agents due to their easy availability, no health risks, and environment-friendly features, especially against *Callosobruchus maculatus* under storage conditions.

Conclusion

The agricultural status and economies of different nations have been enhanced and improved by the efficient and healthy production of stored grains. The main limitation in the production of stored grains, especially pulses, is the outbreak of insect pests, especially *Callosobruchus maculates*, which cause heavy losses in open as well as warehouses. Chemical and synthetic pesticides play a significant role in soil development, especially at the time of grain storage in post-harvest conditions. However, synthetic chemicals cause health hazards and also negatively influence the environment. Insect repellant and killing properties of plants have attracted researchers to formulate powder and extract formulations that are proving highly fruitful in pest control and management. Locally available plant species may be screened for the presence of insecticidal properties against *Callosobruchus maculates* and other pests.

References

Abdelgaleil, S. A., Gad, H. A., Hamza, A. F., & Al-Anany, M. S. (2021). Insecticidal efficacy of two inert dusts and *Trichoderma harzianum*, applied alone or in combination, against *Callosobruchus maculatus* and *Callosobruchus chinensis* on stored cowpea seeds. *Crop Protection*, 146, 105656.

Adhikary, P., & Barik, A. (2012). Effect of temperature on biology of *Callosobruchus maculatus* (F.). *Indian Journal of Entomology*, 74(3), 261-266.

Ahmed, S., & Grainge, M. (1986). Potential of the neem tree (*Azadirachta indica*) for pest control and rural development. *Economic botany*, *40*, 201-209.

Allotey, J., & Oyewo, E. O. (2004). Some aspects of the biology and control of *Callosobruchus maculatus* (F.) on some stored soybean *Glycine max* (L.) Merr varieties. *African Journal of Food, Agriculture, Nutrition and Development*, *4*(2).

Altieri, M. A., Ponti, L., & Nicholls, C. I. (2005). Manipulating vineyard biodiversity for improved insect pest management: case studies from northern California. *The International Journal of Biodiversity Science and Management*, *1*(4), 191-203.

Arora, G. L. (1977). Taxonomy of the Bruchidae (Coleoptera) of Northwest India. Part I. Adults. *Oriental insects*, *11*(sup7), 1-132.

Ashamo, M. O., Ileke, K. D., & Ogungbite, O. C. (2021). Entomotoxicity of some agro-wastes against cowpea bruchid, *Callosobruchus maculatus* (Fab.) [Coleoptera: Chrysomelidae] infesting cowpea seeds in storage. *Heliyon*, *7*(6), e07202.

Atwal, A. S., & Dhaliwal, G. S. (2008). Agriculture pests of South India and their management, Kalyani Publishers, Ludhiana.

Augustine, N., & Balikai, R. A. (2019). Biology of pulse beetle, *Callosobruchus chinensis* (Linnaeus) on cowpea variety DC-15. *Journal of Entomology and Zoology Studies*, *7*(1), 513-516.

Badii, K. B., Asante, S. K. & Sowley, E. N. K. (2013). Varietal susceptibility ofcowpea (*Vigna unguiculata* L.) to the storage beetle, *Callosobruchus maculatus* F. (Coleoptera: Bruchidae). 2 (4), 82-89.

Bandara, K. A. N. P., & Saxena, R. C. (1995). A technique for handling and sexing *Callosobruchus maculatus* (F.) adults (Coleoptera: Bruchidae). *Journal of Stored Products Research*, *31*(1), 97-100.

Barde, A. A., Misari, S. M. & Dike, M. C. (2014). Observations on the biology of *Callosobruchus maculatus* (Fab.) (Coleoptera: Bruchidae) under ambient laboratory conditions. *Afrrev Stech: An International Journal of Science and Technology*. 3 (3), 27-33.

Bharathi, T. D., Krishnayya, P. V. & Madhumathi, T. (2017). Developmental response of *Callosobruchus maculates* F & *C. chinensis* L. on different pulse host-grains. *Chemical Science Review Letter*. 6 (22),786-792.

Bhargava, M. C., Choudhary, R. K. & Yadav, S. R. (2008). Biology and host preference of pulse Beetle (*Callosobruchus chinesis* L.) on different pulses. *Journal of Maharashtra Agriculture University*. 33 (1), 44-46.

Boeke, S. J., Baumgart, I. R., vanLoon, J. J. A., Van Huis, A., Dicke, M. & Kossou, D. K. (2004). Toxicity and repellence of African plants traditionally used for the protection of stored cowpea against *Callosobruchus maculatus*. *Journal of Stored Products Research*.40(4),423-438.

Borikar, P. S., & Pawar, V. M. (1996). Life fecundity tables for pulse beetle, *Callosobruchus chinensis* (Linnaeus) infesting mung bean, *Vigna radiata* (L.) Wilczek. *Journal of Entomological Research*, *20*(1), 59-65.

Borude, D. M., Sonkamble, M. M. & Mutkule, D. S. (2012). Screening of different pulses for biology of *Callosobruchus chinesis* Linn. *Journal of Eco-friendly Agriculture*.7 (1), 98-99.

Bridwell, J. C. (1929). The Cowpea Bruchid (Coleoptera) under another Name.-A Plea for one Kind of Entomological Specialist. *Proceedings of the Entomological Society of Washington, 31*(2).

Chakraborty, S., Mondal, P. & Senapati, S. K. (2015) Studies on biology of the pest beetle, *Callosobruchus chinensis* Linn. infesting green gram. *Current Biotica.* 9 (1),93-97.

Collard, B. C., & Mackill, D. J. (2008). Marker-assisted selection: an approach for precision plant breeding in the twenty-first century. *Philosophical Transactions of the Royal Society B: Biological Sciences, 363*(1491), 557-572.

Cotton, R. T. (1960) *Pest of stored grain on grain products.* United State, Burgescomp. Minniapolis, Minn, p.289.

Credland, A. G. (1994). 1398. The city of london imperial volunteers embarking on the ariosto. *Journal of the Society for Army Historical Research, 72*(291), 187-187.

Dent, D. (1991). Insect pest management Wallingford: CAB International. *Natural enemies*, 295-372.

Devi, M. B., & Devi, N. (2013). Effect of some botanicals on adult mortality, oviposition and development of pulse beetle, *Callosobruchus chinensis* L. on gram. *Indian Journal of Entomology, 75*(4), 295-297.

Dias, C. A. R. (1986). *Ecological studies on Callosobruchus maculates Fab. Callosobruchus chinensis and Callosobruchus analis K. infesting legumes in India* (Doctoral dissertation, Ph. D. thesis, PG school, IARI, New Delhi).

Dina, S. O. (1971). *The reaction of various types of pulses on the life process of Bruchids Callosobruchus chinensis (L.) and C. maculatus (F.)* (M.Sc. thesis, Post Graduate School, Indian Agricultural Research Institute, and New Delhi).

Dobie, P., Haines, C. P. Hodges, R. J. & Prevett, P. F. (1984). *Insect and Arachnids of Tropical Stored Product: Their Biology and Identification (A Training Manual).* Slough, United Kingdom, Tropical Products Institute.

Ekeh, F. N., Oleru, K. I., Ivoke, N., Nwani, C. D. & Eyo, J. E. (2013). Effects of *Citrussinensis* peel oil on the oviposition and development of cowpea beetle, *Callosobruchus maculates* (Coleoptera: Chrysomelidae) in some legume grains. *Pakistan Journal of Zoology.* 45 (4), 967-974.

Eliopoulos, P. A. (2006). Lifetables of *Venturia canescens* (Hymenoptera: Ichneumonidae) parasitizing the Mediterranean flour moth (Lepidoptera: Pyralidae). *Journal of Economic Entomology.* 99 (1), 237-243.

Fabricius, J. C. (1775). *Systema Entomologiae*, p.659.

Fox, C. W. (1993). Multiple mating, lifetime fecundity and female mortality of the bruchid beetle, *Callosobruchus maculatus* (Coleoptera: Bruchidae). *Functional Ecology*,7, 203-208.

Fox, C. W., & Reed, D. H. (2011). Inbreeding depression increases with environmental stress: an experimental study and meta-analysis. *Evolution, 65*(1), 246-258.

Gahukar, R. T. (2014). Factors affecting content and bio efficacy of neem (*Azadirachta indica* A. Juss.) phytochemicals used in agricultural pest control: a review. *Crop Protection.* 62, 93-99.

Gahukar, R. T. (1994). Storage of food grains and insect control in developing countries. *International Journal of Tropical Insect Science.* 15(4-5), 383-400.

Gupta, S., Apte, S. D. & Wasi, N. (2016). Influence of storage containers on *Callosobruchus chinensis* (L.) infesting green gram. *Archives of Applied Science Research.* 8, 21-23.

Hagstrum, D. W., & Phillips, T. W. (2017). Evolution of stored-product entomology: protecting the world food supply. *Annual review of entomology, 62*, 379-397.

Hajam, Y. A., & Kumar, R. (2022). Management of stored grain pest with special reference to *Callosobruchus maculatus*, a major pest of cowpea: A review. *Heliyon, 8*(1).

Hoffmann, A. (1945). *Fauna De France 44. Coleopteres Bruchides Et Anthribides.* Paris, Paul Lechevalier, p. 184.

Hosamani, G. B., Jagginavar, S. B. & Karabhantanal, S. S. (2018). Evaluation of insecticides on field carryover of pulse beetle *Callosobruchus chinensis* (L.) (Coleoptera: Bruchidae) in green gram and cowpea varieties. *Journal of Entomology and Zoology Studies*, 6 (6), 75-77.

Jacobson, M. (1982). Plants, insects, and man their inter-relationships. *Economic Botany*.36 (3), 346-354.

Jaiswal, D. K., Raju, S. V. S., Kumar, D. & Vani, V. M. (2018). Studies on biology of pulse beetle, *Callosobruchus chinensis* (L.) on stored chickpea under laboratory conditions. *Journal of Pharmacognosy and Phytochemistry*.7 (6), 464-467.

Joshi, V. B., & Ghorpade, S. A. (2001). Life-table studies on pod borer, *Caryedons erratus* (Olivier), a pest of groundnut in storage.

Kabeh, J. D., & Lale, N. E. S. (2008). Role of host plant resistance and intercropping in reducing field infestation of cowpeas (*Vigna unguiculata* (l.) Walp.) by bruchids in the Nigerian Sudan Savanna. *Academic Journal of Entomology*, *1*(1), 12-18.

Kalpna, Younis, H. A., & Kumar, R. (2022). Management of stored grain pest with special reference to Callosobruchus maculatus, a major pest of cowpea: A review. *Heliyon*, *8(1)*, e08703,https://doi.org/10.1016/j.heliyon.2021.e08703.

Karimzadeh, R., Javanshir, M. & Hejazi, M. J. (2020) Individual and combined effects of insecticides, inertdusts and high temperatures on *Callosobruchus maculates* (Coleoptera: Chrysomelidae). *Journal of Stored Products Research*. 89,101693.

Kaur, M. & Ramzan, M. (2001). Effect of physical characters of soybean on oviposition and development of *Callosobruchus maculatus* (Fabricius). *Journal of Research of Punjab Agriculture University*. 38 (3-4), 207-211.

Kavitha, G., Mahalakshmi, M. S., Reddy, K. B., Reni, Y. P & Radhika, K. (2018). Development of pulse bruchid, *Callosobruchus chinensis* (L.) on different genotypes of green gram under no choice storage conditions. *Journal of Entomology and Zoology Studies*. 6 (5), 975-980.

Kazemi, F., Talebi, A., Fathipour, Y. & Farahani, S. (2009). A comparative study on the effect of four leguminous species on biological and population growth parameters of *Callosobruchus maculatus* (Fab.) (Col: Bruchidae). *Advances in Environmental Biology*.3(3),226-232.

Kergoat, G. J., Delobel, A., Le RuB. & Silvain, J. F. (2008). Seed beetles in the age of the molecule: recent advances on systematics and host-plant association patterns In: Jolivet P, Santiago-Blay J, Schmitt M, (eds.) *Research on Chrysomelidae*. Leiden, Brill, pp. 59-86.

Khanvilkar, S. V., & Dalvi, C. S. (1984). Carryover of pulse beetle, *Callosobruchus maculatus* (Fabr) infestation from field and its control. *Bulletin of Grain Technology*, *22*(1), 54-61.

Kishor, K., & Tiwari, B. M. R. (2021). Effect of cow dung smoke on pulse beetle, *Callosobruchus chinensis* (Linn.) and quality parameters of stored pulses at Pantnagar, Uttarakhand. *Journal of. Entomology and Zoology Studies*, *9*(1), 1288-1294.

Lal, D., & Raj, D. V. (2012). Mating, oviposition, fecundity and longevity of *Callosobruchus maculatus* (Fab.) on different pigeon pea varieties. *Bulletin of Environment: Pharmacology and Life Sciences*, *1*(11), 12-15.

Lioi, L., Sparvoli, F., Galasso, I., Lanave, C. & Bollini, R. (2003) Lectin-related resistance factors against bruchids evolved through several duplication events. *Theoretical and Applied Genetics*.107 (5),814-822.

Maina, Y. T., & Lale, N. E. S. (2004). Integrated management of *Callosobruchus maculatus* (F.) Infesting cowpea seeds in storage using varietal resistance, application of neem (*Azadiracht aindica* A. Juss) seed oil and solar heat. *International Journal of Agriculture & Biology*, 1560–8530/2004/06–3–440–446.

Mehta, V., & Negi, N. (2020). Biology of pulse beetle *Callosobruchus chinensis* (L.) on three pulses. Messina, F. J. (1991). Life-history variation in a seed beetle: adult egg-laying vs. larval competitive ability. *Oecologia*. 85(3),447-455.

Mishra, S. K., Macedo, M. L. R., Panda, S. K., & Panigrahi, J. (2018). Bruchid pest management in pulses: past practices, present status and use of modern breeding tools for development of resistant varieties. *Annals of Applied Biology*, *172*(1), 4-19.

Mohan, S., & Subbarao, P. V. (2000). First second of *Callosobruchus maculatus* infesting blackgram and greengram under field conditions of Tamil Nadu. *Bulletin of Grain Technology*, *33*(1), 73-74.

Mohapatra, D., Giri, S. K., & Kar, A. (2014). Effect of microwave aided disinfestation of *Callosobruchus maculatus* on green gram quality. *Int J Agric Food Sci Technol*, *5*(2), 55-62.

Moreno, R. A. P., Duque, G, A., Cruz, J. & Trochez, P. A. (2000) Life cycle and hosts of *Callosobruchus maculatus* (Coleoptera: Bruchidae). *Revista Colombianade Entomologia*. 26 (3/4), 131-135.

Moumouni, D. A., Doumma, A. & Seyni, I. S. (2014). Effect of solar drying on the biological parameters of the cowpea weevil, *Callosobruchus maculatus* Fab. (Coleoptera- Bruchinae) in Sahelian area. *Journal of Applied Biosciences*. 84, 7723-7729.

Omar, Y. M., & Mahmoud, M. A. (2020). Effects of three constant temperature ranges to control *Callosobruchus chinensis* (Coleoptera: Bruchidae): a serious pest of pulses in Egypt. *International journal of tropical insect science*, *40*, 1013-1020.

Onyido, A. E., Zeibe, C. C., Okonkwo, N. J., Ezugbo-Nwobi, I. K., Egbuche, C. M., Udemezue, I. O. & Ezeanya, L. C. (2011). Damage caused by the bean Bruchid, *Callosobruchus maculatus* (Fabricius) on different legume seeds on sale in Awka and Onitsha Markets, Anambra State, South Eastern Nigeria. *African Research Review.*5(4), 116–123.

Parr, M. J., Tran, B. M., Simmonds, M. S. & Credland, P. F. (1996). Oviposition behavior of the cowpea seed beetle, *Callosobruchus maculatus*. *Physiological Entomology.*21 (2),107-117.

Patel, V. K. (2002). *Studies on characterization of genotypes and seed storability against pulse beetle (Callosobruchus chinensis) in horsegram.* (M.Sc. thesis, UBKV, Pundibari̅, Cooch Behar, West Bengal).

Patel, V. K., Chaudhuri, N., & Senapati, S. K. (2005). Biology of pulse beetle (*Callosobruchus chinensis* Linn.) as influenced by feeding of different grain pulses. *Agricultural Science Digest*, *25*(4), 254-256.

Patil, D. R. (2007). Population build-up, varietal screening and management of pulse beetle (Callosobruchus chinensis L.) through botanical materials in stored chickpea (M.Sc. thesis, Junagadh Agricultural University).

Paul, A., Radhakrishnan, M., Anandakumar, S., Shanmugasundaram, S., & Anandharamakrishnan, C. (2020). Disinfestation techniques for major cereals: A status report. *Comprehensive Reviews in Food Science and Food Safety*, *19*(3), 1125-1155.

Perez-Mendoza, J., Flinn, P. W., Campbell, J. F., Hagstrum, D. W & Throne, J. E. (2004). Detection of stored-grain insect infestation in wheat transported in rail road hopper-cars. *Journal of Economic Entomology*. 97(4),1474-1483.

Peumans, W. J., & Van Damme, E. J. (1995). Lectins as plant defense proteins. *Plant Physiology.*109 (2), 347.

Phillips, T. W. & Throne, J. E. (2009). Biorational approaches to managing stored-product insects. *Annual Review of Entomology*. 55 (1), 375-397.

Pic, M. (1902). Presumed new beetles of Rhodesia. *Review Entomology.*21,4-7.

Pimentel, D. (1991). Diversification of biological controls strategies in agriculture. *Crop Protection*. 10 (4), 243-253.

Prakash, A., Rao, J., Mukherjee, A. K., Berliner, J., Pokhare, S. S., Adak, T., Munda, S. & Shashank, P. R. (2014). *Climate change: impact on crop pests*. Odisha, Applied Zoologists Research Association (AZRA), pp.1-46.

Radha, R., & Susheela, P. (2014). Efficacy of plant extracts on the toxicity, ovipositional deterrence and damage assessment of the cowpea weevil, *Callosobruchus maculatus* (Coleoptera: Bruchidae). *Journal of Entomology and Zoology Studies*, *2*(3), 16-20.

Raghwani, B. R. (1998). *Management of pulse beetle (Callosobruchus maculatus F.) in stored pigeonpea* (Ph.D. thesis, Gujarat Agricultural University, Sardar Krushinagar).

Raina, A. K. (1970). *Callosobruchus* spp. Infesting stored pulses in India and comparative study of their biology. *Indian Journal of Entomology*. 32, 303-310.

Rajendran, S., & Sriranjini, V. (2008). Corrigendum to "Plant products as fumigants for stored-product insect control." *Journal of Stored Products Research* 44 (2008) 126–135]. *Journal of Stored Products Research*, *3*(44), 304.

Ramzan, M. (1994). Efficacy of edible oils against pulse beetle, *Callosobruchus maculatus* (Fab.). *Journal of Insect Science (India)*. 7 (1), 37-39.

Shankar, U., & Abrol, D. P. (2012). Integrated Pest Management in Stored Grains. In: *Integrated pest management: principles and practice*, CABI, p.386.

Sharma, K., Kumar, R., & Hajam, Y. A. (2022). The laboratory assessment of host specific developmental study of *Callosobruchus maculatus* reared on stored grain pulses. *Journal of Entomological Research*, *46*(suppl), 1118-1123.

Sharma, M., Pareek, B. L. & Sharma, K. P. (2007). Studies on the biology of pulse beetle, *Callosobruchus maculatus* (F.) (Coleoptera: bruchidae). *Soyabean Research*. 5,38-42.

Sharma, R., Devi, R., Soni, A., Sharma, U., Yadav, S., Sharma, R. & Kumar, A. (2016). Growth and developmental responses of *Callosobruchus maculatus* (F.) on various pulses. *Legume Research-An International Journal*. 39(5),840-843.

Sharma, S. S. (1989). Review of literature of the losses caused by *Callosobruchus* species (Bruchidea: Coleoptera) during storage of pulses. *Bulletin of Grain Technology*.22, 62-68.

Sindhura, K. A. & Godhani, P. H. (2020). Comparative biology of pulse beetle *Callosobruchus maculatus* (F.). *Indian Journal of Entomology*. 82(2),232-234.

Sindhura, K. A., & Godhani, P. H. (2020). Comparative biology of pulse beetle.

Singh, D. P., & Ahlawat, L. P. S. (2005). Greengram (*Vigna radiata*) and blackgram (*V. mungo*) improvement in India: Past, present and future prospects. *Indian Journal of Agricultural Sciences*.75 (5), 243-250.

Singh, S. C., & Rina Kumari, R. K. (2000). A study of the biology of *Callosobruchus chinensis* (Linn.) infesting stored pulses (grain legumes) in India.

Solanki, D. K., & Mittal, D. K. (2018). Biology of pulse beetle *Callosobruchus chinensis* in storage condition in gram. *International Journal of Agriculture Sciences, 10(7),* 5682-5686.

Soundararajan, R. P., Chitra, N., Geetha, S. & Poorani, J. (2012). Biological control of bruchid *Callosobruchus maculatus* (F.) in black gram. *Journal of Biopesticide*. 5, Southgate, B. J. (1979) Biology of the bruchidae. *Annual Review of Entomology*, 24 (1), 449-473.

Srivastava, K. M., & Pant, J. C. (1989). Growth and developmental response of *Callosobruchus maculatus* (Fabr.) to different pulses.

Sulehrie, M. A. Q., Golob, P., Moss, C. & Tran, B. M. D. (1998). The oviposition and development of a Pakistani biotype of *Callosobruchus maculatus* (F.) (Coleoptera: Bruchidae) on different host legumes. In: Jin, Z.; Liang, Q.; Liang, Y.; Tan, X.; Guan, L. (eds.), *Proceedings of the 7th International Working Conference on Stored-Product Protection,14-19October1998, Beijing.* China, Sichuan Publishing House of Science and Technology. pp.43-50.

Tabu, D., Selvaraj, T., Singh, S. K & Mulugeta, N. (2012). Management of Adzukibean beetle (*Callosobruchus chinensis* L.) using some botanicals, inert materials and edible oils in stored chickpea. *Journal of Agricultural Technology*. 8 (3), 881-902.

Talukder, F. (2009). Pesticide resistance in stored-product insects and alternative biorational management: a brief review. *Journal of Agriculture and Marine Science*.14, 9-15.

Tsedeke, A. (1985). Evaluation of some insecticides for thecontrol of the beanbruchids in stored haricot bean. *Ethiopian Journal of Agricultural Sciences*. 7, 112-117.

Upadhyay, R. K., & Ahmad, S. (2011). Management strategies for control of stored grain insect pests in farmer stores and public ware houses. *World Journal of Agricultural Sciences*, *7*(5), 527-549.

Valizadegan, O., Pourmirza, A. A., & Safaralizadeh, M. H. (2009). Combination of microwaves radiation and cold storage for control of *Oryzaephilussurin amensis* (L.) (Col. Silvanidae). *Journal of Biological Sciences*, *9*(3), 231-236.

Varma, S., & Anandhi, P. (2010). Biology of pulse beetle (*Callosobruchus chinensis* Linn., Coleoptera: Bruchidae) and their management through botanicals on stored mung grains in Allahabad region. *Legume Research-An International Journal*, *33*(1), 38-41.

Vyas, H. H. (2004). *Biology, varietal screening and management of C. chinensis (Linnaeus) in stored cowpea* (M.Sc. thesis, Junagadh Agricultural University, Junagadh).

Wanderley, M. J. A., Costa, N. P. D., Silva, T., Borges, M., Cruz, G. R. B. D. & Melo, T. D. S. (2020) Use of vegetable oils to control bean weevil development on cowpea grains. *Revista Caatinga*. 32,1117-1124.

Wijenayake, D. U. S., & Karunaratne, M. M. S. C. (1999). Ovipositional preference and development of the cowpea beetle *Callosobruchus chinensis* on different stored pulses.

Xu, W. (1999). Experiments on epidemiology and life habit of the cowpea weevil (*Callosobruchus maculatus* F.) *Zhejiang Nongye Kexue*. (5),222-224.

Yadav, U., & Tiwari, R. (2018). Effect of smoke on insect mortality and quality parameters of stored wheat at Pantnagar, Uttarakhand. *Journal of Entomology and Zoology Studies, 6*(3), 1661-1666.

Zalavadia, R. V. (1971). Some studies on bionomics and control of pulse beetle (Callosobruchus maculatus fabricius) under storage conditions of Saurashtra area. (M.Sc. thesis, Gujarat Agricultural University, Junagadh, Gujarat, India).

Chapter 8

Insect Pests of Forest Plants and Their Management

Anshuman Semwal[1,*]
Rajender Kumar[2]
Ojas Chauhan[1]
Chander Singh[1]
Priyanka Sharma[1]
Nikita Chauhan[1]
Lalit Kalia[1]
VGS Chandel[1]
Ankit Uniyal[3]
Aakriti Gusain[3]
and Ravindra Chauhan[4]
[1]Department of Entomology, Dr. YSP, UHF, Nauni, Solan, HP, India
[2]Department of Tree Improvement and Genetic Resources· Dr. YSP, UHF, Nauni, Solan, HP, India
[3]Department of Entomology, GBPUA&T, Pantnager, Uttarakhand, India
[4]Department of Entomology, VCSG, UUHF, Bharsar, Pauri Garhwal Uttarakhand, India

Abstract

According to the Food and Agriculture Organisation of the United Nations (FAO), land with tree crown cover (or stand density) of more than about 20% of the total area is defined as 'forest' in 'developed regions', whereas in 'developing countries', ecosystems with a minimum of only 10% tree (and/or bamboo) crown cover can be referred to as 'forest'. Insects serve important functions in the ecosystem of forests, both from a beneficial or neutral perspective and as "pests." Insects interact with trees and the surrounding abiotic and biotic conditions on multiple, complex levels. Trees are just as vulnerable to insect destruction as agricultural produce. It is only possible to control insects and diseases if their precise identification, biology, most susceptible stage, peak activity period, and environmental relationships are known. The objective of forest pest management should be to keep all system components in a state of equilibrium that increases the chance of long-term sustainability. Effective tree pest management requires careful consideration of ecological and economic concerns. Identification of the pest, its biology, and the kind of damage are among the elements that influence the selection of appropriate management

* Corresponding Author's Email: anshumanentomology@gmail.com.

In: Insects as Humans' Frenemies: From Friends to Foes
Editor: Istkhar Rao
ISBN: 979-8-89530-320-7

tactics and treatments, if any. In the end, judgements on pest management reflect a compromise between the value of the product, the level of pest damage, the relative efficacy and expense of control methods, and the influence on the environment. Therefore, forest pest management has a solid conceptual, methodological, and technical basis for incorporation into forest resource management planning and operations. To make pest control an integrated element of forest resource management, the "wait and see" strategy must be replaced with a "look ahead" policy. Strategies for pest monitoring and control must be completely integrated into the whole planning and decision-making process. Conventional control methods continue to be the primary instrument for the majority of pest situations and the foundation of integrated pest management systems. As a result, forest managers and aspiring entomologists must be well-versed in a variety of pest control techniques and alternatives for using them separately or in combinations as a component of forest pest management.

Keywords: forest, insect pests, integrated control methods, efficacy, biological control

1. Introduction

Forests are intricate ecosystems that not only contribute to the well-being of rural communities by providing essential goods like lumber, fuelwood, fibre, and other non-wood forest products but also supply several other important goods. Aside from providing ecosystem services like preventing desertification, protecting watersheds, enhancing carbon sequestration, and maintaining biodiversity, they also play an important role in preserving social and cultural values. Fire, pollution, invasive species, insects, and diseases pose a serious threat to the protection of these valuable resources (Bajpai, 2001). It is well known that forest trees are host to many kinds of insect pests. Apart from insects, other types of animals might act as tree pests, including mites, weeds, vertebrates, and nematodes. They can hurt tree growth, vitality, and survival; the yield and quality of wood and non-wood products; wildlife habitat; recreation; and aesthetic and cultural values. Since forestry practices were adopted systematically, pest problems affecting forest trees have been better understood by man. Forest trees are susceptible to pest attacks in all parts, including roots, stems, leaves, shoots, and terminal leaders. Depending on the severity of the damage, the harvestable product can be negatively impacted (FAO 2009). In some cases, the effect will be too slight to affect the value of the harvested product, while in other cases, the damage will cause the trees to be stunted or killed (Tomlinson et al., 2010). In response to these concerns, methods for controlling pests were developed. Therefore, forest protection refers to shielding (preventing it from being destroyed by danger) forests from the dangers that could either destroy or damage them enough so that they do not yield the expected benefits. It can also be defined as forestry activities that focus on preventing and controlling damage to forests by human action, animals, insects, fungi, and adverse climate conditions. (Gupta & Sankaran, 2021) Forests are severely damaged by insects. There is no specific stage of a plant's life in which insects cannot cause injuries; from seed to final plant product, it can be observed throughout the forest plant growth cycle. Weevils, pulse beetles, and several species of moths can attack the seeds even before they are collected. Forest plant nurseries may suffer damage from leaf eaters like cutworms, grasshoppers, and crickets or root eaters like termites and white grubs. Plants often suffer severe damage from defoliators, bark borers, cambium-feeding insects, and sap-sucking insects when they reach the sapling and pole stages. (Verma et al., 1996) Thus, to protect forest trees against insect injuries, it is important to detect and assess the attack as soon as possible because, by delaying the action, insects are more likely to multiply, causing more damage to the forest. An effective tree pest management strategy should incorporate ecological and economic factors (Khan et al., 1988). Pest biology, damage type, and climatic conditions for its development are all factors that determine which control methods and strategies are most appropriate for pest management. An effective pest management plan is based on a balance between product value, pest damage, the effectiveness of control measures, and environmental impact. The goal of management should be to keep the balance between the pests and hosts at a level where an economically significant injury won't happen (Kumar et al., 1994).

2. Detection and Assessment of Damage

Monitoring and sampling are two important factors influencing insect pest management in forest areas that need to be discussed in detail (Dhaliwal et al., 2013). A detection and assessment management programme for any insect population should begin with population monitoring, which is the basic requirement for damage assessment while sampling this population provides an estimate of the total population of this pest in a given area.

Insect pest monitoring and their natural enemies are vital tools in IPM—the process of making management decisions based on that information. The term "monitoring" refers to the practice of inspecting your farm, forest, or structure for the presence of pests, how many they are or how much devastation they have wreaked. The likelihood of an infestation and the most effective method of control depends on the identification of the pest; in other words, "monitoring can serve as a means to ascertain the spatial dispersion of pests or to evaluate the efficacy of pest management interventions" (Dhaliwal et al., 2013). The following objectives need to be considered for monitoring the damage assessment:

A. Assessment of variations in insect populations and their spread in forests.
B. Details regarding the development and biology of insects in forests.
C. Impact of biotic and abiotic variables on the forest pests' population.

Surveying insect pest populations is an important aspect. An insect pest survey refers to a comprehensive gathering of data on the insect population within a specific geographical location at a particular point in time. The forest pest survey may be divided into qualitative and quantitative surveys. The primary objective of a qualitative survey is typically focused on the detection of pests, and this includes a list of pest species along with density references such as common, abundant, and uncommon. They are often used to determine the extent of an outbreak of newly imported pests in a forest region. These surveys are usually carried out at international borders, where goods from the forestry industry are examined to prevent the introduction or invasion of new species of pests. In a quantitative survey, the study of pests provides a numerical definition of the density of the pest population throughout both time and geography. It gives information about the destructive capacity of a species, and the data collected from it may be used to anticipate future patterns of population growth. This examination provides a framework for decision-making that is necessary to take management measures for a pest that causes economic damage to forest plant species (Dhaliwal & Arora, 2003).

Based on the damage, insect pests can be classified as defoliator insects and plant sap-sucking insects. A defoliator is an insect that consumes leaves or needles, causing the foliage to become chewed, mined, or missing. There are a variety of factors that contribute to defoliating insect damage, including tree species, feeding intensity, and the time of the year when feeding occurs. Most defoliating insects cause little damage to trees, but a few may cause significant damage. A small number of defoliators are significant forest pests that may cause widespread tree mortality during outbreaks. The larval stages of moths, butterflies, and sawflies are the most important defoliating insects in conifers. The defoliators may be:

i. *Leaf miners:* Chlorophyll-bearing tissues are consumed by these organisms between the epidermal layers. Example: Maple leaf blotch miner (*Cameraria aceriella* Clemens).
ii. *Skeletonizers:* On one side of the leaf, these insects remove the chlorophyll-containing tissues and the epidermis, The epidermis and veins are left intact. Example: Birch leaf skeletonizer (*Bucculatrix canadensisella*).
iii. *Whole leaf feeders:* All leaf tissue is consumed or removed by these insects as they feed on leaf margins. Example: Black-back prominent moth (*Clostera anastomosis*).
iv. *Shot-hole feeders:* The larvae do not feed on the leaf margins because they consume all leaf tissues by forming chew holes in the leaves. Example: Maple leafcutter larvae (*Paraclemensia acerifoliella*).

Sap-sucking insects consume the sweet sap generated by the leaves and transmitted via the phloem tissue under the bark. Certain insect feeding alters plant growth hormones, generating galls, which are distinguishable malformations of plant leaves or shoots. Examples include black pine needle scalea (Nuclaspis californica) and giant conifer aphids (Cinara spp.). A broad description of the insects damaging forests is given in the following section.

3. Insects Damaging the Forest

3.1. Defoliators

3.1.1. Teak Defoliator, Hyblaea Puera (Cramer) (Lepidoptera: Hyblaeidae)

This defoliator commonly grows along the southern and western ghats of India, Maharashtra, and Uttarakhand. This is an oligophagous pest, although its primary host is teak (Dhobe & Naik, 2015).

3.1.1.1. Identification Characteristics

The moth's forewings are greyish brown, while the hindwings are marked by black and orange-yellow colour. The wingspan is three to four centimetres in an adult moth. The larvae are defoliators. Pupae have a brownish colour and have an opaque nature (Seervi et al., 2022).

3.1.1.2. Developmental Process

The fertilised female deposits her eggs singly along the veins on the underside of the young leaves. A single female may produce around 400 eggs. The egg hatches in about two days, and the larvae consume the foliage of the teak tree. The larval stages have a duration of 10-12 days in summer and in winter it extends up to 20 days. The mature larva falls to the earth with the aid of silken threads and pupates there (Kabade et al., 2015).

3.1.1.3. Nature of Damage

Only the caterpillars are harmful to the teak plants. They consume the young and fragile leaves of the teak tree, which stunts the plant's development. The pest results in around 44% of the damage (Tripathy et al., 2018).

3.1.1.4. Management Practices

1) Collection of egg masses, larvae, and pupae, followed by destruction, is one of the management practices (Klapwilk et al., 2016).
2) Clean cultivation and field excavation to expose pupae to environmental mortality factors aid in minimising the pest's prevalence.
3) In the larval stage of this teak defoliator, several parasitoids such as *Brachymeraia lasus*, *Eriborus gardneri*, and *Symplensis* sp. account for around 70% of mortality.
4) Pheromone traps may be used to attract adults of pests.
5) The use of 0.1% carbaryl is effective against this pest's larvae.

3.1.2. Teak Skeletonizer Eutectona Machaeralis (Walker) (Lepidoptera: Pyralidae)

The teak skeletonizer is common in the western ghats, Maharashtra, Uttarakhand, Andhra Pradesh, Tamil Nadu, and Karnataka (Dhobe & Naik, 2015).

3.1.2.1. Identification Characteristics

The pest affects teak and several *Callicarpa* species. On the white forewings of the brilliant yellow moths are fulvous or pink transverse markings in a zigzag pattern or serrate lines. The hindwings are white with a marginal line or band that is ochreous or reddish. The temperature and the humidity both have an effect on the design and colour of the wings (Alam et al., 2004).

3.1.2.2. Developmental Process

The fertilised female deposits her eggs on the plant's sensitive leaves or in the vicinity of tender leaves. There are about 250 eggs laid on the underside of leaves by a single female. There are five larval instars present in the pest life cycle. The last instar larvae will pupate along with the leaves as they fall. During winters, the pest hibernates from January to March. The developmental processes (egg-adult) are completed within one month (Mukesh et al., 2022).

3.1.2.3. Nature of Damage

The introduction of rain in July marked the beginning of the pest infestation, which had been ongoing since July. The larvae consume all of the leaf's green substance while leaving the veinlet network unharmed, effectively removing the leaf from the plant. The final instar can feed on older leaves of teak plants. Due to the lower amount of chlorophyll, the growth may be suppressed (Nayak et al., 2002).

3.1.2.4. Management Practices

1) Collect the egg masses as well as the larvae and pupae, and then destroy them.
2) Clean cultivation and field excavation to expose pupae to environmental mortality factors aid in minimising the pest's prevalence (Dhaliwal & Arora, 2003).
3) (3). The egg parasitoid *Tricogramma raoi* was reported effective when it was released at a rate of 1.50 lakhs per hectare, which resulted in a significant reduction in the population of leaf skeletonizer larvae, pupae, and adults (Shukla et al., 2003).
4) Pheromone traps may be used to attract pest adults (Dhaliwal & Arora, 2003).
5) The use of 0.1% carbaryl is effective against this pest's larvae (Bajpai, 2001).

3.1.3. Pink Gypsy Moth Lymantria Mathura (Moore) (Lepidoptera: Lymantriidae)

The pink gypsy moth is widespread in western ghats, Assam, the Himalayan area, Uttarakhand, and Sikkim. This pest is found in climates ranging from cold temperate to warm and in permanent forests, temperate coniferous, tropical and subtropical broadleaf dry and wet forests. It is a polyphagous pest that causes extensive plant damage. Significant pink gipsy moth host plants include Sal, Asna, Australian red cedar, zelkowa, cherry, willow, peach, apple, ash, walnut, oak, pine chestnut, and wax tree, among others (Nair, 2007).

3.1.3.1. Identification Characteristics

Male moths have black patches on their vertices. The base colour of the forewing is lighter, while the dorsal side is black and marked with zigzags. The hindwings are orange, with a black mark at the end of each cell and a curving band of submarginal dots. Females are distinguished by their white frons and fuscus on the head and thorax and their two black dots on the vertex. The abdomen is crimson, with tiny black markings on the head. White forewings with red and black dots and characteristic zigzag markings. The male wing span is 40–51 mm, and the female wing span is 96–112 mm. The larva is dark and covered with bristles (Zlotina et al., 1999).

3.1.3.2. Developmental Process

Eggs are laid by the female on the leaves of the host plants. A single female may lay hundreds of eggs, which she does in clusters (Deepa & Remadevi, 2006). The entire time spent as a larva is around eighty-five days, during which it goes through five or six instars. Pupation takes place in the soil, and it might take anywhere from 10 to 15 days for the adult to emerge. Each year, just one generation is produced (Nair, 2007).

3.1.3.3. Nature of Damage

In extreme infestations, the caterpillar feeds en masse on flowers and foliage and skeletonises the whole plant. Young shoots are annihilated, and utter devastation is not unusual. (Zlotina et al., 1999).

3.1.3.4. Management Practices

1) collect and destroy egg aggregates as well as gregarious and solitary larvae. (Dhaliwal & Arora, 2003).

2) Dig up the ground in the summer to expose pupae to natural causes of death.
3) The moths can be caught and killed with pheromone traps.
4) Adults are mostly active at night; thus, a light trap is the best way to get rid of pests via adults (Klapwilk et al., 2016).
5) 0.03% azadirachtin should be sprayed on the crop.

3.1.4. Indian Gypsy Moth Lymantria Obfuscate (Walker) (Lepidoptera: Lymantriidae)

The Indian gipsy moth is found in Himachal Pradesh, Uttarakhand, Northwest Himalaya, and Jammu & Kashmir. The insects eat willow, popular, apricots, apple, and walnut, among other hosts (Showket et al., 2020).

3.1.4.1. Identification Characteristics

The male moth is brownish-brown with a 32-mm wing span. There are post-medial double lines on the forewings more regular and with a black lunule at the end of each cell on the hind wings. The larvae have a light brown colour with short dorsal and lengthy lateral hair tufts. In addition, the larva has a dark brown dorsal band with central and lateral lines. The body length of mature larvae is around 40 to 50 mm (Inouea et al., 2019).

3.1.4.2. Developmental Process

In June and July, the female deposits her light greyish brown, ovoid, and shiny eggs on the bark in clusters. Each batch contains 200 to 400 eggs and is coated in yellowish-brown hair. The winter is spent in the egg stage, and hatching occurs in March or early April. The mature larvae collect on the tree bark, under the surface of branches, behind debris and stones, and pupate in the soil inside the debris. The pupal phase lasts between 9 and 21 days. Each year, just one generation is produced. Males live 4–10 days, whereas females live 11–31 days (Showket et al., 2020).

3.1.4.3. Nature of Damage

The sole damaging stage of the insect is the caterpillar, which consumes leaves voraciously and skeletonises the plant in severe infestations. Caterpillars are voracious eaters that feed throughout the night. It is a significant pest of willow in the Kashmir valley and the north-western Himalayas (Dowden, 1935).

3.1.4.4. Management Practices

1) Collect and destroy caterpillar egg masses and gregarious forms.
2) Provide refuge for mature larvae at the base of a tree, then remove, gather, and kill the larvae.
3) The light trap is an economic-friendly management practice in the crop.
4) Utilising clean cultivation to expose larvae to natural causes of mortality is beneficial.
5) Dig the crop to expose pupae to natural causes of death (Klapwilk et al., 2016).
6) The pathogens, particularly *Entomophaga maimaiga* and the nucleopolyhedrovirus, have been discovered to show promising results in the conventional biological control of the harmful gipsy moth in the forest (Showket et al., 2020).
7) Apply dimethoate (0.03%), phosphamidon (0.03%), or carbaryl (0.15%) to the infested crop.

3.1.5. Deodar Defoliator Ectropis Deodarae (Prout) (Lepidoptera: Geometridae)

In Uttar Pradesh, Kerala, and Punjab, the deodar defoliator has been discovered feeding on cedar (*Cedrus deodara*). It has been recorded at altitudes above 1500 meters (Singh et al., 1989).

3.1.5.1. Identification Characteristics

Moths are medium-sized insects with underdeveloped fore and hind wings and many irregular black spots. The pupa is brownish and of the obtact type (Thakur, 2000).

3.1.5.2. Developmental Process

In the spring, diapausing pupae emerge as moths, which mate shortly after emergence. After mating, females begin climbing the deodar tree to deposit their eggs on its needles. In spring, eggs are deposited in large numbers and hatch within 2–5 days. Pupation is done in the soil under a layer of humus. Hibernation occurs in the pupal stage (Singh et al., 1989).

3.1.5.3. Nature of Damage

Caterpillars are the sole harmful stage of the insect problem. The caterpillar is the most damaging to the deodar tree because it feeds on the needles. In extreme infestations, they skeletonise the whole plant (Singh et al., 1989).

3.1.5.4. Management Practices

1) Retrieve and eliminate the egg masses as well as the caterpillar.
2) Dig a field under the shadow of the tree to expose the pupae to the natural elements that cause their demise.
3) Proper culture, which includes exposing pupae to both biotic and abiotic stimuli, contributes to the successful management of the pest (Klapwilk et al., 2016).
4) To combat the adult larvae and pupae of this pest, shrews should be released into the forest.
5) The defoliator's assault can be mitigated by implementing grease bands on the primary trunk. This measure obstructs the wingless female moths from ascending the trunk and depositing eggs on the needles (Beeson, 1941).
6) Dust the crop with carbaryl 5% at a rate of 25 kg per hectare, or spray the crop with azadirachtin at a concentration of 0.03% (Verma & Chander, 1995).

3.2. Sapsuckers

Insects such as suckers, aphids, thrips, coccids, and psyllids, which feed on the sap of leaves and severely affect the trees' health, may be found in forests (Tyagi et al., 2014). Honeydew is produced by some of the species, which, when covered with sooty mould, prevents the process of photosynthesis from taking place (Dhaliwal et al., 2013).

3.2.1. Subabul Psyllid Heteropsylla Cubana (Crawford) (Hemiptera: Psyllidae)

The subabul psyllid is distributed throughout India, Sri Lanka, Indonesia, Thailand, Taiwan, and the Philippines. It is an important pest of *Leucaena leucocephala* (Misra et al., 2001).

3.2.1.1. Identification Characteristics

The adult psyllid is light yellow with 4-5 antennal segments and measures about 0.32 mm in body length and 0.15 mm in width. Nymphs resemble adults except that they are smaller in size and devoid of wings (Mishra, 2020).

3.2.1.2. Developmental Process

Eggs are laid by females on the fragile leaves of subabul twigs, and they hatch after two to three days. The freshly emerging nymphs move around the plant aimlessly in search of more fragile parts to feed on, and once they find them, they begin to sip the cell sap from the more tender leaves. There are five nymphal instars, and the duration of time spent as a nymph is between eight and ten days. Within ten to thirteen days, the life cycle is completed, and adults have a maximum lifespan of seven days. On subabul, up to eight or ten generations may be accomplished in a single calendar year. (Mishra, 2020).

3.2.1.3. Nature of Damage

The nymphs and the adults all feed on the cell sap that is extracted from the most vulnerable parts of the plant, such as the leaves, buds, stems, blooming bodies, and fruiting bodies. By sucking the sap from the cells, psyllids inject toxins into the plant's body. Curled and yellowed leaves and dropping flowers are the results of this. On the leaves and other plant parts, nymphs and adults secrete honeydew, which attracts sooty moulds, causing pathogens. Plants are affected by sooty moulds on their leaves, which affect photosynthesis and growth (Mishra, 2020).

3.2.1.4. Management Practices

1) Dispose of the infested plant parts along with the pests.
2) The white muscardine diseases caused by the fungus *Beauria bassiana* cause 64 to 100 per cent mortality of the pest population (Misra et al., 2001).
3) Ladybird beetles (*Menochilus sexmaculatus*) feed on nymphs and adults of the psyllid, which can be manipulated to control the pest (Klapwilk et al., 2016).
4) Varieties K6, K8, K500, and K636 are resistant to this pest (Misra et al., 2001).
5) Spray 0.10% carbaryl on the infested plant part can also be applied for successful control.

3.2.2. Gamhar Lace Bug Tingis Bessoni (Drake) (Hemiptera: Tingidae)

In India, the gamhar lace insect is an occasional threat to young gamhar (*Gmelina arborea*) seedlings. It is common in Uttarakhand, Uttar Pradesh, Kerala, Maharashtra, and Madhya Pradesh. The many little black lace bugs (4.5 x 1.7 mm) gather on the stems and branches of young trees and graze voraciously (Meshram & Bhowate, 2017).

3.2.2.1. Developmental Process

Eggs are placed by females in a vertical row and put into the fragile shoot tissue of the plant. The nymphs concentrate at the point where the lamina of the leaf meets the axilla of the leaf. There are five stages, or instars, of the nymphal life cycle, and the whole process may take anywhere from 11 to 40 days between April and October, depending on the temperature. There is a significant amount of overlap between generations; in fact, one year may see the completion of as many as seven generations. The eggs that were deposited in the cold temperatures survive the winter and don't hatch until March of the following year (Meshram & Bhowate, 2017).

3.2.2.2. Nature of Damage

At the base of the leaf blade, where the bigger veins are located, both the adults and the nymphs eat in enormous numbers by sucking the sap from the plant. The leaves get speckled and discoloured and eventually fall off as a direct consequence of the insect's eating on them. The shoots on the saplings wither and die, and they develop epicormic branching. If the infestation is severe, the plant might lose all of its leaves. The insect assault may be followed by an infestation of fungus, which is characterised by necrotic lesions that are black and appear at the bases of the leaves. This may be followed by the defoliation and drying out of the new shoots. The first step in management is to collect and kill all of the nymphs and adults (Meshram et al., 2001).

3.2.2.3. Management Practices

1) Removing and killing the insect-infested sections of the plant, as well as the nymph and adult stages of the pest, is one of the common practices for management (Kumar & Chandra, 2017).
2) Apply a 0.02% concentration of monocrotophos or a 0.005% concentration of deltamethrin. Spraying a combination of the insecticide and the fungicide, carbendazim (0.1%), should be done in the event when there is an additional fungal infestation (Meshram and Tewari, 2003).

Conclusion

The two ways of forest pest management are artificial and natural. Natural control involves climatic conditions, nutrition, host resistance, and biotic variables, while the artificial control embraces silvicultural, biological, microbial, and chemical controls. Forest ecology consists of several components, all of which are essential for the long-term health of the forest. Roots, stems, leaves, shoots, and terminal leaders are all susceptible to attack by pests. Minor pest damage does not influence the value of harvested crops; however, severe pest damage stunts or kills trees and diminishes their market value. Insects and mites, illnesses, weeds, vertebrates, and nematodes are examples of tree pests. Destructive insects are a vital component of forest ecosystems and have substantial economic, social, and ecological implications for forest production and values. The negative effects should be avoided or maintained at bearable levels using ecologically sound methods that are consistent with forest management practices. Forest-damaging insects are described more precisely in terms of their economic impact and disruptive effect on forest pest control plans and operations. Today, forest pest management makes extensive use of quantitative methodologies to measure and analyse forest growth and development as well as the dynamics of forest insect populations. There is a great deal of diversity and complexity in forest ecosystems. We may be impacted by the health of forests since so many people depend on them. For this reason, it's critical to maintain the forest and shield it from threats like pests.

References

Alam, T., Ali, M. S. & Chaturvedi O. O. P. (2004). Biology and seasonal history of Teak foliage skeletonizer in north Bihar', *Indian Journal of agroforestry*, 85–87.

Bajpai, R. (2001). Major insect pest menance to dominant multipurpose trees of central India: Status and control measures. *Indian Journal of agroforestry*, 3(2), 178-179.

Beeson, C. F. C. (1941). *The Ecology and Control of Forest Insects of India and Neighbouring Countries.* Bishen Singh Mahendra Pal Singh, Dehradun.

Deepa, B., & Remadevi, OK., (2006). Ovicidal activity of the wood of *Pterocarpus marsupium* Roxb. (Fabales:Fabaceae) against the teak defoliator, *Hyblaea puera* Cramer. *Current Biotica*, 5(3), 370-372.

Dhaliwal, G. S., & Arora, R. (2003). *Integrated Pest Management: Concepts and Approaches*. CABI.

Dhaliwal, G. S.., Singh, R., & Jindal, V. (2013). *A Textbook of Integrated Pest Management.* Kalyani Publishers, Ludhiana.

Dhobe, N. S., & K. V. Naik. (2015). Bionomics of Teak leaf skeletonizer, *Eutectona machaeralis* Walker (Pyrallidae: Lepidoptera) under laboratory conditions. *The Indian Forester*, 141(5),554–560.

Dowden, P. B. (1935). *Brachymeria intermedia* (Nees), a primary parasite, and *B. compsilurae* a secondary parasite, of the gypsy moth. *Journal of Agricultural Research*, 50, 495-523.

FAO (Food and Agriculture Organisation), (2009). *Global review of forest pests and diseases*. Rome.

Gupta, K., & Sankaran, K. V. (2021). Forest Biosecurity Systems and Processes: An Indian Perspective. *Frontiers in Forests and Global Change*, 4,1-12.

Inouea, M. N., Ohnob, Y. S., Hagaa, Y., Aaraia, H., Sanoa, T., Martemyanovc, V. V., & Kunimi. Y. (2019). Population dynamics and geographical distribution of the gypsy moth, *Lymantria dispar* in Japan. *Forest Ecology and Management*, 434, 154-164.

Kabade, K. H., & Patel, Z. P. (2015) Biology of teak defoliator (*Hyblaea puera* Cramer, Hyblaeidae: Lepidoptera). *Bioinfolet.* 2015, 12(1c), 343-346.

Khan, H. R., Kumar, S., & Prasad, L. (1988). Studies on seasonal activity of some agro-forestry insect pests by light-trap. *Indian Forester*, 114, 215-229.

Klapwilk, M., Bylund, H., Schroeder., & Bjorkman. (2016). Forest management and natural biocontrol of insect pests. *Forestry: An International Journal of Forest Research.* 89 (3), 253-262.

Kumar, A., & Chandra, G. 2017. Insect pest infestation on *Gmelina arborea* Roxb. in different agroclimatic zones of Jharkhand, India. *Arthropods*, 2017, 6(1), 8-20.

Kumar, M., Gangwar, B., Kumar, P., Chaudhary, A. K., & Seervi S. (2022). Biology of teak skeletonizer (*Eutectona machaeralis* W.) reared in laboratory conditions on teak (*Tectona grandis* Linn.) leaves in semiarid region of Central India. *Journal of Experimental Zoology*. India, 25, 1659-1664.

Kumar, S., Ahmed, S. I., & Kumar S. (1994) Annotated list of insect pests of forest trees of arid and semi-arid regions Rajasthan and Gujarat. *Oikoassay*, 11(12), 5-9.

Meshram, P. B., & Bhowate, S. (2017). Role of lace bug *Tingis beesoni* and control measures in top dying of *Gmelina arborea* plantations. *Journal of Entomology and Zoology Studies*, 5(3),389-393.

Meshram, P. B., Pande, P. K., & Banerjee, S. K. (2001). Impacts of pest problem in *Gmelina arborea* plantations in western Maharashtra. *Indian Forester*, 127(12), 1377-1386.

Meshram, P. B., & Tewari, C. K. (2003). *Lace bug Tingis* beesoni (Drake) causing top dying of *Gmelina arborea* (Linn.) Hi Tech plantation and evaluation of certain pesticide against the bug. *Journal of Applied Zoological Research*, 14(2), 200-252.

Mishra, R. K., & Roychoudhury. (2020). *Leucaena psyllid, heteropsylla cubana* and its control measures. *Vaniki Sandesh*, 11(4),10-15.

Misra, R. M., Mishra R. K., & Bhandari, R. S. (2001). Biological agents of *Heteropsylla cubana* on *Leucaena leucocephala* in Kalsi, Dehra Dun, Uttaranchal. *Indian Journal of Forestry*, 24,403-408.

Nair, K. S. S. (2007). *Tropical Forest Insect Pests: Ecology, Impact, and Management*. Cambridge University Press.

Nayak, B. G., Hanumantha, M., Ganigera, B. S., & Patil, S. K. (2002) Infestation status of teak skeletonizer, *Eutectona machaeralis* Walker (Lepidoptera: Pyralidae) in different location of Utter Kannada District. *My Forest*, 38 (4), 373-377.

Seervi, S., Gangwar, B., Kumar, P., Kumar, M., Lavania, P., & Venkatesh, Y. N. (2022). Biology of teak defoliator (*Hyblaea puera* C.) reared on teak (*Tectona grandis* L.) leaves under laboratory conditions in semi: Arid region of Central India. *The Pharma Innovation Journal*, 11(9),3164-3168.

Showket, A. D., Owais, M., Sajad, H. W., Ejaz, A. D., Kounser, J., Sajad, H. M., Munazah, Y., Showkat, A., Ajaz, A. K., & Rohie H. (2020). Bio-agents (Pathogens, Parasitoid and Predators) of Gypsy moth *Lymantria obfuscata* Walker (Lepidoptera: Lymantriidae) and their mechanism of action and Natural mortality. *The Pharma Innovation Journal*, 9(11), 23-29.

Shukla, P. K., Srivastava, R., Jalil, P. & Sharama, A. (2003). Tree improvement of teak in Madhya Pradesh (India). *Vanki Sandesh*. 27(2&3): 1-9.

Singh, P., Bhandari, R. S., Fasih, M., Prasad, G., Tyagi, B. D. S., Mishra, B. M. & Mishra, R. K. (1989). Epidemic of deodar defoliator, *Ectropis deodarae* Prout (Lepidoptera: Geometridae) in Kamraj Forest Division (J&K) and its control by aerial spraying of Fenitrothion. *Indian Forester*, 115(6): 459-473.

Thakur, M. L. (2000). *Forest entomology ecology and control*. sai publishers.

Tomlinson, P. B., Tomlinson, P. B., & Zimmerman, M. (2010). *Tropical Trees as Living Systems*. Cambridge University Press.

Tripathy, M. K., Rout, M., & Bhol, N. K. (2018). Biology of *Hyblea purea* (Cramer) reared in laboratory with teak (*Tectona grandis*, Linn) leaves at coastal Odisha, India. *Journal of Entomology and Zoology Studies*. 2018, 6(4), 485-491.

Tyagi, B., Veer, V., & Prakash, S. (2014). *Pests of Forest Importance and Their Management*. Scientific Publishers.

Varma, A., Sudhakar, B., Beena, K., & Bhaskar. (1996) Insect pests associated with nurseries of selected tree crops in Kerala. *Kerala Forest Research Institute, Peechi*, pp 468–473.

Verma, T. D., & Chander, R. (1995). Epidemic of deodar defoliator *Ectropis deodarae* Prout (Lepidoptera: Geometridae) in north- west Himalaya. *Indian Forester*, 12 (7), 663-664.

Zlotina, M. A., Mastro, V. C., Elkinton, J. S., & Leonard, D. E. (1999). Dispersal tendencies of neonate larvae of *Lymantria mathura* and the Asian form of *Lymantria dispar* (Lepidoptera: Lymantriidae). *Environmental Entomology*, 28(2), 240-245.

Chapter 9

An Account of "Threat to Humans by Insects: The Culicine Rivalry"

Anushree Singha Ray[1]
and Kuntal Bhattacharya[2,*]
[1]Department of Zoology, Michael Madhusudan Memorial College, Durgapur, Paschim Bardhaman, West Bengal, India
[2]Department of Zoology, Durgapur Government College, Durgapur, Paschim Bardhaman, West Bengal, India

Abstract

In our lives, the most worrisome insects are mosquitoes. Over 3000 discovered species of mosquitoes and 770 species of *Culex* have been documented, which act as potent insect vectors for many dreadful ailments like Japanese encephalitis (JE), West Nile fever, filariasis, etc. Generally, vectors account for 17% of infections and 7 lakh deaths in humans worldwide. The pathogens that are transmitted by the *Culex* are intensely dependent on the ecosystem, environment, and hosts of the mosquito (upon which species it feeds). There is a chance of infestation by *Culex* in properties that are near marshes, ponds, or lakes, as these mosquitoes are fascinated by temperate and moist environments because these places are ideal for laying their eggs. These common house mosquitoes prefer to bite at dawn and dusk. It detects the human's bite by body heat and the exhaled CO_2. The disease burden is highest in tropical and subtropical areas, where mainly the poorest populations get infected. However, the epidemiological hearsay of various mosquito-borne maladies in assorted parts of Europe proved fascinating evidence that vector-borne diseases are not restricted to tropical and subtropical regions anymore. Among the Culex mosquito-borne diseases, outbreaks of JE, lymphatic filariasis, West Nile fever, and yellow fever claimed a lot of lives and caused chronic suffering, disability, life-long morbidity, disfiguration, and stigmatization. It is known that lymphatic filariasis affects about 100 million people, mainly those with low socio-economic profiles, worldwide every year. Ninety per cent of filarial cases are transmitted by *Culex* mosquitoes, and the prime responsible vector species are *Cx. pipiens*, *Cx. molestus*, *Cx. quinquefasciatus*, and *Cx. plens.* Another deadly disease is Japanese encephalitis (JE), or brain fever, caused by flavivirus, which is transmitted throughout Asia via the mosquito vectors *Cx vishnui* and *Cx tritaeniorhynchus*. Besides these two species, there are several other *Culex* species, i.e., *Cx. epidesmusanhyrcanus*, *Cx. bitaeniorhynchus*, *Cx. whitmorei*, *Cx. pseudovishnui*, *Cx. fuscocephala*, and *Cx. gelidus*, responsible for the transmission of this disease. JE also showed yearly deaths of around 13,600 to 20,400 and about 68,000 infections globally. To date, these deadly diseases have no vaccines; therefore, to reduce the disease burden, we have to manage the vector

* Corresponding Author's Email: kbbjoy@gmail.com.

In: Insects as Humans' Frenemies: From Friends to Foes
Editor: Istkhar Rao
ISBN: 979-8-89530-320-7

population. This chapter scientifically reviews the four key factors, i.e., chemical control, biological control, environmental control, and personal protection, to generate public awareness and proper education to reduce the disease burden in the community.

Keywords: *Culex*, filariasis, encephalitis, chemical control, biological control, environmental control

1. Introduction

Among the phylum Arthropoda, insects are the most diverse group present throughout the world. In this group, varieties of insects play a role as disease vectors, like mosquitoes, flies, ticks, mites, sandflies, bugs, etc.; among these, mosquitoes are the most worrisome insects, having the capability of some dreadful disease transmission (Manikandan et al., 2022). The little winged culicine mosquitoes cause terrible biting annoyance. They act as vectors and transmit various life-threatening diseases like filariasis, Japanese encephalitis, West Nile fever, Rift Valley fever, western equine encephalitis, St. Louis encephalitis, and avian malaria (Kimura et al., 2010), etc., in tropical and subtropical countries, as they are an opportunistic feeder on both humans and animals. Half of the human population can get infected with these diseases at any moment (WHO, 2014). Till now, 770 species of Culicine mosquitoes have been identified, of which several are vectors of some deadly diseases (Turrel, 2012; Mark, 1987). Due to its major social hindrance, in 1996, the World Health Organisation (WHO) asserted it as "public enemy number one" (WHO, 1996).

Focusing on the *Culex* mosquito, it is important to describe the *Culex pipiens* complex, one of the most important groups of the genus *Culex*. It consists of six adherents, namely: *Cx. Pipiens* Linneaus, *Cx. quinquefasciatus* Say, *Cx. Molestus* Forskll, *Cx. Australicus* Dobrotworsky & Drummond, *Cx. Globocoxitus* Dobrotworsky, and *Cx. Pallens* Coquillet, respectively (Zittra et al., 2019). Dumas et al. (2016) reported that a seventh mosquito name is proposed under this group called *Cx. juppi* from South Africa. Other important culicine species are *Cx. tritaeniorhynchus*, *Cx. tarsalis*, *Cx. vishnui*, *Cx. gelidus*, *Cx. fuscocephala*, *Cx. fuscanus*, and *Cx. bitaeniorhynchus*, etc. Disease-transmissible *Culex* species are prevalent in the urban habitats of Asia and Africa, mainly in tropical and subtropical areas where rapid unwitting urbanisation, deforestation, and other appropriate environmental situations help to proliferate mosquito populations (Brown et al., 2008). Mosquitoes have a complete life cycle containing eggs, larvae, wrigglers, pupae, and adults, or imagos. The southern house mosquito (*Cx. quinquefasciatus*) is a grey-coloured, medium-sized mosquito (3.96 to 4.25 mm) prevalent in tropical and sub-tropical provinces. Those Culex-borne diseases distress our outdoor skills (Coffinet et al., 2007; Kiehn et al., 2008), generally by separating us from nature, and gradually it affects our perspective towards biodiversity. It can transmit a variety of pathogens like arbovirus, nematode, and protozoa in the human population. The high abundance of these disease-causing mosquito vectors generally depends on climatic weather (Dhanasekaran et al., 2014). Their habitat, distribution, population, breeding, and feeding (Julian et al., 2010) are directly related to different seasons. Our country is very suitable for mosquito breeding without any hindrance due to its suitable climate with a long monsoon. *Cx. pipiens pipiens* is found across more temperate regions (Farajollahi et al., 2011), whereas *Cx. quinquefasciatus* is prevalent around human dwellings and chooses to breed in places with high organic matter contents, especially in drains, septic tanks, polluted or sewage water, soak pits, unused wells, puddles, and ditches (Harvey-Samuel et al., 2021). They are preferably night-biters (12 a.m. to 3 a.m.) and can fly up to 5 km from the breeding place (Verdonschot & Besse-Lototskaya, 2014). *Cx. tritaeniorhynchus* breeds mainly in fishponds, grassy pools, and rice fields (Service, 2012). Indiscriminate urbanisation and global greenhouse gases affect the ecology of mosquitoes. Different culicine groups show different feeding habits. *Cx. pipiens f. molestus* are endophilic in nature, i.e., they prefer to rest indoors, while *Cx. pipiens pipiens* and *Cx. vishnui* prefer to rest outdoors as they are exophilic in nature. A blood meal is very important to female mosquitoes as it marks the completion of their life cycle, and it is the most important step to carry forward the infection. The irksome mosquito bite is accountable for the hypersensitivity reactions in our skin too.

In general, vectors are responsible for more than 17% of all human infections globally, causing more than 700,000 deaths annually. In India, over 40 million people get infected by mosquito bites annually (Ghosh et al., 2012). Among the *Culex*-borne diseases, lymphatic filariasis (LF) damages the lymphatic system and leads to anomalous amplification of body parts, causing pain, social disgrace, and relentless disability (WHO, 2020). Till now, there is a scarcity of preventive drugs or vaccines for most of the mosquito-borne diseases. Vector control is the key to getting rid of those alarming diseases in the future. The same point applies to culicine-borne diseases too; there are no operative vaccines or precise drugs for different diseases, and so, according to El-Kersh et al. (2016), vector management is the primary factor in eradicating the VBDs (vector-borne diseases). To reduce the disease burden, different countries adopt different technologies to control the vector population. Vector management is done as mosquito exclusion processes of adult and wriggler forms in the means of environmental, chemical, and biological control by the personal protection procedures are practised widely. In this chapter, we discussed thoroughly the powerful control strategies to diminish public health problems regarding vector-borne diseases.

2. Major Culicine Borne Disease Incidence

Mosquito-borne diseases are prevalent throughout the world, and their manifestation depends on the communication of several aspects, for instance, biological, ecological, economic, and social aspects. A heavy disease burden is prevailing worldwide due to *Culex*-borne diseases, among which filariasis is the most common.

2.1. Filariasis

Lymphatic filariasis (LF) is prevalent in 80 countries in the tropics and subtropics, of which 863 million people in 47 countries are threatened and require protective chemotherapy. In 2018, WHO recorded that approximately 40 million people were disfigured by the infection of LF, whereas 51 million people were primarily infected. In India, one-third of diseased and disfigured people stay, and other diseased people are found in Africa, Papua New Guinea, the Americas, Southeast Asia, and the Pacific Islands (WHO, 2018). Filariasis, or in advanced cases, elephantiasis, is caused by a nematode parasite found in the human lymphatic system, named *Wuchereria bancrofti* (Cobbold, 1877), *Brugia malayi* (Brug, 1927), and *Brugia timori* (Partono et al., 1977), transmitted by the mosquito *Cx. quinquefasciatus* Say (Das & Pani, 2002; Hati et al., 1989). The mosquito first ingests the nematode during a blood meal from the infected person, then transmits the pathogen to an unaffected human being during feeding. The mosquito itself transmits 99.4% of filarial cases. The vector of LF elects the breeding places with high organic matter contents, mostly in drains, sewage or polluted water, septic tanks, soak pits, unused wells, etc.

In India, the occurrence of Bancroftian filariasis caused by *Wuchereria bancrofti* is 98%, and Brugian filariasis caused by *Brugia malayi* is 2%. Bihar is the most filariasis-prone state, along with Kerala, Uttar Pradesh, Andhra Pradesh, and Tamil Nadu, respectively. The nematode parasite obstructs the lymphatic system and damages it, which eventually develops fever, lymphadenitis, lymphangitis, and anomalous amplification of body parts, causing pain, social disgrace, and relentless disability (Daz-Menéndez et al., 2011). Five countries have successfully implemented management strategies for filariasis, but other countries still need preventive measures like mass drug administration (MDA) (Shriram et al., 2014). Apart from vector control, it is an effective way to stop the spread of parasites to mosquitoes.

2.2. Japanese Encephalitis

The second most important culicine-borne public health disease is Japanese encephalitis (JE), or brain fever, a prevalent viral encephalitis in Asia. Mainly, *Cx. vishnui* and *Cx. tritaeniorhynchus* spread this disease in Asia, and other vector species are *Cx. annulus*, *Cx. annulirostris*, *Cx. fuscocephala*, and *Cx. gelidus* (Hemmerter et al., 2007). Along with humans, JE also affects birds, horses, pigs, and donkeys. Pigs and wild birds act as reservoirs for the virus, which has an incubation period of around 4–14 days. Worldwide, at least 68,000 infections and 13,000–20,000 deaths are recorded annually (Manikandan et al., 2022). JE cases are mainly found among children under 10 years of age. In India, JE cases have been reported on a large scale in different states like Assam, Andhra Pradesh, Bihar, Karnataka, and Uttar Pradesh (Das et al., 2014). There are no antiviral drugs, but some anti-viral vaccines are present with some drawbacks. Probable helpful vaccines are in the advanced stages of development; till now, vector control is the best strategy to combat this ailment.

2.3. West Nile Fever (WNF)

Another brutal disease, West Nile fever, caused by the West Nile virus (WNV) transmitted by culicine mosquitoes (Bertolotti et al., 2008), showed a remarkable number of infections and deaths. About 1 out of 50 infected people develop fever and other symptoms. It is also a zoonotic disease and is found in birds and horses. According to Granwehr et al. (2004), human infection, results in a minor feverish ailment, meningitis, encephalitis, some sort of poliomyelitis-like condition, and, in due course, death. After infection, the virus takes 3–14 days to show symptoms, as this is the incubation period. After that, the symptoms may persist for some days to quite a few weeks. Cuervo-Parra et al. (2016A) showed a positive side of this disease, i.e., roughly 80% of all infected people did not show any symptoms whatsoever. There has been no accurate vaccine for humans till now (Connelly et al., 2014).

2.4. Rift Valley Fever (RVF)

An RNA virus from the Phlebovirus family, named Rift Valley Fever (RVF) virus, of the genus Bunyaviridae, caused the fever in some domestic animals such as cows, buffaloes, camels, goats, and sheep, as well as in humans. It is transmitted by more than 30 species of mosquitoes, including *Culex* species (OIE, 2014; EFSA, 2013). In Eastern and Southern Africa, RFV is an endemic disease, but it is found in epidemics in Saudi Arabia, Yemen, and Egypt (CSFPH, 2006). By this zoonotic disease, persons who are in direct contact (veterinarians, laboratory personnel, slaughterhouse workers, etc.) get infected in a much greater ratio than normal people.

2.5. St. Louis Encephalitis (SLE)

The culicine-borne SLE virus affects humans, but its normal host is a bird. Generally, the incubation period is as long as the life of a mosquito, i.e., about 10 days. It is a rare disease that remains unnoticed because of its mild symptoms in most cases, but at some point, it can develop some clinical manifestations, such as encephalitis, and eventually death if it remains untreated for a long time (Reisen, 2003). The death rate is between 5% and 20%, where children and the elderly are most susceptible. From Almirón and Crocco (2007), we came to know that the SLE is found in Central America, South America, Mexico, and Canada, i.e., throughout the American continent. To date, there is no effective vaccine against this virus (Connelly et al., 2014), so it is better to avoid mosquito bites for protection from it.

2.6. Western Equine Encephalitis (WEE)

WEE infection is caused by an arbovirus from the family Togaviridae, transmitted by the culicine mosquito *Cx. tarsalis*, the main vector of WEE virus in North America (Mahmood et al., 2004). It is a febrile disease with some neurological severity, e.g., it can be a headache to meningitis or it can turn to encephalitis. Walton (1981) found that Central and South America, Canada, Mexico, and the Western United States are prevalent for this disease. Just like SLE, the susceptible groups are children and elderly people, while adults showed mild symptoms. The mortality rate is also low, ranging from 5% to 15%. Among the infected children, 50% of the infants who survived the disease have enduring brain impairment. As there is no drug of choice or vaccine against the virus (MDH, 2006), people can limit their open-air events at the time of feeding of *Culex* species, especially at dawn and dusk, to avoid getting an infection.

2.7. Venezuelan Equine Encephalitis (VEE)

VEE is caused by the New World alphavirus, which infects equines (horses, zebras, and donkeys) and humans. It causes febrile illness with high viremia having quite a few degrees of severity in vertebrates, which leads to encephalitis. In affected individuals, it showed fever, headache, and mild to severe influenza-like indications. De la Monte et al. reported in 1985 that there are around 1% of reported death cases, with approximately 4–14% of reported neurologically complicative cases. The VEE virus complex is comprised of viruses belonging to the genus Alphavirus from the Togaviridae family. This virus is transmitted more or less to all continents by some infected Culex mosquitoes, such as *Cx. vomerifer*, *Cx. pedroi*, and *Cx. Adamesi* (Brown & Condreay, 2003). Though Aedes is the main vector of it, Culex species transmitted the enzootic virus strains of VEE (Zacks & Paessler, 2010). The life cycle of the virus is also maintained in rodents.

3. Socioeconomic Impact

The widespread vector-borne diseases (VBDs) challenge the public health system, having a wider socioeconomic impact. Suffering from VBDs is a global problem, but mostly in developing countries where proper knowledge is not available to people. Knowledge is the primary criterion for disease control, and most people, especially in rural areas, are deprived of proper knowledge regarding filariasis, JE, and other VBDs. For example, if the general public is unaware of filariasis or JE, prompt treatment would not be possible, and consequently, they would develop a serious problem. Social and economic conditions are much more relatable to VBDs when the government takes measures to control them. Climate change is the main driver of mosquito-borne diseases (MBDs). As mosquitoes are ectothermic, temperature change strongly influences the biting rate, pathogen development within the body (Mordecai, 2017; Paaijmans, 2012), and disease transmission. In addition to temperature, precipitation is one of the most important criteria for mosquito breeding. Episodic rain is responsible for mosquito breeding at unnatural times as well. Besides these causes, it has been noticed that, over the last seven decades, the emergence of different vectors and VBDs has greatly increased due to rapid unplanned urbanisation, disproportionate deforestation, massive use of insecticides and pesticides, ill management of livestock and agro-ecosystems, etc.

4. Vector Control

During the study of those culicine-borne diseases, we found that there is a huge disease burden due to a major drawback, i.e., not having proper knowledge and guidance regarding those illnesses (García-Gutiérrez, 2012). Common people are not aware much about the grievances of those diseases, and many times they remain unreported only because of a lack of knowledge. We saw that, till now, a large number of VBDs do not have any effective vaccines or choiceable drugs to control them. So, to reduce the disease burden, improve the public health system, and, at last, permanently eradicate or prevent those diseases from entering society, we have to control the vector populations. Simply put, we have to manage the mosquito populations to get rid of the ailments they carry, so the risk of infection will be diminished.

Vector management is a great deal, and many countries worldwide have applied so many different strategies to control them. For example, mosquito exclusion in adult form and personal protection are widely used in different countries across the world. But besides these two ways, these vectors can be diminished in their birthplace as well as in their larval stage, and this process is much more useful than adult control because larval confinement is very much necessary to control it (Howard et al., 2007), or it will apply to those areas where adulticidal measures are either costly or undesirable to society. Larval control procedures are preferred in those locations where frequent outbreaks of VBDs are noticed or in those conditions where disease management has failed due to technical reasons. It is a broad area of research, and researchers are now trying to invent an effective mosquito-controlling agent. For example, several larvicidal, pupicidal, and adulticidal agents are used thoroughly along with other methods of control. To target any mosquito species of any group, proper education is necessary to identify the species and its nature regarding the application of such chemicals or other killing agents.

In general, many authorities have termed this control procedure integrated vector management (IVM), which incorporates both chemical and non-chemical control as well as environmental control. The IVM scheme, a sound method for vector control, was first operated in 2004 to manage all VBDs, and it is based on the thought that mosquito control has to depend on an elementary understanding of the target mosquito group's biology, ecology, and behaviour. The relationship of the vector species with their host and its milieu is as important as previously said. Here we picked four important control measures to discuss, like chemical control, biological control, environmental control, and personal protection, used to get protected from VBDs globally (Figure 1).

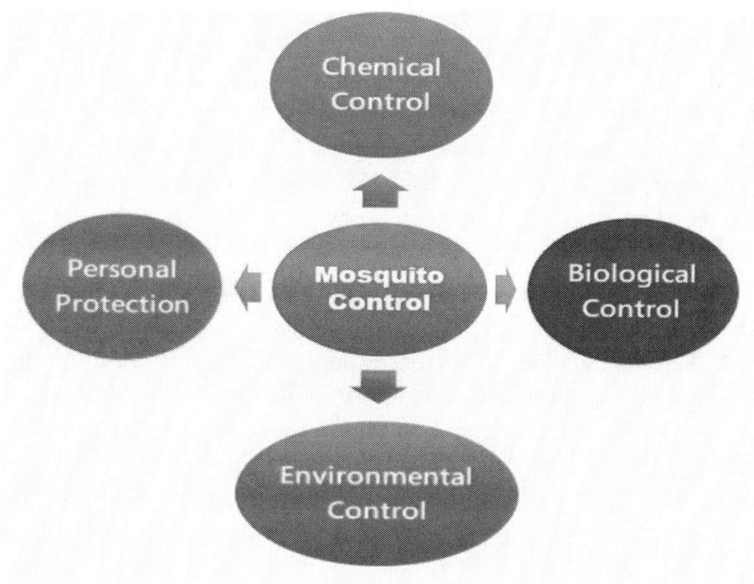

Figure 1. Scheme of Mosquito Reduction.

4.1. Chemical Control

Chemical control measures have been in action since ancient eras; it is the most widely used procedure for mosquito control. As a part of the mosquito control programme in different tropical or subtropical areas, governments or health authorities used diverse kinds of synthetic insecticides. Due to the advancement of chemical science, several mosquitocidal or insecticidal agents were developed; for example, some chlorinated hydrocarbon insecticides like DDT

(Dichloro-diphenyl-trichloroethane), BHC, methoxychlor, etc. were in function from the primaeval period. These are found in different forms, like dust, powder, or liquid, used as larvicidal weapons to combat the VBDs for decades. Besides this, some other chemical insecticides, like organochlorines, organophosphates, diflubenzuron, methoprene, pyrethroids, pyriproxyfen, etc. have been used to manage the vector population (Fathy, 2012). These are contact poisons that cause nervous system breakdown within minutes. Some indoor spraying insecticides are also prescribed by the WHO. Recently, 12 insecticides of four different chemical classes were recommended by the WHO Pesticide Evaluation Scheme (WHOPES) for indoor residual spraying (IRS) purposes.

However, the development of resistance in mosquitoes (Severini et al., 1993; Zhang et al., 2011) against chemical insecticides and some detrimental effects of those insecticides on the environment shifted the curve of chemical insecticide use downwards. Mosquito species now have developed resistance against those insecticides which triggers some special reactions in the mosquito's body and shows several behavioural changes in adults (Jagdish, 2003; Devine, 2007). For instance, pyrethroids and organochlorines like DDT target the voltage-dependent sodium channels in the vector's body, and the enzyme acetylcholinesterase is the target of organophosphate and carbamate insecticides. However, it is reported that vectors can alter the target site or they can bring some changes in the target site to become fully resistant against those insecticides (Ware & Whitacre, 2004). In the last few decades, there has been a drastic diminution seen in many countries regarding chemical insecticides as vector control agents.

Now coming to the second part of the disadvantage of chemical insecticides, i.e., the environmental portion, in which the vastly toxic and nonbiodegradable insecticides, after deposition through biomagnification in the environment, turn into a waste product and smoothly blend in our soil, water, agricultural fields, and food (Wattal et al., 1981; Koul et al., 2008). To date, whichever synthetic pesticide has been used in the environment globally, and those are not biodegradable, deposited in lower to higher trophic levels via biomagnification. It affects public health badly, so there is a point to think about any alternative to chemical insecticides that we can rely upon. Today, we live in an era of unwarranted use of chemical insecticides. Koul et al. (2008) estimated that approximately 2.5 million metric tonnes of pesticides are used per year, which is a matter of concern now. Moreover, several chemical insecticides are not fully safe for non-target organisms, i.e., they don't possess selective lethality. Some highly used synthetic pyrethroids, such as biphenthrin, lambda-cyhalothrin, and cypermethrin, are potent larvicidal agents, but they are not selectively lethal; they should have target specificity to use more; otherwise, they destroy the beneficial species with the targeted one.

Today, it is a challenge for the local government to control the spreading of VBDs. Local governmental bodies and private and non-governmental establishments are all generating mammoth energy to reduce the VBDs and the population of mosquito vectors of those diseases. The insecticides that gave positive results in the past, but due to massive use, are now useless products in the long term. So, due to the urgent need for an alternative form of vector management process, here researchers develop a potent mosquitocidal agent only from plant origin called phytochemicals. Green synthesised phytochemicals have some important characteristics, like being cost-effective, biodegradable, not toxic, environment-friendly, and target-specific (Ansari et al., 2000). Hazardous effects of phytochemicals are too minimal compared to synthetic ones. Due to its biodegradable nature, it doesn't deposit in the environment for too long. It is selectively lethal, so different non-target organisms like fish and other non-harmful insects are more or less safe than chemical ones. Furthermore, it has no ill effect on the human population, and there is little or no information regarding the development of resistance among the vector population against it.

From the ancient period, several medicinal plants were used in medication in India. During that time, when there were no such man-made chemical insecticides, some plant products were used as natural insecticides (Kantamreddi et al., 2010) just after extraction from the plant, such as pyrethrums from the flower *Chrysanthemum cinererifolium* (Hartzell & Willcoxon, 1941), nicotine from the tobacco leaves of *Nicotiana tabacum*, rotenone from *Derris eliptica* (Zubairi et al. 2004), and *Anabasis aphylla*, a Russian weed extract (Campbell et al. 1993), namely anabasine and lupinine (alkaloids), etc. So, researchers started to screen many plants to have mosquitocidal properties in search of alternative measures of vector control.

So, a question may arise: in these phytochemicals, which particular part is responsible for mosquito control? Here the answer is the plant-produced secondary metabolites, which are accumulated in plants for their protection, colour, texture, flavour, aroma, etc. A few secondary metabolites that have remarkable insecticidal effects, as stated previously, are steroids, alkaloids, flavonoids, terpenoids, saponins, glycosides, and tannins from miscellaneous plant species (Shaalan et al. 2005). These chemicals of botanical origin can be used in so many forms against the mosquito population, like larvicidal, pupicidal, and adulticidal agents. At first, crude extracts of any plant parts, i.e., leaves, roots, fruits, flowers, barks, etc., are applied to mosquito larvae, pupae, or adults, and then they are processed to get the pure form of the active component by using different types of polar and non-polar solvents. It is now a global topic of research, and some pure forms of active compounds are now used to control the vector population, which means they are now incorporated into the vector control strategy.

There is huge research available which shows that phytochemicals (Table 1) are very effective in controlling the vector population (ICMR Bulletin, 2003; Murugan, 2007; Ghosh et al., 2012; Bhattacharya & Chandra, 2014), here are some examples of the plants with remarkable mosquitocidal properties are leaves of *Solanum nigrum* (Rawani et al., 2017), *Ravenala madagascariensis* (Bhattacharya et al., 2014a; 2014b), *Acyranthes aspera* (Bhattacharya & Chandra, 2013), *Cayratia trifolia* (Chakraborty et al., 2013), *Nicotiana plumbaginifolia* (Singh et al., 2016), *Capparis zeylanica* (Singha Ray et al., 2016), *Andrographis echiodes* (Das et al., 2019), *Glochidion lanceolarium* (Singh et al., 2019), root of *Tragia involucrate* (Bhattacharya & Chandra, 2014), rhizome of *Curcuma aromatica* (Mallick et al., 2014), seed coat of *Nelumbo nucifera* (Singha Ray et al., 2014), edible pods of *Moringa oleifera* (Mukherjee et al., 2016), fruit peel of *Citrus maxima* (Mallick et al., 2016) etc. Along with this, some traditional plants like neem, nicotine, garlic, and cinnamon possess mosquitocidal activities and show numerous behavioural changes, including loss of appetite, arrested growth, moulting disarrays, fertility reduction, structural defects (Wandscheer et al., 2004; Singha Ray & Bhattacharya, 2020), neuronal effects, etc.

The major mosquitocidal properties of diverse plants may vary with different geographical distributions of them. Some other variations in the insecticidal properties are dependent on the extraction method and the polarity of the chosen solvent during the purification of the phytochemicals. Some other minor factors are also responsible for it, like seasons, weather, climate, nutrition, predators, etc. So, the activity of the active principle may vary a little bit from place to place worldwide.

Table 1. Some important plants having a remarkable mosquitocidal effect

Sl. No.	Plant Species	Plant Parts	Solvent extract	Mosquito Species	References
1.	*Solanum nigrum*	Leaves	Chloroform: methanol (1:1 v/v)	*Cx. quinquefasciatus*	Rawani et al., (2010)
2.	*Ravenala madagascariensis*	Leaves	Chloroform: methanol (1:1 v/v)	*Cx. quinquefasciatus*	Bhattacharya et al., (2014)
3.	*Acyranthes aspera*	Leaves	Acetone	*Cx. vishnui*	Bhattacharya & Chandra, (2014)
4.	*Cayratia trifolia*	Leaves	Petroleum ether Benzene Ethyl acetate	*Cx. quinquefasciatus*	Chakraborty et al., (2013)
5.	*Capparis zeylanica*	Leaves	Ethyl acetate	*Cx. quinquefasciatus*	Singha Ray et al., (2016)
6.	*Andrographis paniculata*	Leaves	benzene, hexane, ethyl acetate, methanol and chloroform leaf	*Cx. quinquefasciatus*	Govindarajan, (2011)
7.	*Citrus maxima*	Fruit peel	n-hexane	*Cx. quinquefasciatus*	Mallick et al., (2016)
8.	*Tragia involucrate*	Root	Chloroform: methanol (1:1 v/v)	*Cx. quinquefasciatus*	Bhattacharya & Chandra, (2014)
9.	*Cassia tora*	Leaves	Methanol	*Cx. quinquefasciatus*	Amerasan et al., (2012)
10.	*Citrullus vulgaris*	Leaves	Methanol	*Cx. quinquefasciatus*	Prabakar & Jebanesan (2004)
11.	*Thymus capitatus*	Leaves	Petroleum ether	*Culex pipiens*	Mansour et al., (2000)
12.	*Citrus aurantium*	Fruit peel	Petroleum ether	*Cx. quinquefasciatus*	Kassir, (1989)
13.	*Pistacia lentiscus*	Leaves	Petroleum ether	*Cx. pipiensmolestus*	Traboulsi et al., (2002)
14.	*Jatropha curcas*	Leaves	Petroleum ether	*Cx. quinquefasciatus*	Rahuman et al., (2007)
15.	*Cleome icosandra*	Leaves	Acetone	*Cx. quinquefasciatus*	Saxena et al., (1992)
16.	*Daucus carota*	Leaves	Hexane	*Cx. annulirostris*	Shaalan et al., (2005)
17.	*Nyctanthesarbortristis*	Flower	Chloroform	*Cx. quinquefasciatus*	Khatune et al., (2001)
18.	*Moringa oleifera*	Bark	Methanol	*Cx. gelidus*	Kamaraj & Rahman (2010)
19.	*Piper retrofractum*	Ripe fruit	Ethanol	*Cx. quinquefasciatus*	Chansang et al., (2005)
20.	*Erythrina indica*	Leaves	Methanol	*Cx. quinquefasciatus*	Govindarajan & Shivakumar, (2014)

4.2. Biological Control

Synthetic or chemical control systems have dominated the control stratagem over the years; their monotonous use disrupts the natural ecosystems, and eventually, mosquitoes get resistant to them (Severiniet al., 1993; Zhang et al., 2011). So, society needs a new approach to controlling the vector population, and if it is the biological one, then the control strategy will be much more acceptable than the previous one (Das et al., 2007). Along with the vector resistance, due to some public health problems and disadvantageous effects on non-target populations (WHO, 1995), an alternate approach to control systems was highly successful. So, the viable alternative, biological control, took over the place of chemical control within a few days and served as a potential technique for mosquito control. Different mosquitocidal biocontrolling agents were assessed, along with their effectiveness, safety of non-target organisms, and environmental standards. Along with bacteria and viruses, different larvivorous fishes (Chatterjee et al., 1997; Ghosh et al., 2005; Chandra et al., 2008; Bhattacharjee et al.,

2009), arthropods (Chatterjee et al., 2007; Mandal et al., 2008), protozoa (Murugesan et al., 2009), fungi (*Entomophaga*, *Entomophtora*, *Neozygites*, *Verticillium*, *Hirsutella*, *Metarhizium*, etc.) (Kaya & Gaugler, 1993; Van Driesche & Bollows, 1996), and nematodes (Mathur, 2003) have been used as potent mosquitocidal agents as these have promising results in mosquito control.

4.2.1. Bacteria

Two important strains of *Bacillus* species, e.g., *Bacillus thuringiensis israelensis* and *Bacillus sphaericus*, are highly endorsed for their potent mosquitocidal activity in the field. Researchers were highly successful in effectively controlling the *Culex* population with the help of the virulence of *Bacillus sphaericus*. It is highly effective to control the 100% larvae of *Cx. quinquefasciatus* in field conditions after seven days of application (Das et al., 2017); no larval emergence was seen afterwards up to the 10th day. During sporulation, these bacteria produce protein crystals with major insecticidal properties, which makes them important to use as a potential mosquito control agent. If the mosquito larvae ingest the bacterium, it produces toxic crystals in the gut of the mosquito larvae, so the mechanism of peristalsis becomes hampered due to the paralysed condition of the gut. Eventually, the larvae can't feed, and death is inevitable due to starvation. Along with it, the bacterium invades the larval body by breaking the epithelium layer of the gut of the insect. Some other effective mosquito-controlling bacteria are *Bacillus cereus*, *Bacillus alvei*, *Bacillus brevis*, *Brevibacillus leterosporous*, *Pseudomonas fluorescens*, *Pseudomonas aeruginosa*, *Clostridium bifermentans*, *Bacillus circulans*, *Bacillus amyloliquefaciens*, and *Bacillus subtilis*, etc. (Poopathi et al., 2013, 2014; Lalithambika et al., 2016).

4.2.2. Viruses

The use of viruses is also very important in controlling mosquitoes. Some entomopathogenic viruses can spread from one generation to the next, which paralyses the mosquito and results in death afterwards (Primo, 1991). A mosquito-specific virus, also known as Densovirus (DNV), from the family Parvoviridiae, came to know that it can reduce the mosquito vector population via pathogenic interactions (Federici, 1985). It can infect so many mosquito species, along with *Cx. pipiens* and *Cx. Pipiens pallans* (Carlson et al., 2006; Zhai et al., 2008). Some highly pathogenic DNVs are capable of controlling the mosquito population by having a high larval death rate.

4.2.3. Fishes

For more than 100 years, fish have been used to control mosquito populations. In India, in 1908 and 1928, two larvivorous fish, namely *Poecilia reticulata* (Guppy) and *Gambusia affinis* (Gambusia), were imported from Texas. At first, it was used to control malaria vectors only, but afterwards, it was also used to control *Culex* mosquitoes as well. Some other biocontrol agents are represented in Table 2, of which *Carassius auratus*, *Xenentodon cancila*, *Oreochromis mosambica*, etc. have more importance.

4.2.4. Insects

Arthropods, protozoa, nematodes, and fungi are biocontrol agents with a lesser degree of success than bacteria and fish. Some examples of these organisms are dragonfly nymphs, adult dragonflies, predator mosquitoes, Toxorhynchites, and other crustacean animals like cyclops, copepods, etc.

4.2.5. Limitations

Though biocontrol is a huge success, it also has some limitations and disadvantages. For example, the fish are voracious eaters and can compete with other fish species in the same area. Some adult fish can eat up the fry of other fish species. *Gambusia* is an opportunistic feeder and can feed on zooplankton, algae, amphibia, eggs, and young of other fish, so there will be an ecological loss. So, it can be used temporarily; it is not a permanent solution, so an alternate method is still needed.

4.3. Environmental Control

The major burden of controlling the vector population relies upon chemical and biological control but as the filarial vector *Cx. quinquefasciatus* is resistant to organochlorine and other chemical insecticides, we have to control some environmental factors to cope with the vector management programme. Environmental management mainly depends on the complete understanding of the ecology and dynamics of the mosquito population as well as of disease epidemiology. Because without proper knowledge, we can't make any environmental changes. Environmental management is classified by environmental modification and manipulation, which consists of topographical, hydrological, ecological, and biological pattern modification of mosquito habitats so those places become unsuitable for mosquito breeding.

We can discuss environmental control under several headings:

i. *Mosquito Population Monitoring:* Both adult and larval mosquitoes can be monitored, where the adult population is observed using mechanical traps and their landing rate (Gebru et al., 2018; Petrić et al., 2014). In the case of landing rate counts, one person should visit per day a particular number of spots, and he or she should count the number of female mosquitoes that rest on a particular body part within a given time. Another way to monitor the mosquito population is to use the mechanical trap, which blows adult mosquitoes into a collection bag. To collect the mosquitoes, some chemical attractants are used, namely, ammonia, carbon dioxide, octanol, lactic acid, etc. To monitor mosquito larval populations, one should use a dipper to collect mosquito larvae from standing water. Ovitraps, or artificial breeding spots, can be used to monitor the larvae at fixed intervals. Mosquito population monitoring is vital to know about the presence of mosquito species, so there will be an indication about any disease they carry, so precautions should be taken against those VBDs.

ii. *Source Reduction:* Under environmental control, source reduction may be an important point to be maintained by ourselves. As it was mentioned earlier, proper education is very important, through which common people came to know about the vectors surrounding their dwellings. As culicine mosquitoes breed on standing water, sewage, or polluted water, we have to clean those areas surrounding our house where they can breed. So, reducing mosquito breeding places would be very effective in controlling it; we just have to eliminate unused water containers, ditches, open drains, etc.; if needed, a community clean-up event may aid in this.

iii. *Open Water Marsh Management:* In the United States, this process is very important now regarding vector control. Here, the use of shallow ditches generates a flow of water inside the marshes and links it with a pond. A network of several ditches helps to drain the habitation; it already reduces the number of vector populations and helps to enter the fish, which feed on the remaining larvae. This process already lessens the chance of using chemical insecticides in the long term.

4.4. Personal Protection

Different control measures are taken by the governments and public health authorities of many countries to diminish the disease burden as well as to control the vector population to eradicate the disease from human life, but until then, we have some responsibilities towards our health that cannot go unnoticed. To be safe from mosquito bites, we should take some precautions that won't lead to any serious health diseases. At first, we have to educate ourselves and our population regarding our environment, diseases, and vectors, then we can take measures against them. As we know, *Culex* mosquitoes prefer to bite, especially at dawn and dusk, so we have to be very cautious during these two times of the day if it is not possible for the whole day. Some effective points of personal protection are as follows:

i) Wear full-sleeve clothing and long pants during the peak biting time.
ii) Use a mosquito net regularly.
iii) Use some ointments all over the exposed body in densely mosquito-populated areas.
iv) Avoid outdoor activity when needed.
v) Take the preventive drug beforehand, if possible.
vi) Clean your surroundings and garden where the adult mosquitoes can rest.
vii) Never unnoticed any symptoms if you are in a VBD-prone area.
viii) Use EPA-approved repellents on the body and clothing.
ix) Use a net on doors and windows.
x) Change the standing water once a week in your home.

Conclusion

Mosquito control is one of the important objectives regarding public health, as we are surrounded by several mosquito-borne diseases with a higher infection and death rate. This is a threat to humanity as the number of cases is rising globally due to numerous causes, including urbanisation, climate change, etc. So, the reduction of this disease burden can be achieved by controlling the vector population by any means, especially by the application of phytochemicals and biopesticides against the disease-causing mosquito population. In this review, we mentioned four types of control systems, each of which has its limitations and disadvantages. In these situations, novel phytochemicals or bacterial formulations are highly dependable among researchers, given the invention of a more appropriate course of vector control. On the other hand, during or after the COVID-19 pandemic, we have already lost many people, so we must lessen the burden of other diseases and deaths as well, especially all the VBDs. We must combat this culicine rivalry shortly to avoid further loss of ourselves.

References

Almirón, W., & Crocco, L. (2007). Mosquitos Urbanos Transmisores de Denguey Encefalitis de San Luis, Manual de capacitación Docente. *Editorial Científica Universitaria*. Córdoba, Argentina, 14–51.

Ansari, M. A., Vasudevan, P., Tandon, M., & Razdan, R. K. (2000). Larvicidal and mosquito repellent action of peppermint (*Mentha piperita*) oil. *Bioresource Technology, 71*(3), 267–71.

Bertolotti, L., Kitron, U. D., Walker, E., Ruiz, M., Jeffrey, B., Scott. L., Gabriel, H., & Tony, G. (2008). Fine-scale genetic variation and evolution of West Nile Virus in a transmission "hot spot" in suburban Chicago, USA. *Virology,374*, 381-9.

Bhattacharjee, I., Aditya, G., & Chandra, G. (2009). Laboratory and field assessment of the potential of larvivorous, air-breathing fishes as predators of culicine mosquitoes. *Biological control, 49*(2), 126-133.

Bhattacharya, K., & Chandra, G. (2013). Bioactivity of *Acyranthes aspera* (Amaranthaceae) Foliage against the Japanese Encephalitis Vector *Culex vishnui* Group. *Journal of Mosquito Research, 3*(13), 89- 96.

Bhattacharya, K., & Chandra, G. (2014). Phagodeterrence, larvicidal and oviposition deterrence activity of *Tragia involucrata* L. (Euphorbiaceae) root extractives against vector of lymphatic filariasis *Culex quinquefasciatus*. *Asian Pacific Journal of Tropical Diseases,4*, S226-S232.

Bhattacharya, K., Burman. S., Nandi, S., Roy, P., Chatterjee, D., & Chandra, G. (2014a). Phytochemical extractions from the leaves of *Ravenala madagascariensis* from Sundarban area and its effect on southern house mosquito *Culex quinquefasciatus* Say (1823) larvae. *Journal of Mosquito Research, 4*(12), 1-6.

Bhattacharya, K., Chandra, I., Kundu, P., Ray, S., Halder, D., & Chandra, G. (2014b). Larval control of *Culex vishnui* group through bio-active fraction of traveller's tree, *Ravenala madagascariensis Sonn.* (Strelitziaceae). *Journal of Mosquito Research, 4*(15), 1-6.

Brown, D. T., & Condreay, L. D. (2003). Replication of alphaviruses in mosquito cells. *The Togaviridae and Flaviviridae.* Plenum Press, New York, p. 171–207.

Brown, H. E., Childs, J. E., Diuk-Wasser, M. A., & Fish, D. (2008). Ecologic factors associated with West Nile virus transmission, North eastern United States. *Emerging Infectious Disease, 14*(10), 1539–1545.

Brug, S. L. (1927). *Filariamalayi*, n.sp., parasitic in man in the Malay Archipelago. *VIIth Congress of Far Eastern Association of Tropical Medicine, 3*: 279.

Campbell, F. L., Sullivan, W. W., & Smith, L. N. (1993). The relative toxicity of nicotine, anabasine, methyl anabasine and lupine for Culicine mosquito larvae. *Journal of Economic Entomology, 26*: 500.

Carlson, J., Suchman, E., & Buchatsky, L. (2006). Densoviruses for control and genetic manipulation of mosquitoes. *Advances in Virus Research, 68*, 361–392.

Centre for Food Security and Public Health (CFSPH). (2006). Rift Valley Fever. *Institute for International Cooperation in Animal Biologics. Iowa State University.* 1–5.

Chakraborty, S., Singha, S., Bhattacharya K., & Chandra, G. (2013). Control of human filarial vector, *Culex quinquefasciatus* Say 1823 (Diptera: Culicidae) through bioactive fraction of *Cayratiatrifolia*leaf. *Asian Pacific Journal of Tropical Biomedicine, 3*(12): 980-84.

Chandra, G., Bhattacharjee, I., Chatterjee, S. N., & Ghosh, A. (2008). Mosquito control by larvivorous fish. *Indian Journal of Medical Research, 127*, 13-27.

Chansang, U., Zahiri, N. S., Bansiddhi, J., Boonruad, T., Thongsrirak, P., Mingmuang, J., & Mulla, M. S. (2005). Mosquito larvicidal activity of aqueous extracts of long pepper (Piper retrofractum Vahl) from Thailand. *Journal of Vector Ecology, 30*(2), 195.

Chatterjee, S. N., Das, S., & Chandra, G. (1997). Gold fish (*Carrasius auratus*) as a strong larval predator of mosquito. *Transaction of the Zoological Society of India,1*, 112-4.

Chatterjee, S. N., Ghosh, A., & Chandra, G. (2007). Eco-Friendly Control of Mosquito Larvae by *Brachytron pratense* Nymph. *Journal of Environmental Health,69*(8), 44-49.

Cobbold, T. S. (1877). Discovery of the adult representative of microscopic filariae. *Lancet, 2*: 70-71.

Coffinet, T., Mourou, J. R., Pradines, B., Toto, J. C., Jarjaval, F., Amalvict, R., & Pages, F. (2007). First record of *Aedes albopictus* in Gabon. *Journal of the American Mosquito Control Association, 23*(4), 471-472.

Connelly, C. R., Bolles, E., Culber, D., De Valerio, J., Donahoe, M., Gabel, K., & Walter, J. (2014). Florida Resident's Guide to Mosquito Control. *Integrated Pest management for Mosquito Reduction around Homes and Neighborhoods. University of Florida, USDA-NIFA*, 22-37.

Cuervo-Parra, J. A., Cortés, T. R., & Ramirez-Lepe, M. (2016). Mosquito-borne diseases, pesticides used for mosquito control, and development of resistance to insecticides. *Insecticides Resistance*, doi: 10.5772/61510.

Das, B. P., Lal, S., & Saxena, V. K. (2004). Outdoor resting preference of *Culex tritaeniorhynchus*, the vector of Japanese encephalitis in Warangal and Karim Nagar districts, Andhra Pradesh. *Journal of vector borne diseases*, *41*(1/2):32.

Das, D., Banerjee, S., Singha Ray, A., & Chandra, G. (2017). Characterization of a Strain of *Bacillus sphaericus* Isolated from Local Rice Field and its Virulence against Mosquito Larvae. *Molecular Entomology,8*(2), 1-9.

Das, M., Singh, A., Bhattacharya, K., Mukherjee, A. & Chandra, G. (2019). Mosquito Larvicidal efficacy of *Andrographis echioides* (Acanthaceae) foliages against vector of lymphatic filariasis *Culex quinquefasciatus* Say (1823). *Proceedings of Zoological Society,* https://doi.org/10.1007/s12595-018-0273-z.

Das, N. G., Goswami, D., & Rabha, B. (2007). Preliminary evaluation of mosquito larvicidal efficacy of plant extracts. *Journal of Vector Borne Diseases, 44*(2), 145-148.

Das, P. K., & Pani, S. P. (2002). Filariasis in India; Epidemiology and control in: helminthology in India Ed. ML Sood. *International Book Distributor*.

De la Monte, S., Castro, F., Bonilla, N. J., Gaskin de, U. A., Hutchins, & G. M. (1985). The systemic pathology of Venezuelan equine encephalitis virus infection in humans. *American Journal of Tropical Medicine and Hygiene*, *34*: 194–202.

Devine, G. J., & Furlong, M. J. (2007). Insecticide use: Contexts and ecological consequences. *Agriculture and Human Values*, *24,* 281-306.

Dhanasekaran, D., & Thangaraj, R. (2014). Microbial secondary metabolites are an alternative approach against insect vector to prevent zoonotic diseases. *Asian Pacific Journal of Tropical Disease*, *4*(4), 253-261.

Díaz-Menéndez, M., Norman, F., Monge-Maillo, B., Pérez-Molina, J. A., & López-Vélez, R. (2011). Filariasis in clinical practice. *Enfermedades infecciosas y microbiologia clinica*, *29*, 27-37.

Dumas, E., Atyame, C. M., Malcolm, C. A., Le Goff, G., Unal, S., Makoundou, P., Pasteur, N., Weill, M., & Duron, O. (2016). Molecular data reveal a cryptic species within the *Culex pipiens*mosquito complex. *Insect Molecular Biology*, *25*(6), 800-809.

El-Kersh, T. A., Ahme, A. M., Al-Sheikh, Y. A., Tripet, F., Ibrahim, M. S., & Metwalli, A. A. (2016). Isolation and characterization of native *Bacillus thuringiensis* strains from Saudi Arabia with enhanced larvicidal toxicity against the mosquito vector *Anopheles gambiae*. *Parasites & vectors*, *9*(1): 1-14.

European Food Safety Authority (EFSA), EFSA Panel on Animal Health and Welfare (AHAW). (2013). Scientific Opinion on Rift Valley fever. *EFSA Journal.11*: 1–48.

Farajollahi, A., Fonseca, D. M., Kramer, L. D., & Kilpatrick, A. M. (2011). "Bird biting" mosquitoes and human disease: A review of the role of *Culex pipiens* complex mosquitoes in epidemiology. *Infection, Genetics and Evolution*, *11*(7), 1577–1585.

Fathy, K. H. (2012). Ecosmart biorational insecticides: alternative insect control strategies. *Insecticides—Advances in Integrated Pest Management,* p. 17–60.

Federici, B. A. (1985). Viral pathogens of mosquito larvae. *Bull Am Mosq Control Association, 6*, 62–74.

García-Gutiérrez, C., Gómez-Peraza, R. L., Aguilar, C. E. L., & León-Váldez, A. (2012). Insecticidas biorracionales para el control de mosquitos y moscas negras en Sinaloa. *Ra Ximhai*, *8*(3), 47-55.

Gebru, A., Jansson, S., Ignell, R., Kirkeby, C., Prangsma, J. C., & Brydegaard, M. (2018). Multiband modulation spectroscopy for the determination of sex and species of mosquitoes in flight. Journal of Biophotonics, 11(8), e201800014.

Ghosh, A., Chowdhury, N., & Chandra, G. (2012). Plant extracts as potential mosquito larvicides. *Indian Journal of Medical Research, 135*(5), 581-598.

Ghosh, A., Mandal, S., Bhattacharjee, I., & Chandra, G. (2005). Biological control of vector mosquitoes by some common exotic fish predators. *Turkish Journal of Biology, 29*(3), 167-171.

Govindarajan, M. (2011). Evaluation of *Andrographis paniculata*Burm.f. (Family: Acanthaceae) extracts against *Culex quinquefasciatus* (Say.) and *Aedes aegypti* (Linn.) (Diptera: Culicidae). Asian Pacific Journal of Tropical Medicine,*4*(3), 176-81.

Govindarajan, M., & Sivakumar, R. (2014). Larvicidal, ovicidal, and adulticidal efficacy of *Erythrina indica* (Lam.) (Family: Fabaceae) against *Anopheles stephensi, Aedes aegypti*, and *Culex quinquefasciatus* (Diptera: Culicidae). *Parasitology Research,113*(2), 777-91.

Granwehr, B. P., Lillibridge, K. M., Higgs, S., Mason, P. W., Aronson, J. F., Campbell, G. A., & Barrett, A. D. (2004). West Nile virus: where are we now? *Lancet Infectious Diseases*, *4*: 547–556.

Hartzell, A., & Wilcoxon, F. (1941). A survey of plant products for insecticidal properties. Contributions from Boyce Thompson Institute, *12*, 127-141.

Harvey-Samuel, T., Thomas, A., Jolene, S., Chris, N. N., Samoa, A., Patricia, P., Steven, S., & Luke, A. (2021). *Culex quinquefasciatus*: status as a threat to island avifauna and options for genetic control. *CABI Agriculture and Bioscience*, *2*:9.

Hati, A. K., Chandra, G., Bhattacharyya, A., Biswas, D., Chatterjee, K. K., & Dwibedi, H. N. (1989). Annual transmission potential of *Bancroftian filariasis* in an urban and a rural area of West Bengal, India. *American Journal of Tropical Medicine and Hygiene, 40*(4), 305-307.

Hemmerter, S., Slapeta, J., van den Hurk, A. F., Cooper, D. R., Whelan, I. P., Russell, C. R., Johansen, A. C., & Beebe, W. N. (2007). A curious coincidence: mosquito biodiversity and the limits of the Japanese encephalitis virus in Australasia. *BMC Evolutionary Biology*, *7*: 1–11.

Howard, A. F. B., Zhou, G., & Omlin, F. (2007). Malaria mosquito control using edible fish in western Kenya: Preliminary findings of a controlled study. *BMC Public Health,7*, 199-204.

ICMR Bulletin. (2003). Prospects of using herbal products in the control of mosquito vectors. *ICMR Bulletin, 33*(1), 1-10.

Jagdish, K. (2003). An inventory of *Culicidae* diversity in Haryana state. *Journal of Vector Borne Diseases, 40*, 112-114.

Julian, E. G. R., Bradley, J. B., Jose, A. F. A., Maria, A. L. P., Wilberth, A. C. C., Luis, F. F. F., & Barry, J. B. (2010). Host-feeding preference of the mosquito, *Culex quinquefasciatus*, in Yucatan State, Mexico. *Journal of Insect Science, 10*(1), 32.

Kamaraj, C., & Rahuman, A. A. (2010). Larvicidal and adulticidal potential of medicinal plant extracts from south India against vectors. *Asian Pacific Journal of Tropical Medicine,3*(12), 948-953.

Kantamreddi, V. S. S. N., Lakshmi, Y. N. & Kasapu, V. V. V. S. (2010) Preliminary phytochemical analysis of some important Indian plant species. *International Journal of Pharma and Bio Sciences, 1* (4), 351 - 357.

Kassir, J. T., Mohsen, Z. H., & Mehdi, N. S. (1989). Toxic effects of limonene against *Culex quinquefasciatus* Say larvae and its interference with oviposition. *Anzeiger für Schädlingskunde, Pflanzenschutz, Umweltschutz,62*(1), 19-21.

Kaya, H. K., & Gaugler, R. (1993). Entomopathogenic nematodes. *Annual Review of Entomology, 38*, 181-206.

Khatune, N. A., Haque, M. E., & Mosaddik, M. A. (2002). Laboratory evaluation of *Nyctanthesarbor-tristis* Linn. flower extract and its isolated compound against common filarial vector, *Culex quinquefasciatus* Say (Diptera: Culicidae) larvae. *Pakistan Journal of Biological Science,4*(5), 585-587.

Kiehn, L., Murphy K. E., Yudin, M. H., & Loeb, M. (2008). Self-reported protective behaviour against West Nile Virus among pregnant women in Toronto. *Journal of Obstetrics and Gynaecology,30* (12),1103-1109.

Kimura, M., Darbro, J. M., & Harrington, L. C. (2010). Avian malaria parasites share congeneric mosquito vectors. *Journal of Parasitology*, *96*(1), 144-52.

Koul, O., Walia, S., & Dhaliwal, G. S. (2008). Essential oils as green pesticides: potential and constraints. *Biopesticide International,4*, 63–84.

Lalithambika, B., & Vani, C. (2016). *Pseudomonas aeruginosa* KUN2, extracellular toxins-A potential source for the control of dengue vector. *Journal of Vector Borne Diseases, 53*(2), 105.

Mahmood, F., Reisen, W. K., Chiles, R. E., & Fang, Y. (2004). Western equine encephalomyelitis virus infection affects the life table characteristics of *Culex tarsalis* (Diptera: Culicidae). *Journal of Medical Entomology, 41*: 982–986.

Mak, J. W. (1987). Epidemiology of lymphatic filariasis. *Ciba Foundation Symposium,* (Vol. 127, No. 1, pp. 5-14).

Mallick, S., Bhattacharya, K., & Chandra, G. (2014). Mosquito larvicidal potentiality of wild turmeric, *Curcuma aromatica* rhizome, extracts against Japanese Encephalitis vector *Culex vishnui* group. *Journal of Mosquito Research, 4*(19), 1-6.

Mallick, S., Mukherjee, D., Singha Ray, A. & Chandra, G. (2016). Larvicidal efficacy of fruit peel extracts of *Citrus maxima* against *Culex quinquefasciatus*. *Journal of Mosquito Research, 6*(20), 1-8.

Mandal, S. K., Ghosh, A., Bhattacharjee, I., & Chandra, G. (2008). Biocontrol efficiency of odonate nymphs against larvae of the mosquito, *Culex quinquefasciatus* Say, 1823. *Acta tropica,106*(2), 109-114.

Manikandan, S., Mathivanan, A., Bora, B., Hemaladkshmi, P., Abhisubesh, V. & Poopathi, S. (2022). A review on vector borne disease transmission: current strategies of mosquito vector control. *Indian Journal of Entomology*, Ref. No. e22410 DoI. No.: 10.55446/IJE.2022.593.

Mansour, S. A., Messeha, S. S., & el-Gengaihi, S. E. (2000). Botanical biocides. 4. Mosquitocidal activity of certain *Thymus capitatus* constituents. *Journal of Natural Toxins,9*(1), 49-62.

Mathur, S. J. N. (2003). Prospects of using Herbal Products in the Control of Mosquito Vectors. *Indian Council of Medical Research,1*(33), 1-10.

Minnesota Department of Health (MDH). (2006). Western Equine Encephalitis Fact Sheet. Minnesota Department of Health. p. 1.

Mordecai, E. A., Cohen, J. M., Evans, M. V., Gudapati, P., Johnson, L. R., Lippi, C. A., & Weikel, D. P. (2017). Detecting the impact of temperature on transmission of Zika, dengue, and chikungunya using mechanistic models. *PLoS Neglected Tropical Diseases*, *11*(4), e0005568.

Mukherjee, D., Singha Ray, A., Bhattacharya, K., & Chandra, G. (2016). Strobilus extractives of *Thuja orientalis* as novel antibacterial agent against some pathogenic bacteria. *International Journal of Pharma and Bio Sciences, 7*(1), (B) 156-160.

Murugan. K., Murugan, P., & Noortheen, A. (2007). Larvicidal and repellent potential of *Albizzia amara* Boivin and *Ocimumbasilicum* Linn against dengue vector, *Aedes aegypti* (Insecta: Diptera: Culicidae). *Bioresource Technology,98*, 198-201.

Murugesan, A. G., Sathesh, P. C., & Selvakumar, C. (2009). Biolarvicidal activity of extracellular metabolites of the keratinophilic fungus *Trichophyton mentagrophytes* against larvae of *Aedes aegypti*– a major vector for chikungunya and dengue. *Folia Microbiology,54*(3), 213-216.

OIE—World Organisation for Animal Health. Rift Valley Fever. OIE Terrestrial Manual 2014. OIE Reference Laboratories, p. 1–20.

Paaijmans, K. P., Blanford, S., Chan, B. H. K., & Thomas, M. B. (2012). Warmer temperatures reduce the vectorial capacity of malaria mosquitoes. *Biology Letters,8,* 465–8.

Partono, F., Aennis, D. T., Atmosoedjono, S., Oemijati, S., & Cross, J. H. (1977). *Brugia timori* sp.n. (Nematoda: Filarioidea) from Flores Island, Indonesia. *Journal of Parasitology, 63*(3): 540-546.

Petrić, D., Bellini, R., Scholte, E. J., Rakotoarivony, L. M., & Schaffner, F. (2014). Monitoring population and environmental parameters of invasive mosquito species in Europe. *Parasites & Vectors, 7*(1), 187.

Poopathi, S., Mani, C., & Rajeswari, G. (2013). Potential of sugarcane bagasse (agro-industrial waste) for the production of *Bacillus thuringiensis israelensis*. *Tropical Biomedicine, 30*(3), 504-515.

Poopathi, S., Thirugnanasambantham, K., Mani, C., Ragul, K., & Sundarapandian, S. M. (2014). Isolation of mosquitocidal bacteria (*Bacillus thuringiensis*, *B. sphaericus* and *B. cereus*) from the excreta of arid birds. *Indian Journal of Experimental Biology*, *52*, 739-747.

Prabhakar, K., & Jebanesa, A. (2004). Larvicidal efficacy of some cucurbitacious plant leaf extracts against *Culex quinquefasciatus*. *Bio resource Technology, 95*, 113-4.

Primo, Y. E. (1991). Mundi prensa ed. Ecologíaquímica. *Nuevosmétodos de lucha contra insectos.* Madrid; p. 191.

Rahuman, A. A., Gopalakrishnan, G., Venkatesan, P., & Geetha, K. (2008). Larvicidal activity of some Euphorbiaceae plant extracts against *Aedes aegypti* and *Culex quinquefasciatus* (Diptera: Culicidae). *Parasitology Research,102*(5), 867-73.

Rawani, A., Ghosh, A., & Chandra, G. (2010). Mosquito larvicidal activities of *Solanumnigrum* L. leaf extract against *Culex quinquefasciatus* Say. *Parasitology Research, 107*, 1235–1240.

Rawani, A., Singha Ray, A., Ghosh, A., Sarkar, M. & Chandra, G. (2017). Larvicidal activity of phytosteroid compounds from leaf extract of *Solanum nigrum* against *Culex vishnui* group and *Anopheles subpictus*. *BMC Research Notes,10*(135), 1-8.

Reisen, W. K. (2003). Epidemiology of St. Louis encephalitis virus. *Advances in Virus Research,* p. 139–183.

Roberts, D. (2014). Mosquito larvae change their feeding behavior in response to kairomones from some predators. *Journal of Medical Entomology, 51*(2), 368-74.

Saxena, R. C., Dixit, O. P., & Sukumaran, P. (1992). Laboratory assessment of indigenous plant extracts for anti-juvenile hormone activity in *Culex quinquefasciatus. The Indian Journal of Medical Research,95*, 204-206.

Service, M. (2012). Medical entomology for students (5th edition). *Cambridge University Press*, Cambridge. 55-57.

Severini, C., Rom, R., Marinucci, M., & Rajmond, M. (1993). Mechanisms of insecticide resistance in field populations of *Culex pipiens*from Italy. *Journal of the American Mosquito Control Association, 9*: 164-168.

Shaalan, E. A. S., Canyonb, D., Younesc, M. W. F., Abdel-Wahaba, H., & Mansoura, A. H. (2005). A review of botanical phytochemicals with mosquitocidal potential. *Environment International*, *31*, 1149–66.

Shaalan, E. A., Canyon, D., Younes, M. W., Abdel-Wahab, H., & Mansour, A. H. (2005). A review of botanical phytochemicals with mosquitocidal potential. *Environment International, 31*(8), 1149-66.

Shriram, A. N., Krishnamoorthy, K., Sivan, A., Saha, B. P., Kumaraswami, V., & Vijayachari, P. (2014). Impact of MDA and the prospects of elimination of the lone focus of diurnally sub periodic lymphatic filariasis in Nicobar Islands, India. *Acta Tropica*, *133*: 93-97.

Singh, A., Bhattacharya, K., & Chandra, G. (2019). Larvicidal potentiality of the leaf extracts of *Glochidionlanceolarium* (Phyllanthaceae) against the Japanese encephalitis vector *Culex vishnui* (Culicidae). *Oriental Insects*, doi: 10.1080/00305316.2018.1487887.

Singh, A., Bhattacharya, K., Singha Ray, A. & Chandra, G. (2016). Larvicidal efficacy of mature leaf extract of *Nicotiana plumbaginifolia* Viv. (Solanaceae) against southern house mosquito. *International Journal of Pharma and Bio Sciences, 7*(2), (B) 162 – 167.

Singha Ray, A. & Bhattacharya, K. (2020). An overview on nano-materials as mosquito controlling agents against *Anopheles stephensi. Nanoscience and Nanotechnology*, *4*(3), 45-49.

Singha Ray, A., & Bhattacharya, K. (2021). COVID-19 or SARS-CoV-2 pandemic and its management: A Review. *Annals of the Romanian Society for Cell Biology,25*(6), 20057-20073.

Singha Ray, A., & Bhattacharya, K. (2022). An overview on the zoonotic aspects of COVID-19. *Techno World*, p: 109-122.

Singha Ray, A., Bhattacharya, K., & Chandra, G. (2015). Target specific larvicidal effect of *Capparis zeylanica* L. (Capparaceae) foliages against filarial vector *Culex quinquefasciatus* Say (1823). *International Journal of Pharma and Bio Sciences, 6* (3), (B) 139-148.

Singha Ray, A., Bhattacharya, K., Singh, A. & Chandra, G. (2014). Larvicidal Activity of *Nelumbo nucifera* Gaertn. (Nymphaeaceae) against *Anopheles stephensi* (Liston 1901) and its Effect on Non-target Organisms. *Journal of Mosquito Research, 4*(10), 1-7.

Traboulsi, A. F., Taoubi, K., el-Haj, S., Bessiere, J. M., & Rammal, S. (2002). Insecticidal properties of essential plant oils against the mosquito *Culex pipiensmolestus* (Diptera: Culicidae). *Pest Management Science,58*(5), 491-5.

Turell, M. J. (2012). Members of the *Culex pipiens* complex as vectors of viruses1. *Journal of the American Mosquito Control Association*, *28*(4s), 123.

Van Driesche, R. G., & Bollows, T. S. (1996). Biological Control. Chapman & Hall. p. 539.

Verdonschot P. F. M., & Besse-Lototskaya A. A. (2014). Flight distance of mosquitoes (Culicidae): A metadata analysis to support the management of barrier zones around rewetted and newly constructed wetlands. *Limnologica*, *45*: 69-79.

Walton, T. E. (1981). Venezuelan, eastern, and western encephalomyelitis. *Virus Diseases of Food Animals: A World Geography of Epidemiology and Control. Disease Monographs.* Vol. 2. Academic Press, New York. p. 587–625.

Wandscheer, C. B., Duque, J. E., da Silva, M. A. N., Fukuyama, Y., Wohlke, J. L., Adelmann, J., & Fontana, J. D. (2004). Larvicidal action of ethanolic extracts from fruit endocarps of *Melia azedarach* and *Azadirachta indica* against the dengue mosquito *Aedes aegypti*. *Toxicon*, *44*, 829–835.

Ware, G. W., & Whitacre, D. M. (2004). Introducción a losInsecticidas. *The Pesticide Book*, 6th ed. Willoughby.

Wattal, B. L., Joshi G. C., & Das M. (1981). Role of agriculture insecticides in precipitating vector resistance. *Journal of Communicable Diseases*, 13, 71-73.

WHO. (2014). A global brief on vector-borne diseases. Geneva: *World Health Organization*.

World Health Organization. (1995). Vector control for malaria and other mosquito borne diseases. *WHO Technical Report series, 857*, 1-91.

World Health Organization. (1996). Report of the WHO informal consultation on the evaluation on the testing of insecticides. *CTD/WHO PES/IC/96.1. Geneva,* 69.

World Health Organization. (2020). Vector Borne Diseases. https://www. who.int/news-room/fact-sheets/detail/vector-borne-diseases.

Zacks, M. A., & Paessler, S. (2010). *Encephalitic alphaviruses. Veterinary Microbiology*, *140*: 281–286.

Zhai, Y. G., Lv, X. J., Sun, X. H., Fu, S. H., Gong, Z. D., Fen, Y., & Liang, G. D. (2008). Isolation and characterization of the full coding sequence of a novel densovirus from the mosquito *Culex pipiens* pallens. *Journal of General Virology*, *89*(1), 195-199.

Zhang, W. J., Jiang, F. B., & Ou, J. F. (2011). Global pesticide consumption and pollution: With China as a focus. *Proceedings of the International Academy of Ecology and Environmental Sciences,1*(2): 125-144.

Zittra, C., Moog, O., Christian, E., & Fuehrer, H. P. (2019). DNA-aided identification of *Culex* mosquitoes (Diptera: Culicidae) reveals unexpected diversity in underground cavities in Austria. *Parasitology Research, 118*(5) 1385–1.

Zubairi, S. I., Sarmidi, M. R., & Aziz, R. A. (2004). A study into the effect of concentration process on the yield of rotenone from the extract of local plant species. *Proceeding of 'Seminar Sebatian Smulajadi Malaysia Ke- 20', Unimas, Sarawak.*

Chapter 10

The Mosquito as the Archnemesis of Humanity: The Anopheline Saga

Mou Singha Ray[1]
and Kuntal Bhattacharya[2,*]
[1]Department of Zoology, Visva Bharati, Santiniketan, Bolpur, Birbhum, West Bengal, India
[2]Department of Zoology, Durgapur Government College, Durgapur, Paschim Bardhaman, West Bengal, India

Abstract

Anopheles, a medically important vector of malaria, filariasis, and other arbovirus infections, has undoubtedly had the most socioeconomic influences on human beings since the dawn of civilisation. Integrated information owing to vector biology, bionomics, developmental rates (time and space determinants of disease propagation), and survival in an ever-changing climate is of paramount importance for understanding many aspects of the vectorial capacity of anopheline groups, and this knowledge is of utmost need for new vector control discoveries as well. As most anopheline mosquitoes are dead-end hosts for malarial parasites, only a few out of the five hundred recognised species of anopheles can transmit the pathogen to humans. The evolutionary pressures affecting the vector and the pathogen have been a topic of research in recent years to unravel the dynamics of vectorial aptitude and competency. It is now evident that the selection pressures exerted by the pathogen on the vector are in favour of its augmented transmission, and even the parasite is eligible to influence the vector for its benefit. On the contrary, malaria control programmes are facing major drawbacks as the increasing DDT, HCH, malathion, pyrethroid resistance in adult mosquitoes and chlorpyriphos, and fenthion resistance in anopheline larvae have been well documented. The potentiality of transgenic approaches to create refractory mosquitoes has bestowed the convenience of a relevant genetic drive system for the control of vector-borne diseases. The selection of combinations of interventions targeting vectors in their different stages of the lifecycle will result in maximum reductions in disease burden. Comprehensive knowledge of the present situation of each of the human malaria dominant vector species (DVS) distributions is necessary for integrated vector control, and ecological and behavioural studies of each species will be highly enquired. This chapter scientifically summarises the deadly epic vector Anopheles in terms of its biology, population dynamics, role in fatal disease transmission, vector bionomics, vectorial capacity, evolutionary role in adapting behaviours against an ever-changing climate, insecticide resistance, and probable control strategies. This summarisation is aimed at being used as a detailed reference for modern research on host-

* Corresponding Author's Email: kbbjoy@gmail.com.

In: Insects as Humans' Frenemies: From Friends to Foes
Editor: Istkhar Rao
ISBN: 979-8-89530-320-7

vector interrelationships in varying aspects of biological, ecological, and taxonomical relevance concerning species-specific vector control.

Keywords: *Anopheles*, Dominant Vector Species (DVS), vector biology, insecticide resistance, vector control

1. Introduction

Anopheles, the medically important vectors of malaria, filaria, and several arboviruses, have been arch-rivals of human beings since the dawn of civilization. The World Malaria Report, based on 84 malaria-endemic countries, showed 247 million, 245 million, and 232 million malarial cases globally in the years 2021, 2022, and 2019, respectively. The comparison of death estimations is also following rising trends (WHO, 2022). There are nearly 450 known existing *Anopheles* species, among which only 60 species are eligible enough to be known as actual malaria vectors in the wild to transmit *Plasmodium* to (Sylvie et al., 2008), and 41 dominant vector species/species complexes (DVS) are responsible for malaria transmission at a major public health concern level (Hay et al., 2010; Sinka et al., 2012). These DVS vary greatly in terms of their feeding patterns and habitats. *An. darlingi*, the potent DVS of South America, generally lives a highly opportunistic life, and it readily bites humans indoors and outdoors as and when they come upon it. Vectorial capacity and competence also possess a high impact on various major and minor roles in malaria transmission, even at the species level (Sylvie et al., 2008). The genome sequencing of *An. gambiae* has greatly benefited the understanding of the genetic determinants of vector capacity and competence (Holt et al., 2002). Anthropophagic habit, competence to adopt and develop protozoan parasitic infection, a relatively longer life span, and highly adaptive population growth rates are the hallmarks of an extremely efficient human malaria vector. Malaria transmission rate varies based on vector distribution, vector lifespan, vector abundance in a specific area, host-vector-pathogen contact time, vector-pathogen susceptibility, and host-vector survivorship, all of which influence the ability of the pathogen. Pathogens can influence host-vector survivorship in various degrees depending on regional ecological factors such as regional climatic conditions, regional topographic levels, water levels, larval vector habitat diversity, and mostly on the lifestyle of mankind (Hadis et al., 1997; Rwegoshora et al., 2007). For infection transmission, two consecutive blood feasts are of utmost need for ease in pathogen uptake and ultimate transmission to a liable subject, such as a human. The longing for human blood in *Anopheles* mosquitoes makes them vectors of utmost medical importance for malaria and Bancroftian filariasis, and multiple blood meal intakes from multiple hosts make the *Anopheles* mosquitoes indulge in massive arbovirus transmission (Snow, 1983; Anderson et al., 1995). Host preferences for blood meal intake can vary according to the nutritional needs of the vector body, the availability of the host species, vector density, and the intrinsic host species preferences of the vector as well. (Loyala et al., 1993; Githeko et al., 1994; Burgos et al., 1988; Zimmerman et al., 2006).

1.1. The Importance of Multiple Blood-Feeding

In mosquitoes, cyclic reproduction is initiated by blood meal intake and ends by oviposition, followed by the female vector becoming able to take another blood meal. For anopheline mosquitoes in particular, this cyclical reproduction is considered a gonotrophic concordance (Swellengrebel, 1929; de Buck & Swellengrebel, 1934). A study reported about 3 bloodmeal consumptions of *An. culicifacies* Giles, starting from eclosion to first oviposition (Reisen & Khan, 1976). At least two blood meals were necessary for each gonotrophic phase for *An. gambiae* Giles and *An. funestus* Giles (Hocking & Maclnnes, 1948), with 20% gonotrophic discordance in the population where the ovaries reached the late latent stage only with the first blood meal, and a second meal was critical for complete development (Gillies, 1988). Altogether, multiple blood feeding is considered a widespread requirement and well-planned strategy to complete the anopheline gonotrophic cycle, and multiple feeding becomes possible through the blood protein concentration mechanism, known as the prediuresis process (Briegel & Rezzonico, 1985; Lahondère & Lazzari, 2012). Female anophelines are in urgent need of sequential bloodmeal to make reproduction successful, by getting the required yolk protein and yolk lipid resources.

1.2. Blood Feeding Behaviour and Host Preferences

The selection for host preference and host phenotypic plasticity are processes explaining the existing variations in blood-feeding behaviour. Host unavailability, host accessibility, innate preferences for the host, changes in genetic factors, and environmental change are some issues that control and modify host preference selection. Altering the host selection patterns by keeping the innate host preferences intact is considered the plasticity of the host phenotype. By showing anthropophilic behaviour (by preferring human blood the most), by presenting endophilic behaviour (by resting doors), and by exploiting man-made breeding habits, *An. gambiae* proves itself as a powerful and dangerous malaria vector (Besansky et al., 2004). While *An. arabiensis* is more opportunistic in behaviour (Coluzzi et al., 2002; Besansky et al., 2004; Day, 005), *An. gambiae* shows extremely anthropophilic/anthropophagic behaviour with some temporal variability (variation across different time intervals) and spatial variability (variation across different locations). (Lemassonet al., 1997; Diatta et al., 1998; Costantiniet al., 1998; Bøgh et al., 2001; Sousa et al., 2001; Ducheminet al., 2001; Wanji et al., 2003; Mwangangiet al., 2003; Cohuetet al., 2004; Caputo et al., 2008). Numerous studies on malaria vector host preferences have provided considerable information on the "Human Blood Index" (HBI) and feeding habits of anophelines. The HBI refers to the proportion of human-origin blood in freshly fed *Anopheles* (Garrett-Jones, 1964). HBI is one significant impactful factor in malaria transmission, though it does not prove inherent host preferences, as HBI denotes final host selection mainly, which can be affected by natural factors like host availability and temporal and spatial variation (Costantini et al., 1999). Anophelines differ in malaria transmission by their resting and feeding habits, i.e., by showing endophily (resting in a manmade habitat, indoors), exophily (resting outdoors), endophagy (preferring feeding indoors), exophagy (feeding outdoors), anthropophily (preferring human blood), and zoophily (preferring domestic animal blood, mainly cattle). Still, human and domestic animal availability has a great impact on host preferences (Garrett-Jones et al., 1980; Hackett, 1931). When livestock became unavailable, zoophilic vectors also showed signs of adapting to human blood (Bruce-Chwatt & Zulueta, 1980).

2. Human Host Preference for Desired Bloodmeal

Mosquitoes, which are in close contact with human hosts, show extreme human blood preference. Ayala and Coluzzi (2005) reported the extreme anthropophilic nature of *An. gambia* in the prolonged human-host contact scenario, and vectorial capacity is also reported to change massively in that case. The main reason for this human host preference is the easy availability of the preferred breeding site and the desired blood meal, both provided by humans. All members of the *An. gambiae* complex differs in host predilection because of the obvious genomic discrepancy, which proves that behind every trophic behaviour, genetic information plays a vital role. Various chromosomal rearrangements (inversion) play an important role in putting a hold on coadapted alleles from undergoing chromosomal recombination (Coluzzi, 1982; Takken & Costantini, 2006). Lacroix et al. (2005) reported that humans infected with gametocytes are preferable. Due to various factors like body weight and skin bacteria that produce volatiles, mosquitoes strongly prefer pregnant women's blood (Brakset et al., 1999; Port et al., 1980; Anselle et al., 2002). Lindsay et al. (2000) reported that, around 15 metres away, pregnant women are more preferable than nonpregnant women. For better malaria control during pregnancy, this area of research should be looked into thoroughly. A study showed the strong aggressive behaviour of infected vector mosquitoes during biting (Wekesa et al., 1992; Koella et al., 1999). Cellular and molecular studies of olfactory components of vectors showed very interesting updates that can clear the mystery of choosing the preferable human host for bloodmeal, and thus, by gaining this knowledge, alteration of vector trophic behaviour can be possible, which in turn can limit malaria transmission (Luet al., 2007; Halem et al., 2006).

2.1. Vector Bionomics

2.1.1. Vector Aquatic Stages

Anopheles aquatic stages are strongly influenced by water temperature, salinity, water mineral level, water pollution level, predators present in the water, and vector density. Vector growth rate is directly proportional to the temperature, up to a certain limit; after that, vector mortality increases with the rising temperature. Temperature shows an inversely proportional relationship between vector size and the number of eggs released by female mosquitoes (Bates, 1939). Water salinity has a high impact by influencing vector larval mortality rates (Eckstein, 1936). The properly balanced ratio of different minerals provides a favourable environment for larval vector survival (Bates, 1939). Water pollution levels prevent the larval growth of *An. messeae* (Artemiev, 1980). Van Seventer (1969) reported industrial pollution as the cause of the larval mortality of *An. maculipenniss.* And *An. atroparvus*. Various fish, amphibia, reptiles, birds, and insect species predate on immature anopheline larvae (Service, 1976). *Gambusia affinisin*, one fish species, predates on immature anopheline stages voraciously; many fungi, protozoa, and mites (hydrachnid mites) predate on vector larvae and possess a threat against larval growth (Bates, 1949). A study reported the significant effect of vectorial density on the larval and pupal growth and size of *An. atroparvus*. Larval mortality showed a directly proportional relationship with larval density; density-dependent cannibalism was also noticed (Jetten & Takken, 1994). Detinova et al. (1962) showed an inversely proportional relationship between mean fecundity and the vector gonotrophic cycle. In the meantime, vector larvae distribution partially depends on oviposition site selection by females, with the help of visual and chemical stimuli (Bates, 1949).

2.1.2. Vector Adult Stages

An. sacharovi, *An. messeae*, *An. labranchiae, An. Atroparvus*, and *An. maculipenniss*. reportedly show swarming behaviour (Cambournac & Hill, 1940; Horsfall, 1955; Erel, 1973). Bates (1949) reported light as the most important stimulus of dusk, and other factors like temperature, humidity, sound, CO_2 level, air movements, and odours heavily impact the swarming behaviour. Post-fertilisation, bloodmeal is necessary for the maturation of eggs (Muir, 1988). Yoeli and Mer (1938) reported the need for two bloodmeals for complete egg maturation in the case of *An. sacharovi*, but the demand for a second blood meal could be cut down if the criteria of high temperature (24–30 °C) pupal development are fulfilled, or it can also be done by providing fruit juice before the first blood meal. Despite being crepuscular insects, anopheline daylength response varies due to the temperature too (Vinogradova, 1960; Kasap, 1987).

2.2. Vector Dispersal

Adult anophelines disperse according to their feeding, resting, and ovipositional requirements, whereas species specificity, season, meteorology, and the site available for the previously mentioned requirements control the dispersal direction, distance travelled, and flight density as well. (Horsfall, 1955). Dispersal is reported to be active (local dispersal events in breeding site proximity) and passive (dispersed by the help of vertical or horizontal air currents or any kind of transport) (Boyd, 1949).

2.3. Vector Density

Anopheline vector density relies upon the aquatic and adult stage of life parameters, which are vector fecundity, growth rate, and mortality. These parameters, in turn, get altered by chemical, biological, and physical factors of the vector habitat. According to the study report of Leeson (1939), in a highly humid condition at 25°C, the male and female anopheline mosquito ratio is nearly 1:1.

3. Transmitting Pathogens, Gut-Microbiota and Paratransgenesis

Anopheles mosquitoes transmit pathogens including the malaria parasite *Plasmodium* (Yassine & Osta, 2010), filarial worms *Wuchereria* and *Brugia* microfilariae (Manguin et al., 2010), and the arbovirus O'Nyong Nyong virus (ONNV) (Brault et al., 2004; Waldock et al., 2012). These pathogens infect the mosquito midgut during bloodmeal taking time and reach other body tissues through haemolymph and in the next bloodmeal-taking event, the pathogens are transmitted to a new uninfected human host and, thus the infection spreads. The developmental time taken by the pathogens in the mosquitoes, their intermediate host, is termed the extrinsic incubation period (EIP). Thus, during the developing period of EIP, pathogens interact with the gut microbiota of mosquitos, and it is reported that these mosquito gut microbiotas provide resistance against these deadly pathogens, either by direct interaction or by bacteria-mediated secondary interaction. (Cirimotich et al., 2011; Meister et al., 2009; Dong et al., 2009). The diversity of the gut-bacterial community remains less in normal conditions, whereas bacteria massively proliferate (10-900-fold increase in 24 h to 30 h) after consuming bloodmeal after a blood meal event (Pumpuni et al., 1996; Kumar et al., 2010; Wang et al., 2011). The Anopheline gut-microbiota detection report by 16S rRNA sequencing, showed the evident presence of *Pseudomonas* (most abundant) followed by *Aeromonas*, *Asaia*, *Comamonas*, *Elizabethkingia*, *Enterobacter*, *Klebsiella*, *Pantoea* and *Serratia* (Jadin et al., 1967; Pumpuni et al., 1993; Straif et al., 1998; Gonzalez-Ceron et al., 2003; Lindh et al., 2005; Favia et al., 2007; Terenius et al., 2008; et al., 2008; Lindh et al., 2008; Rani et al., 2009; Dong et al., 2009; Kambris et al., 2009; Muller et al., 2010; Kumar et al., 2010; Chouaia et al., 2010; Damiani et al., 2010; Kajla et al., 2010; Manguin et al., 2010; Noden et al., 2011; Wang et al., 2011; Kampfer et al., 2011; Dinparast et al., 2011; Boissiere et al., 2012; Chavshin et al., 2012; Osei-Poku et al., 2012). The *Anopheles* immune system controls the gut-microbiota population and keeps itself safe from infection by applying resistance and tolerance modes of protection, where tolerance modes of protection allow bacterial and *Plasmodium* load enhancement (Meister et al., 2009; Kumar et al., 2010). Anopheline senses for breeding waterbody bacteria and human skin bacteria are significantly impressive, and they help in altering the rate of oviposition and bloodmeal-taking behaviour (Lindh et al., 2008; Verhulst et al., 2009). The microbiota composition also influences social, reproductive, and feeding preferences and may be important for vector control. One easy and straight way to diminish the vector competence is to use bacterial strains, i.e., natural microbiota (e.g., *Enterobacter* espZ strain), causing *Plasmodium* resistance (Cirimotich et al., 2011) or artificially introduced bacteria (e.g., *Wolbachia*) and granting wide spectrum resistance (Moreira et al., 2009), because of their natural incompatibility with pathogen development and replication. In the paratransgenesis technique, pathogen resistance takes place by introducing genetically modified bacteria into the vector. Natural *Anopheles* symbiont *Pantoea agglomerans* is reported to have been engineered successfully, followed by expressing and secreting desired proteins to prevent mosquito midgut invasion by *Plasmodium* in *An. stephensi* and *An. gambiae* without putting any effect on the mosquito lifespan, by EPIP4 (*Plasmodium* enolase-plasminogen interaction peptide), which comes into competition with *Plasmodium* EPIP for plasminogen binding, or by aiming the parasite directly, which is seen in the case of the anti-plasmodial scorpine protein (Bisi et al., 2011; Wang et al., 2012). *Pantoea* and *Asaia* are the two merging potential bacterial candidates for successful para-transgenesis in the field population as well (Favia et al., 2007; Boissiere et al., 2012).

3.1. Vector Susceptibility to Malaria Parasites and Refractoriness to Infection

Anopheline susceptibility for *Plasmodium* infection has been proven to be highly affected by the genomic basis of host-parasite interrelationships. *P. vivax*, *P. falciparum*, and P. *malariae* show successful transmission by *An. superpictus* and *An. maculipenniss* (Weyer, 1939; Barber & Rice, 1935). *An. sacharovi*, *An. Labranchiae,* and *An. atroparvus* show susceptibility to the European variety of *P. falciparum*. From recent study reports, it is known that *An. sacharovi*, *An. labranchiae*, and *An. atroparvus* no longer show susceptibility to the *Plasmodium* tropical strains. Anophelines show distinguished levels of quantitative susceptibility for human malarial parasites, with incomplete to complete refractoriness. The study shows poor experimental infection reports of anophelines, which normally are not susceptible to *P. falciparum* (Habtewold et al., 2008; de Zulueta et al., 1975). The level of infection depends upon the mosquito susceptibility and power of evasion by the parasite; being identified as non-self, infectious elements become targeted by the immune system of the mosquitoes, and thus, to some extent, infection efficiency becomes limited (Sindenet et al., 2004). But in some rare cases, unnatural host-parasite interaction shows high vector susceptibility (Collins & Roberts, 1991; Nace et al., 2004). The anopheline genome has thus always been of great importance for the search of refractory genetic determinants and vector susceptibility genetic factors; though the parasite genome bears high significance as well, the most important factor should be considered as the ever-changing environment itself and the relationship of the host and parasite with it (Boëte, 2009).

3.2. Vectorial Capacity (VCAP) and Basic Reproductive Rate

Vectorial capacity, or VCAP, is considered the daily measure at which prospective future inoculations can arise from the present infection status. Vectorial capacity is an excellent way to determine malaria transmission risk in an area, and thus it is highly helpful for the eradication programme as well (Garrett-Jones, 1964; Smith et al., 2012). The total number of malaria cases generated from one infective case is the basic reproductive rate, which must be equal to 1 for malaria to spread further, and if the basic reproductive rate becomes zero, then the disease will decline. Vector longevity is one of the most vital factors in the decline or upsurge of the basic reproductive rate (Macdonald, 1957).

3.3. Sporogonic Development and Infection Transmission

Sporozoites secure a great opportunity of being transmitted as early as they can reach the salivary gland, and this in turn imparts strong evolutionary and selective pressure to make the sporogonic development period short, which is one of the most important factors for vectorial capacity determination. In the relatively lengthy sporogonic development period of *P. falciparum* (12–23 days) (World Health Organisation, 1975), anophelines are allowed to be infectious for a very small period only; in their relatively short life span, for example, in the case of *An. gambiae*, it is 2–4 weeks only (Sylvie et al., 2008). After being infectious, the high parasite burden leads to impairment and alterations in the blood-feeding efficiencies, and a significant upsurge in the biting rate is seen as a result of vector behavioural manipulation. (Koella,1999). For high sporozoite growth in a very short development time, optimal temperature is necessary to carry out the cycle, as temperature and development time are in an inversely proportional relationship with each other (World Health Organisation, 1975). Sinden et al. (2007) reported density-dependent parasite development during sporogony.

3.4. Evolutionary Selection Pressures, Longevity, Age-Dependent Mortality and Virulence

Irrational uses of broad-spectrum insecticides in public and agricultural domains lead to some serious problems in society, where the main is insecticide resistance, and by the strong evolutionary selection pressure of nature, resistant strains continue to propagate successfully, causing more harm to mankind and proving to be beneficial for the parasite and vector, as mosquitoes got high reproductive success and parasites got guaranteed transmission (Schwartz & Koella, 2002; Chouaibou et al., 2008). Mosquito longevity-related factors affecting the transmission potential are EIP survival and getting the opportunity to have the required blood meal (Smith et al., 2014). However, as EIP and mosquito survivability chances are inversely proportional to each other, the infection transmission possibilities increase with the young age of the infectious mosquito (Smith et al., 2012; Smith et al., 2014). The vectorial capacity hugely relies upon the age-dependent mortality rate of the mosquito (Styer et al., 2007; Dawes et al., 2009; Bellan, 2010; Novoseltsevet et al., 2012; Ryan et al., 2015; Brand et al., 2016). For successful infection transmission and infectious host encounters, the daily mortality rate of mosquitoes plays a huge role, as the fundamental reproduction quantity (R0) relies upon vector lifespan and duration of infectivity (Dye, 1992; Garrett-Jones, 1964). Thus, knowledge about vector survival factors could pave the way for the design of brand-new malaria control strategies. Though many anopheline species exhibit age-dependent mortality in cases of infection (Clements & Paterson, 1981), some indicate a similar survival rate in both cases of infection and non-infection (De Buck, 1936; Boyd MF, 1940; Robert et al., 1990; Chege & Beier, 1990; Gamage-Mendis et al., 1993; Hogg & Hurd, 1997), and some anophelines are reported to show poor survivability (Gad et al., 1979; Klein et al., 1982; Klein et al., 1986; Hogg & Hurd, 1995a; Hogg & Hurd, 1995b; Ferguson & Read, 2002). Ferguson and Read (2002) reported an overall anopheline survival reduction in vector-parasite association in a laboratory experimental study, but it failed to be proved in natural field conditions and laboratory experiments for a longer period (Ferguson & Read, 2002) because anophelines will face the detrimental effect of the parasite-vector association only in the case of high parasite burden (De Buck, 1936; Hogg & Hurd, 1995b), which is not normally the case, as most infected anophelines in nature carry 2-3 parasite oocysts on average, which is far less than being harmful (Haji et al., 1996; Pringle, 1966; Graves et al., 1988; Rhosenberg et al., 1990). One of the major impactful factors in the case of anopheline mortality could be the sporozoite load, due to the obvious anopheline mortality differences in lengthy anopheline studies, while sporozoites reside in the salivary glands of the mosquito (Ferguson & Read, 2002); and Sinden et al. (2007) reported about the density-dependent parasite development during sporogony, which supports the fact as well.

The parasite causes host epithelial cell damage, midgut cell apoptosis and ovary follicular epithelial cell damage, resulting in reduced egg production; thus, affecting vector fecundity rather than vector survival (Hurd & Carter, 2004; Vlachou et al., 2004; Hurd et al., 2005) and massive post-infection immune response generation affects vector fecundity or longevity (Schwartz & Koella, 2004; Mendes et al., 2008; Armitage et al., 2003; Dong et al., 2006). However, due to energy resource competition, anopheline behaviour gets changed as parasites are also competing for the resources which leads to vector mortality (Rivero & Ferguson, 2003; Lambrechts et et al., 2006). Due to the high biting rate but inefficient blood meal intake, the *An. gambiae-P. falciparum* association makes feeding-linked mortality evident (Anderson et al., 2000). Lefevre et al. (2008) noticed that the parasite showed a well-developed manipulating strategy of exploiting compensatory behaviour. Data Meta-analysis revealed fewer incidents of infection-induced mortality in nature than in laboratory experiments (Ferguson & Read, 2002), which explains the existence of evolutionary forces that modulate coevolutionary strategies between the hosts and the parasites. As vector survival is beneficial for both the parasite and the vector, low virulence is gradually obtained by the parasites, while, in parallel, mosquitoes less dented by infection are proved to be selectively advantageous for the increasing chances of parasite transmission. Here, it is noticed by Ferguson et al. (2003a, 2003b) and Ferguson and Read (2002) that successful infection is driven by the genetics of both parasite and mosquito. But Lambrechts et al. (2006) showed that different environmental factors are responsible for the evolution of virulence. The prospect of malaria transmission control with the help of genetically modified mosquitoes (GMMs) is an excellent way out. However, infection virulence, prevalence, and anopheline cost-benefit balance highly affect the success of the GMM strategy in nature (Marrelli et al., 2007).

4. Increased Infection Due to Insecticide Resistance

Using mutation and natural selection, mosquitoes counteract the detrimental effects of insecticides by propagating the resistant strains, which in turn leads to insecticide resistance (Weillet et al., 2004; Diabate et al., 2004; Ranson et al., 2011). Invertebrate host infection levels are extremely affected by genetic insecticide resistance. Insecticide-sensitive mosquitoes are proven to be more prone to being infected than the resistant ones, in the case of filaria. By altering the potential oxidation-reduction reaction in the mosquito tissues, insecticide resistance enormously influences parasite transmission and, against infection, offers safety. (McCarroll & Hemingway, 2002).

4.1. Using Delayed Action Insecticides and Restraining Insecticide Resistance

Late-life-acting (LLA) insecticides can be used to hinder the evolution of insecticide resistance, cut short vector longevity, and then limit malaria transmission by selectively targeting aged malaria-transmitting mosquitoes only. As a result, resistance-approving selection pressure diminishes (Read et al., 2009).

4.2. Integrated Vector Management (IVM)

The universal strategic planning framework for integrated vector management (IVM) demands a confirmative-revolutionary perspective involving familiarisation with modern plans of action and carrying out 24/7 surveillance of local vector ecology, epidemiology, and resources (World Health Organisation, 2004). IVM is of paramount importance, to achieve the highest eradication and maintain the mosquito control impact (World Health Organisation, 2007). WHO's Position Statement for Controlling Malaria and lymphatic filariasis endorses IVM for more cost-friendly approaches (WHO, 2011; Weekly Epidemiological Record, http://www.who.int/wer). The Global Malaria Programme (GMP) provides both short- and long-term goals, where cutting down the malaria burden to the point of losing the capability of being a public health problem is the short-term goal; on the other hand, in endemic countries, making the worldwide incidence rate zero through a gradual elimination process is the long-term goal (WHO, 2011)). Proper vector control and malaria case control using proper diagnosis and remedy are the two head pillars of GMP (WHO, 2011; ANVR Quarterly Newsletter (2007)). Using LLINs (long-lasting insecticide nets) and/or ITNs (insecticide-treated nets) and utilising IRS (indoor residual spraying) along with insecticide treatment are the two different but equally effective approaches of GMP (WHO, 2011). Malaria control, from the perspective of vectorial capacity, proves the disproportionately high benefits of eliminating adult mosquitoes (Macdonald, 1957). Multi-pronged approaches to malaria control have been proven to be highly effective. By using insecticide-impregnated bed nets, reducing and delaying mosquito bites and enhancing the adult mosquito mortality rate is highly effective. On the other hand, practical limits exist on the continuously upsurging mortality of the mosquito, and the hike in expenditure and the benefits may not increase linearly in a simultaneous manner as well. The global DDT-based malaria control campaign certainly didn't prove to be successful in malaria eradication all over the world of the infamous insecticide resistance evolution, but larval control is evident for being more effective due to its fixed localization, while adult mosquitoes have the advantage of long-distance flight; though, the relative benefit differs along with the ecology of the vector larvae (Killeen et al., 2001).

Conclusion

In an ever-changing environment, constantly evolving mosquitoes' interactions with parasites and hosts (mainly humans) determine their vectorial capacity and malaria system competence. *Plasmodium* is proven to be highly successful in main vector selection because *Anopheles* has been modifying itself for the benefit of the parasite, for desired longevity, immune evasion, and mastery in anthropophily by responding to the selection pressures exerted by the parasite. Selective forces exerted by the parasites are distinct and specific for each vector, and for gaining knowledge about the specific selective forces, comparing closely related vector species genomes and molecular evolution concerning changing vectorial capacity and competence is a good way to go, as strong evolutionary forces influence vectorial capacity and important parameters (host choice, longevity) the most, and lower evolutionary selective forces affect vector immunity more. This kind of research is extremely helpful for exploring new malaria control targets. To plan the perfect malaria control strategy, evolutionary ecological parameters should be considered, and to understand malaria epidemiology and eradicate it, vector bionomics, evolutionary selective forces, vector ecology influencing vectorial capacity, and vectorial competence are extremely important factors to consider. In the field of mosquito immunity and olfaction genetics research, rapid progression is seen (Christophideset al., 2002; Luet al., 2007; Hallemet al., 2006), but very little progression is evident in the field of mosquito adaptation concerning its anthropophilic nature and relation concerning parasitic virulence genetic factors. Moreover, infectious parasite availability, vector characteristics, malaria vector density, longevity, extrinsic incubation period duration, vector susceptibility to parasites, vector feeding preferences, human behaviour and habitat in the case of anthropophilic, and vector dispersal behaviour all influence the malaria incidence rate in some specific area. Thus, to apply the appropriate malaria control strategy, understanding the type of existing anopheline species in that particular area and their behaviour with host and parasite alike is a vital issue.

References

Anderson, R. A., & Brust, R. A. (1995). Field evidence for multiple host contacts during blood feeding by *Culex tarsalis, Cx. restuans*, and *Cx. nigripalpus* (Diptera: Culicidae). *Journal of Medical Entomology*, *32*(5), 705-710.

Ansell, J., Hamilton, K. A., Pinder, M., Walraven, G. E. L., & Lindsay, S. W. (2002). Short-range attractiveness of pregnant women to *Anopheles gambiae* mosquitoes. *Transactions of the Royal Society of Tropical Medicine and Hygiene*, *96*(2), 113-116.

Armitage, S. A. et al. (2003) Examining costs of induced and constitutive immune investment in Tenebrio molitor. J. Evol. Biol. 16, 1038–1044.

Artemiev, M. M. (1980). Anopheles mosquito—main malaria vector in the USSR. *International scientific project on ecologically safe methods for control of malaria and its vectors. International scientific project on ecologically safe methods for control of malaria and its vectors*, 45-71.

Ayala, F. J., & Coluzzi, M. (2005). Chromosome speciation: humans, Drosophila, and mosquitoes. *Proceedings of the National Academy of Sciences*, *102*(suppl_1), 6535-6542.

Barber, M. A., & Rice, J. B. (1935). Malaria studies in Greece: the malaria infection rate in nature and in the laboratory of certain species of *Anopheles* of East Macedonia. *Annals of Tropical Medicine & Parasitology*, *29*(3), 329-348.

Bates, M. (1939). The Use of Salt Solutions for the Demonstration of Physiological Differences between the Larvae of certain European Anopheline Mosquitoes. *American Journal of Tropical Medicine*, *19*(4).

Bates, M. (1949). Ecology of anopheline mosquitoes. *MF Boyd, Malariology*, *1*, 302-330.

Bellan, S. E. (2010). The importance of age dependent mortality and the extrinsic incubation period in models of mosquito-borne disease transmission and control. *PLoS One*, *5*(4), e10165.

Besansky, N. J., Hill, C. A., & Costantini, C. (2004). No accounting for taste: host preference in malaria vectors. *Trends in parasitology*, *20*(6), 249-251.

Bisi, D. C., & Lampe, D. J. (2011). Secretion of anti-Plasmodium effector proteins from a natural *Pantoeaagglomerans* isolate by using PelB and HlyA secretion signals. *Applied and environmental microbiology*, *77*(13), 4669-4675.

Boëte, C. (2009). Anopheles mosquitoes: not just flying malaria vectors… especially in the field. *Trends in parasitology*, *25*(2), 53-55.

Bøgh, C., Clarke, S. E., Pinder, M., Sanyang, F., & Lindsay, S. W. (2001). Effect of passive zooprophylaxis on malaria transmission in The *Gambia. Journal of Medical Entomology*, *38*(6), 822-828.

Boissière, A., Tchioffo, M. T., Bachar, D., Abate, L., Marie, A., Nsango, S. E., & Morlais, I. (2012). Midgut microbiota of the malaria mosquito vector *Anopheles gambiae* and interactions with *Plasmodium falciparum* infection. *PLoS pathogens*, *8*(5), e1002742.

Boyd, M. F. (1940). On the Correlation between the Incidence of Stomach and Gland Infection in *Anopheles quadrimaculatus* infected with *Plasmodium vivax. American Journal of Tropical Medicine, 20*(1).

Boyd, M. F. (1949). Malariology. A comprehensive Survey of all Aspects of this Group of Diseases from a global Standpoint. *Malariology. A Comprehensive Survey of all Aspects of this Group of Diseases from a Global Standpoint., 2.*

Braks, M. A. H., Anderson, R. A., & Knols, B. G. J. (1999). Infochemicals in mosquito host selection: human skin microflora and *Plasmodium* parasites. *Parasitology Today, 15*(10), 409-413.

Brand, S. P., Rock, K. S., & Keeling, M. J. (2016). The interaction between vector life history and short vector life in vector-borne disease transmission and control. *PLoS Computational Biology, 12*(4), e1004837.

Brault, A. C., Foy, B. D., Myles, K. M., Kelly, C. L. H., Higgs, S., Weaver, S. C., Powers, A. M. (2004). Infection patterns of o'nyongnyong virus in the malaria-transmitting mosquito, *Anopheles gambiae. Insect molecular biology, 13*(6), 625-635.

Briegel, H., & Rezzonico, L. (1985). Concentration of host blood protein during feeding by anopheline mosquitoes (Diptera: Culicidae). *Journal of Medical Entomology, 22*(6), 612-618.

Briones, A. M., Shililu, J., Githure, J., Novak, R., & Raskin, L. (2008). *Thorselliaanophelis* is the dominant bacterium in a Kenyan population of adult *Anopheles gambiae* mosquitoes. *The ISME journal, 2*(1), 74-82.

Bruce-Chwatt, L. J., & De Zulueta, J. (1980). *The rise and fall of malaria in Europe: a historico-epidemiological study*. Published for the Regional Office for Europe of the World Health Organization by Oxford University Press, Walton Street, Oxford OX2 6DP.

Burkot, T. R., Graves, P. M., Paru, R., & Lagog, M. (1988). Mixed blood feeding by the malaria vectors in the *Anopheles punctulatus* complex (Diptera: Culicidae). *Journal of Medical Entomology, 25*(4), 205-213.

Cambournac, F. J. C., & Hill, R. B. (1940). Observation on the swarming of *Anopheles maculipennis*, var. atroparvus. *American Journal of Tropical Medicine, 20*(1).

Caputo, B., Nwakanma, D., Jawara, M., Adiamoh, M., Dia, I., Konate, L., della Torre, A. (2008). *Anopheles gambiae* complex along The Gambia River, with particular reference to the molecular forms of *An. gambiae* ss. *Malaria Journal, 7*(1), 1-17.

Chavshin, A. R., Oshaghi, M. A., Vatandoost, H., Pourmand, M. R., Raeisi, A., Enayati, A. A., Ghoorchian, S. (2012). Identification of bacterial microflora in the midgut of the larvae and adult of wild caught *Anopheles stephensi*: a step toward finding suitable paratransgenesis candidates. *Acta tropica, 121*(2), 129-134.

Chege, G. M., & Beier, J. C. (1990). Effect of Plasmodium falciparum on the survival of naturally infected afrotropical*Anopheles* (Diptera: Culicidae). *Journal of Medical Entomology, 27*(4), 454-458.

Chouaia, B., Rossi, P., Montagna, M., Ricci, I., Crotti, E., Damiani, C., Bandi, C. (2010). Molecular evidence for multiple infections as revealed by typing of *Asaia* bacterial symbionts of four mosquito species. *Applied and environmental microbiology, 76*(22), 7444-7450.

Chouaïbou, M., Etang, J., Brevault, T., Nwane, P., Hinzoumbé, C. K., Mimpfoundi, R., & Simard, F. (2008). Dynamics of insecticide resistance in the malaria vector *Anopheles gambiae*sl from an area of extensive cotton cultivation in Northern Cameroon. *Tropical Medicine & International Health, 13*(4), 476-486.

Christophides, G. K., Zdobnov, E., Barillas-Mury, C., Birney, E., Blandin, S., Blass, C., & Kafatos, F. C. (2002). Immunity-related genes and gene families in *Anopheles gambiae. Science, 298*(5591), 159-165.

Cirimotich, C. M., Dong, Y., Clayton, A. M., Sandiford, S. L., Souza-Neto, J. A., Mulenga, M., & Dimopoulos, G. (2011). Natural microbe-mediated refractoriness to *Plasmodium* infection in Anopheles gambiae. *Science, 332*(6031), 855-858.

Clements, A. N., & Paterson, G. D. (1981). The analysis of mortality and survival rates in wild populations of mosquitoes. *Journal of Applied Ecology*, 373-399.

Cohuet, A., Simard, F., Wondji, C. S., Antonio-Nkondjio, C., Awono-Ambene, P., & Fontenille, D. (2004). High malaria transmission intensity due to *Anopheles funestus* (Diptera: Culicidae) in a village of savannah–forest transition area in Cameroon. *Journal of Medical Entomology, 41*(5), 901-905.

Collins, W. E., & Roberts, J. M. (1991). *Anopheles gambiae* as a host for geographic isolates of *Plasmodium vivax. Journal of the American Mosquito Control Association, 7*(4), 569-573.

Coluzzi, M. (1982). Spatial distribution of chromosomal inversions and speciation in anopheline mosquitoes. *Progress in clinical and biological research, 96*, 143-153.

Coluzzi, M., Sabatini, A., della Torre, A., Di Deco, M. A., & Petrarca, V. (2002). A polytene chromosome analysis of the *Anopheles gambiae* species complex. *Science, 298*(5597), 1415-1418.

Costantini, C., Sagnon, N. F., della Torre, A., Diallo, M., Brady, J., Gibson, G., & Coluzzi, M. (1998). Odor-mediated host preferences of West African mosquitoes, with particular reference to malaria vectors. *The American journal of tropical medicine and hygiene, 58*(1), 56-63.

Costantini, C., Sagnon, N. F., Torre, A. D., & Coluzzi, M. (1999). Mosquito behavioural aspects of vector-human interactions in the *Anopheles gambiae* complex. *Parassitologia, 41*(1/3), 209-220.

Damiani, C., Ricci, I., Crotti, E., Rossi, P., Rizzi, A., Scuppa, P., Favia, G. (2010). Mosquito-bacteria symbiosis: the case of *Anopheles gambiae* and *Asaia*. *Microbial ecology, 60*, 644-654.

Dawes, E. J., Churcher, T. S., Zhuang, S., Sinden, R. E., & Basáñez, M. G. (2009). Anopheles mortality is both age-and Plasmodium-density dependent: implications for malaria transmission. *Malaria journal, 8*(1), 1-16.

Day, J. F. (2005). Host-seeking strategies of mosquito disease vectors. *Journal of the American Mosquito Control Association, 21*(sp1), 17-22.

De Buck, A. (1936). Some results of six years' mosquito infection work. *American Journal of Hygiene, 24*(1).

De Buck, A., & Swellengrebel, N. H. (1934). Behaviour of Dutch *Anopheles atroparvus* and *messeae* in Winter under artificial Conditions. *Rivista di Malariologia, 13*(4), 404-416.

De Zulueta, J., Ramsdale, C. D., & Coluzzi, M. (1975). Receptivity to malaria in Europe. *Bulletin of the World Health Organization, 52*(1), 109.

Detinova, T. S., Bertram, D. S., & World Health Organization. (1962). *Age-grouping methods in Diptera of medical importance, with special reference to some vectors of malaria.* World Health Organization.

Diabaté, A., Brengues, C., Baldet, T., Dabire, K. R., Hougard, J. M., Akogbeto, M., ... & Chandre, F. (2004). The spread of the Leu-Phekdr mutation through *Anopheles gambiae* complex in Burkina Faso: genetic introgression and de novo phenomena. *Tropical Medicine & International Health, 9*(12), 1267-1273.

Diatta, M., Spiegel, A., Lochouarn, L., & Fontenille, D. (1998). Similar feeding preferences of *Anopheles gambiae* and *A. arabiensis* in Senegal. *Transactions of the Royal Society of Tropical Medicine and Hygiene, 92*(3), 270-272.

DinparastDjadid, N., Jazayeri, H., Raz, A., Favia, G., Ricci, I., & Zakeri, S. (2011). Identification of the midgut microbiota of *An. stephensi* and *An. maculipennis* for their application as a paratransgenic tool against malaria. *PloS one, 6*(12), e28484.

Dong, Y., Manfredini, F., & Dimopoulos, G. (2009). Implication of the mosquito midgut microbiota in the defense against malaria parasites. *PLoS pathogens, 5*(5), e1000423.

Duchemin, J. B., Tsy, J. M. L. P., Rabarison, P., Roux, J., Coluzzi, M., & Costantini, C. (2001). Zoophily of *Anopheles arabiensis* and *An. gambiae* in Madagascar demonstrated by odour-baited entry traps. *Medical and Veterinary Entomology, 15*(1), 50-57.

Dye, C. (1992). The analysis of parasite transmission by bloodsucking insects. *Annual review of entomology, 37*(1), 1-19.

Eckstein, F. (1936). Experimentellebeobachtungenan*Anophelesmaculipennis*. *Arch. f. Schiffs-u. Tropen-Hyg, 40*, 381-395.

Erel, D. (1973). *Anadolu vektörlerivemücadelemetotlari*. AkişBasımevi.

Favia, G., Ricci, I., Damiani, C., Raddadi, N., Crotti, E., Marzorati, M., ... & Daffonchio, D. (2007). Bacteria of the genus *Asaia* stably associate with *Anopheles stephensi*, an Asian malarial mosquito vector. *Proceedings of the National Academy of Sciences, 104*(21), 9047-9051.

Ferguson, H. M., & Read, A. F. (2002). Genetic and environmental determinants of malaria parasite virulence in mosquitoes. *Proceedings of the Royal Society of London. Series B: Biological Sciences, 269*(1497), 1217-1224.

Ferguson, H. M., Mackinnon, M. J., Chan, B. H., & Read, A. F. (2003a). Mosquito mortality and the evolution of malaria virulence. *Evolution, 57*(12), 2792-2804.

Ferguson, H. M., Rivero, A., & Read, A. F. (2003b). The influence of malaria parasite genetic diversity and anaemia on mosquito feeding and fecundity. *Parasitology, 127*(1), 9-19.

Gad, A. M., Maier, W. A., & Piekarski, G. (1979). Pathology of *Anopheles stephensi* after infection with *Plasmodium bergheiberghei*. II. Changes in amino acid contents. *Zeitschrift fur Parasitenkunde (Berlin, Germany), 60*(3), 263-276.

Gamage-Mendis, A. C., Rajakaruna, J., Weerasinghe, S., Mendis, C., Carter, R., & Mendis, K. N. (1993). Infectivity of *Plasmodium vivax* and *P. falciparum* to *Anopheles tessellatus*; relationship between oocyst and sporozoite development. *Transactions of the Royal society of tropical medicine and hygiene*, *87*(1), 3-6.

Garrett-Jones, C. (1964). The human blood index of malaria vectors in relation to epidemiological assessment. *Bulletin of the World Health Organization*, *30*(2), 241.

Garrett-Jones, C., Boreham, P. F. L., & Pant, C. P. (1980). Feeding habits of anophelines (Diptera: Culicidae) in 1971–78, concerning the human blood index: a review. *Bulletin of entomological research*, *70*(2), 165-185.

Gillies, M. T. (1988). Anopheline mosquitoes. Vector behaviour and bionomics. *Malaria: Principle and practices of malariology*, *1*, 453-485.

Githeko, A. K., Service, M. W., Mbogo, C. M., Atieli, F. K., & Juma, F. O. (1994). Origin of blood meals in indoor and outdoor resting malaria vectors in western Kenya. *Acta tropica*, *58*(3-4), 307-316.

Gonzalez-Ceron, L., Santillan, F., Rodriguez, M. H., Mendez, D., & Hernandez-Avila, J. E. (2003). Bacteria in midguts of field-collected *Anopheles albimanus* block *Plasmodium vivax*sporogonic development. *Journal of medical entomology*, *40*(3), 371-374.

Graves, P. M., Burkot, T. R., Carter, R., Cattani, J. A., Lagog, M., Parker, J., & Alpers, M. P. (1988). Measurement of malarial infectivity of human populations to mosquitoes in the Madang area, Papua New Guinea. *Parasitology*, *96*(2), 251-263.

Habtewold, T., Povelones, M., Blagborough, A. M., & Christophides, G. K. (2008). Transmission blocking immunity in the malaria non-vector mosquito *Anopheles quadriannulatus* species A. *PLoS pathogens*, *4*(5), e1000070.

Hackett, L. W., & Missiroli, A. (1931). The Natural Disappearance of Malaria in Certain Regions of Europe. *American Journal of Hygiene*, *13*(1).

Hadis, M., Lulu, M., Makonnen, Y., & Asfaw, T. (1997). Host choice by indoor-resting *Anopheles arabiensis* in Ethiopia. *Transactions of the Royal Society of Tropical Medicine and Hygiene*, *91*(4), 376-378.

Haji, H., Smith, T., Charlwood, J. D., & Meuwissen, J. H. (1996). Absence of relationships between selected human factors and natural infectivity of *Plasmodium falciparum* to mosquitoes in an area of high transmission. *Parasitology*, *113*(5), 425-431.

Hallem, E. A., Dahanukar, A., & Carlson, J. R. (2006). Insect odor and taste receptors. *Annu. Rev. Entomol.*, *51*, 113-135.

Hay, S. I., Sinka, M. E., Okara, R. M., Kabaria, C. W., Mbithi, P. M., Tago, C. C., & Godfray, H. C. J. (2010). Developing global maps of the dominant *Anopheles* vectors of human malaria. *PLoS medicine*, *7*(2), e1000209.

Hocking, K. S., & MacInnes, D. G. (1948). Notes on the bionomics of *Anopheles gambiae* and *A. funestus* in East Africa. *Bulletin of Entomological Research*, *39*(3), 453-465.

Hogg, J. C., & Hurd, H. (1995a). Malaria-induced reduction of fecundity during the first gonotrophic cycle of *Anopheles stephensi* mosquitoes. *Medical and veterinary entomology*, *9*(2), 176-180.

Hogg, J. C., & Hurd, H. (1995b). *Plasmodium yoeliinigeriensis*: the effect of high and low intensity of infection upon the egg production and bloodmeal size of *Anopheles stephensi* during three gonotrophic cycles. *Parasitology*, *111*(5), 555-562.

Hogg, J. C., & Hurd, H. (1997). The effects of natural *Plasmodium falciparum* infection on the fecundity and mortality of *Anopheles gambiae*sl in north east Tanzania. *Parasitology*, *114*(4), 325-331.

Holt, R. A., Subramanian, G. M., Halpern, A., Sutton, G. G., Charlab, R., Nusskern, D. R., ... & Hoffman, S. L. (2002). The genome sequence of the malaria mosquito *Anopheles gambiae. science*, *298*(5591), 129-149.

Horsfall, W. E. (1955). Mosquitoes. Their bionomics and relation to disease. *Mosquitoes. Their Bionomics and Relation to Disease.*

Hurd, H., & Carter, V. (2004). The role of programmed cell death in Plasmodium–mosquito interactions. *International Journal for Parasitology*, *34*(13-14), 1459-1472.

Hurd, H., Carter, V., & Nacer, A. (2005). Interactions between malaria and mosquitoes: the role of apoptosis in parasite establishment and vector response to infection. *Role of Apoptosis in Infection*, 185-217.

Jadin, J. (1967). Role of bacteria in the digestive tube of insects, vectors of plasmodidae and trypanosomidae. *Annales des societesbelges de medecinetropicale, de parasitologie, et de mycologie*, *47*(4), 331-342.

Jetten, T. H., & Takken, W. (1994). Anophelism without malaria in Europe: a review of the ecology and distribution of the genus Anopheles in Europe (No. 94-5). Wageningen Agricultural University.

Kajla, M. K., Andreeva, O., Gilbreath III, T. M., & Paskewitz, S. M. (2010). Characterization of expression, activity and role in antibacterial immunity of *Anopheles gambiae* lysozyme c-1. *Comparative Biochemistry and Physiology Part B: Biochemistry and Molecular Biology*, *155*(2), 201-209.

Kambris, Z., Cook, P. E., Phuc, H. K., & Sinkins, S. P. (2009). Immune activation by life-shortening Wolbachia and reduced filarial competence in mosquitoes. *Science*, *326*(5949), 134-136.

Kämpfer, P., Matthews, H., Glaeser, S. P., Martin, K., Lodders, N., & Faye, I. (2011). *Elizabethkingiaanophelis* sp. nov., isolated from the midgut of the mosquito *Anopheles gambiae*. *International journal of systematic and evolutionary microbiology*, *61*(11), 2670-2675.

Kasap, M. (1987). Effects of photoperiod on some attributes of immature stages of An. sacharovi. TürkHijyenveDeneyselBiyolojiDergisi, 44, 53-61.

Killeen, G. F., McKenzie, F. E., Foy, B. D., Bøgh, C., & Beier, J. C. (2001). The availability of potential hosts as a determinant of feeding behaviours and malaria transmission by African mosquito populations. *Transactions of the Royal Society of Tropical Medicine and Hygiene*, *95*(5), 469-476.

Klein, T. A., BA, H., & RG, A. (1982). Detrimental effects of Plasmodium cynomolgi infections on the longevity of *Anopheles dirus*.

Klein, T. A., Harrison, B. A., Grove, J. S., Dixon, S. V., & Andre, R. G. (1986). Correlation of survival rates of *Anopheles dirus* A (Diptera: Culicidae) with different infection densities of *Plasmodium cynomolgi*. *Bulletin of the World Health Organization*, *64*(6), 901.

Koella, J. C. (1999). An evolutionary view of the interactions between anopheline mosquitoes and malaria parasites. *Microbes and infection*, *1*(4), 303-308.

Kumar, S., Molina-Cruz, A., Gupta, L., Rodrigues, J., & Barillas-Mury, C. (2010). A peroxidase/dual oxidase system modulates midgut epithelial immunity in *Anopheles gambiae*. *Science*, *327*(5973), 1644-1648.

Lacroix, R., Mukabana, W. R., Gouagna, L. C., & Koella, J. C. (2005). Malaria infection increases attractiveness of humans to mosquitoes. *PLoS biology*, *3*(9), e298.

Lahondère, C., & Lazzari, C. R. (2012). Mosquitoes cool down during blood feeding to avoid overheating. *Current Biology*, *22*(1), 40-45.

Lambrechts, L., Chavatte, J. M., Snounou, G., & Koella, J. C. (2006). Environmental influence on the genetic basis of mosquito resistance to malaria parasites. *Proceedings of the Royal Society B: Biological Sciences*, *273*(1593), 1501-1506.

Leeson, H. S. (1939). Longevity of *Anopheles maculipennis* race atroparvus, van thiel, at controlled temperature and humidity after one blood meal. *Bulletin of Entomological Research*, *30*(3), 295-301.

Lemasson, J. J., Fontenille, D., Lochouarn, L., Dia, I., Simard, F., Ba, K., & Molez, J. F. (1997). Comparison of behavior and vector efficiency of *Anopheles gambiae* and *An. arabiensis* (Diptera: Culicidae) in Barkedji, a Sahelian area of Senegal. *Journal of Medical Entomology*, *34*(4), 396-403.

Lindh, J. M., Borg-Karlson, A. K., & Faye, I. (2008). Transstadial and horizontal transfer of bacteria within a colony of *Anopheles gambiae* (Diptera: Culicidae) and oviposition response to bacteria-containing water. *Acta tropica*, *107*(3), 242-250.

Lindh, J. M., Terenius, O., & Faye, I. (2005). 16S rRNA gene-based identification of midgut bacteria from field-caught *Anopheles gambiae*sensulato and *A. funestus* mosquitoes reveals new species related to known insect symbionts. *Applied and environmental microbiology*, *71*(11), 7217-7223.

Loyola, E. G., Gonzalez-Ceron, L., Rodriguez, M. H., Arredondo-Jimenez, J. I., Bennett, S., & Bown, D. N. (1993). *Anopheles albimanus* (Diptera: Culicidae) host selection patterns in three ecological areas of the coastal plains of Chiapas, southern Mexico. *Journal of Medical Entomology*, *30*(3), 518-523.

Lu, T., Qiu, Y. T., Wang, G., Kwon, J. Y., Rutzler, M., Kwon, H. W., & Zwiebel, L. J. (2007). Odor coding in the maxillary palp of the malaria vector mosquito *Anopheles gambiae*. *Current Biology*, *17*(18), 1533-1544.

Macdonald, G. (1957). The epidemiology and control of malaria. *The Epidemiology and Control of Malaria.*

Manguin, S., Bangs, M. J., Pothikasikorn, J., & Chareonviriyaphap, T. (2010). Review on global co-transmission of human *Plasmodium* species and *Wuchereriabancrofti* by *Anopheles* mosquitoes. *Infection, Genetics and Evolution, 10*(2), 159-177.

Marrelli, M. T., Li, C., Rasgon, J. L., & Jacobs-Lorena, M. (2007). Transgenic malaria-resistant mosquitoes have a fitness advantage when feeding on Plasmodium-infected blood. *Proceedings of the National Academy of Sciences, 104*(13), 5580-5583.

McCarroll, L., & Hemingway, J. (2002). Can insecticide resistance status affect parasite transmission in mosquitoes? *Insect Biochemistry and Molecular Biology, 32*(10), 1345-1351.

Meister, S., Agianian, B., Turlure, F., Relógio, A., Morlais, I., Kafatos, F. C., & Christophides, G. K. (2009). Anopheles gambiae PGRPLC-mediated defense against bacteria modulates infections with malaria parasites. *Plos pathogens, 5*(8), e1000542.

Mendes, A. M., Schlegelmilch, T., Cohuet, A., Awono-Ambene, P., De Iorio, M., Fontenille, D., Vlachou, D. (2008). Conserved mosquito/parasite interactions affect development of *Plasmodium falciparum* in Africa. *PLoS Pathogens, 4*(5), e1000069.

Moreira, L. A., Iturbe-Ormaetxe, I., Jeffery, J. A., Lu, G., Pyke, A. T., Hedges, L. M., & O'Neill, S. L. (2009). A Wolbachia symbiont in *Aedes aegypti* limits infection with dengue, Chikungunya, and Plasmodium. *Cell, 139*(7), 1268-1278.

Muir, D. A. (1988). Anopheline mosquitoes: vector reproduction, life-cycle and biotope. *Malaria: principles and practice of malariology, 1*, 431-451.

Müller, G. C., Beier, J. C., Traore, S. F., Toure, M. B., Traore, M. M., Bah, S., Schlein, Y. (2010). Field experiments of Anopheles gambiae attraction to local fruits/seedpods and flowering plants in Mali to optimize strategies for malaria vector control in Africa using attractive toxic sugar bait methods. *Malaria journal, 9*(1), 1-11.

Mwangangi, J. M., Mbogo, C. M., Nzovu, J. G., Githure, J. I., Yan, G., & Beier, J. C. (2003). Blood-meal analysis for anopheline mosquitoes sampled along the Kenyan coast. *Journal of the American Mosquito Control Association, 19*(4), 371-375.

Nace, D., Williams, T., Sullivan, J., Williams, A., Galland, G. G., & Collins, W. E. (2004). Susceptibility of *Anopheles farauti* to infection with different species of *Plasmodium. Journal of the American Mosquito Control Association, 20*(3), 272-276.

Noden, B. H., Vaughan, J. A., Pumpuni, C. B., & Beier, J. C. (2011). Mosquito ingestion of antibodies against mosquito midgut microbiota improves conversion of ookinetes to oocysts for *Plasmodium falciparum*, but not *P. yoelii. Parasitology International, 60*(4), 440-446.

Novoseltsev, V. N., Michalski, A. I., Novoseltseva, J. A., Yashin, A. I., Carey, J. R., & Ellis, A. M. (2012). An age-structured extension to the vectorial capacity model. *PLoS One, 7*(6), e39479.

Osei-Poku, J., Mbogo, C. M., Palmer, W. J., & Jiggins, F. M. (2012). Deep sequencing reveals extensive variation in the gut microbiota of wild mosquitoes from K enya. *Molecular Ecology, 21*(20), 5138-5150.

Port, G. R., Boreham, P. F. L., & Bryan, J. H. (1980). The relationship of host size to feeding by mosquitoes of the *Anopheles gambiae* Giles complex (Diptera: Culicidae). *Bulletin of Entomological Research, 70*(1), 133-144.

Pringle, G. (1966). A quantitative study of naturally-acquired malaria infections in *Anopheles gambiae* and *Anopheles funestas* in a highly malarious area of East Africa. *Transactions of the Royal Society of Tropical Medicine and Hygiene, 60*(5).

Pumpuni, C. B., Beier, M. S., Nataro, J. P., Guers, L. D., & Davis, J. R. (1993). *Plasmodium falciparum*: inhibition of sporogonic development in *Anopheles stephensi* by gram-negative bacteria. *Experimental parasitology, 77*(2), 195-199.

Pumpuni, C. B., Demaio, J., Kent, M., Davis, J. R., & Beier, J. C. (1996). Bacterial population dynamics in three anopheline species: the impact on *Plasmodium*sporogonic development. *The American journal of tropical medicine and hygiene, 54*(2), 214-218.

Ramsdale, C. D., Coluzzi, M., & World Health Organization. (1975). Studies on the infectivity of tropical African strains of *Plasmodium falciparum* to some southern European vectors of malaria (No. WHO/MAL/75.859). World Health Organization.

Rani, A., Sharma, A., Rajagopal, R., Adak, T., & Bhatnagar, R. K. (2009). Bacterial diversity analysis of larvae and adult midgut microflora using culture-dependent and culture-independent methods in lab-reared and field-collected *Anopheles stephensi*-an Asian malarial vector. *BMC Microbiology*, *9*, 1-22.

Ranson, H. N′ Guessan R, Lines J, Moiroux N, Nkuni Z, Corbel V. (2011). Pyrethroid resistance in African anopheline mosquitoes: what are the implications for malaria control, 91-98.

Read, A. F., Lynch, P. A., & Thomas, M. B. (2009). How to make evolution-proof insecticides for malaria control. *PLoS biology*, *7*(4), e1000058.

Reisen, W. K., Aslam khan, M. (1976). Observations on the swarming and mating behaviour of *Anopheles culicifacies* Giles in nature. *Bulletin of the World Health Organization*, *54*(2), 155.

Rhosenberg, R., Andre, R. G., & Somchit, L. (1990). Highly efficient dry season transmission of malaria in Thailand. *Transactions of the Royal Society of Tropical Medicine and Hygiene*, *84*(1), 22-28.

Rivero, A., & Ferguson, H. M. (2003). The energetic budget of *Anopheles stephensi* infected with *Plasmodium chabaudi*: is energy depletion a mechanism for virulence? *Proceedings of the Royal Society of London. Series B: Biological Sciences*, *270*(1522), 1365-1371.

Robert, V., Verhave, J. P., & Carnevale, P. (1990). *Plasmodium falciparum* infection does not increase the precocious mortality rate of *Anopheles gambiae*. *Transactions of the Royal Society of Tropical Medicine and Hygiene*, *84*(3), 346-347.

Rwegoshora, R. T., Simonsen, P. E., Meyrowitsch, D. W., Malecela-Lazaro, M. N., Michael, E., & Pedersen, E. M. (2007). Bancroftian filariasis: house-to-house variation in the vectors and transmission—and the relationship to human infection—in an endemic community of coastal Tanzania. *Annals of Tropical Medicine & Parasitology*, *101*(1), 51-60.

Ryan, S. J., Ben-Horin, T., & Johnson, L. R. (2015). Malaria control and senescence: the importance of accounting for the pace and shape of aging in wild mosquitoes. *Ecosphere*, *6*(9), 1-13.

Schwartz, A., & Koella, J. C. (2002). Melanization of Plasmodium falciparum and C-25 sephadex beads by field-caught *Anopheles gambiae* (Diptera: Culicidae) from southern Tanzania. *Journal of Medical Entomology*, *39*(1), 84-88.

Schwartz, A., & Koella, J. C. (2004). The cost of immunity in the yellow fever mosquito, *Aedes aegypti* depends on immune activation. *Journal of Evolutionary Biology*, *17*(4), 834-840.

Service, M. W. (1976). Mosquito Ecology. Applied Science Publishers, London 583pp. *28*(2), 174.

Sinden, R. E., Alavi, Y., & Raine, J. D. (2004). Mosquito–malaria interactions: a reappraisal of the concepts of susceptibility and refractoriness. *Insect biochemistry and molecular biology*, *34*(7), 625-629.

Sinden, R. E., Dawes, E. J., Alavi, Y., Waldock, J., Finney, O., Mendoza, J., Basanez, M. G. (2007). Progression of *Plasmodium berghei* through *Anopheles stephensi* is density-dependent. *PLoS pathogens*, *3*(12), e195.

Sinka, M. E., Bangs, M. J., Manguin, S., Rubio-Palis, Y., Chareonviriyaphap, T., Coetzee, M., Hay, S. I. (2012). A global map of dominant malaria vectors. *Parasites & vectors*, *5*(1), 1-11.

Smith, D. L., Battle, K. E., Hay, S. I., Barker, C. M., Scott, T. W., & McKenzie, F. E. (2012). Ross, Macdonald, and a theory for the dynamics and control of mosquito-transmitted pathogens. *PLoS pathogens*, *8*(4), e1002588.

Smith, D. L., Perkins, T. A., Reiner Jr, R. C., Barker, C. M., Niu, T., Chaves, L. F., & Scott, T. W. (2014). Recasting the theory of mosquito-borne pathogen transmission dynamics and control. *Transactions of the Royal Society of Tropical Medicine and Hygiene*, *108*(4), 185-197.

Snow, W. F. (1983). Mosquito production and species succession from an area of irrigated rice fields in The Gambia, West Africa. *The Journal of Tropical Medicine and Hygiene*, *86*(6), 237-245.

Sousa, C. A., Pinto, J., Almeida, P. G., Ferreira, C., Do Rosário, V. E., & Charlwood, J. D. (2001). Dogs as a favored host choice of *Anopheles gambiae*sensustricto (Diptera: Culicidae) of Sao Tome, west Africa. *Journal of Medical Entomology*, *38*(1), 122-125.

Straif, S. C., Mbogo, C. N., Toure, A. M., Walker, E. D., Kaufman, M., Toure, Y. T., & Beier, J. C. (1998). Midgut bacteria in *Anopheles gambiae* and *An. funestus* (Diptera: Culicidae) from Kenya and Mali. *Journal of Medical Entomology*, *35*(3), 222-226.

Styer, L. M., Carey, J. R., Wang, J. L., & Scott, T. W. (2007). Mosquitoes do senesce: departure from the paradigm of constant mortality. *The American journal of tropical medicine and hygiene*, *76*(1), 111.

Swellengrebel, N. H. (1929). On the Influence of the Wind in the Spread of *Anopheles maculipennis*. *American Journal of Hygiene*, *10*(2).

Sylvie, M., Pierre, C., & Jean, M. (2008). *Biodiversity of malaria in the world*. John Libbey Eurotext.

Takken, W., & Costantini, C. (2006). The genetics of vector-host interactions: alternative strategies for genetic engineering for malaria control. *Genetically Modified Mosquitoes for Vector Control*.

Terenius, O., De Oliveira, C. D., Pinheiro, W. D., Tadei, W. P., James, A. A., & Marinotti, O. (2008). 16S rRNA gene sequences from bacteria associated with adult *Anopheles darlingi* (Diptera: Culicidae) mosquitoes. *Journal of Medical Entomology*, *45*(1), 172-175.

Thomas, C. J., & Lindsay, S. W. (2000). Local-scale variation in malaria infection amongst rural Gambian children estimated by satellite remote sensing. *Transactions of the Royal Society of Tropical Medicine and Hygiene*, *94*(2), 159-163.

Van Seventer, H. A. (1969). The disappearance of malaria in the Netherlands. *The disappearance of malaria in the Netherlands*.

Verhulst, N. O., Beijleveld, H., Knols, B. G., Takken, W., Schraa, G., Bouwmeester, H. J., & Smallegange, R. C. (2009). Cultured skin microbiota attracts malaria mosquitoes. *Malaria journal*, *8*, 1-12.

Vinogradova, E. B. (1960). An experimental investigation of the ecological factors inducing imaginal diapause in bloodsucking mosquitoes (Diptera, Culicidae). *Ent Obozr*, *39*, 327-340.

Vlachou, D., Zimmermann, T., Cantera, R., Janse, C. J., Waters, A. P., & Kafatos, F. C. (2004). Real-time, in vivo analysis of malaria ookinete locomotion and mosquito midgut invasion. *Cellular microbiology*, *6*(7), 671-685.

Waldock, J., Olson, K. E., & Christophides, G. K. (2012). *Anopheles gambiae* antiviral immune response to systemic O'nyong-nyong infection. *PLoS Neglected Tropical Diseases*, *6*(3), e1565.

Wang, S., Ghosh, A. K., Bongio, N., Stebbings, K. A., Lampe, D. J., & Jacobs-Lorena, M. (2012). Fighting malaria with engineered symbiotic bacteria from vector mosquitoes. *Proceedings of the National Academy of Sciences*, *109*(31), 12734-12739.

Wang, Y., Iii, T. M. G., Kukutla, P., & Yan, G. (2011). Xu J. Dynamic gut microbiome across life history of the malaria mosquito *Anopheles gambiae* in Kenya. PLOS ONE. *Public Libr Sci*, *6*, e24767.

Wanji, S., Tanke, T., Atanga, S. N., Ajonina, C., Nicholas, T., & Fontenille, D. (2003). *Anopheles* species of the Mount Cameroon region: biting habits, feeding behaviour and entomological inoculation rates. *Tropical Medicine & International Health*, *8*(7), 643-649.

Weill, M., Malcolm, C., Chandre, F., Mogensen, K., Berthomieu, A., Marquine, M., & Raymond, M. (2004). The unique mutation in ace-1 giving high insecticide resistance is easily detectable in mosquito vectors. *Insect molecular biology*, *13*(1), 1-7.

Wekesa, J. W., Copeland, R. S., & Mwangi, R. W. (1992). Effect of *Plasmodium falciparum* on blood feeding behavior of naturally infected *Anopheles* mosquitoes in western Kenya. *The American journal of tropical medicine and hygiene*, *47*(4), 484-488.

Weyer, F., & Bock, E. (1939). Experiments on the Transmission of various Species of *Plasmodium* to *Anopheles*. *Archiv fur Schiffs-und Tropenhygiene*, *43*(6).

WHO (2022) World Malaria Report 2022. https://www.who.int/teams/global-malaria-programme/reports/world-malaria-report-2022.

World Health Organization (1975) WHO Manual on Practical Entomology in Malaria, Part II, World Health Organization.

World Health Organization for Africa: Implementation of indoor residual spraying of insecticides for malaria control in the WHO African region report Geneva: World Health Organization; 2007.

World Health Organization. World Malaria Report: 2011. Geneva: World Health Organization, 2011: 259.

Yassine, H., & Osta, M. A. (2010). *Anopheles gambiae* innate immunity. *Cellular microbiology*, *12*(1), 1-9.

Yoeli, M., & Mer, G. G. (1938). The Relation of Blood Feeds to the Maturation of Ova in *Anopheles elutus*. *Transactions of the Royal Society of Tropical Medicine and Hygiene*, *31*(4).

Zimmerman, R. H., Galardo, A. K. R., Lounibos, L. P., Arruda, M., & Wirtz, R. (2006). Bloodmeal hosts of *Anopheles* species (Diptera: Culicidae) in a malaria-endemic area of the Brazilian Amazon. *Journal of Medical Entomology*, *43*(5), 947-956.

Chapter 11

The Generalist vs. The Specialist Predator: Their Roles in Biological Control

Anil Kumar[1,*]
Vishwajeet[2]
and Nishi Keshari[3]
[1]Department of Entomology, Sugarcane Research Institute,
Dr. Rajendra Prasad Central Agricultural University, Pusa, Samastipur, Bihar, India
[2]Department of Entomology, PGCA, Dr. Rajendra Prasad Central Agricultural University, Pusa, Samastipur, Bihar, India
[3]Department of Nematology, PGCA, Dr. Rajendra Prasad Central Agricultural University, Pusa, Samastipur, Bihar, India

Abstract

Predators are animals that prey on other animals; they may be beetles, true bugs, lacewings, wasps, dragonflies, spiders, predatory mites, vertebrates, and snakes. They are usually very active because they must hunt their prey. Predatory insects feed many insect pests, and they are an important component of the natural control of agricultural pests. Understanding the role of predators in the ecosystem will not only help to implement biological control in pest management but also conserve biodiversity. Predators can be exploited in 'biological control', which forms the basis of integrated pest management and can help reduce the use of chemical insecticides. Based on the choice of diet, predators can be broadly classified as generalists and specialists. The distinction between the two will form the structure of this chapter. Specialist predators feed on only one or a few species of prey, and the success of classical biological control mostly relies on the introduction of specialist natural enemies. But generalist predators have a wide range of prey available. They may feed at all stages, including eggs, larvae, nymphs, pupae, and adults. Some of the predators are predatory in larval as well as adult stages (e.g., ladybird beetles), while others predate only in the larval stage (e.g., lacewings). Generalist predators are quick-acting; they are already present and start to feed upon newly arrived pests in a density-dependent manner; on the other hand, specialists may take a longer time to arrive. Pest suppression at the very beginning stage may delay or reduce the intensity of the pest attack, or it may give enough time for the specialist to arrive. The ability of a specialist predator to hunt down its specific prey may help to immediately suppress the pest when released inundatively, in a similar fashion as insecticide spray. However, the integration of both specialists and generalists in the cropping ecosystem can be additive and help suppress the pest population below the economic threshold level. Due to their behaviour, predators are a farmer's best friend.

---

* Corresponding Author's Email: anil.kumar@rpcau.ac.in; agri_anil@rediffmail.com.

In: Insects as Humans' Frenemies: From Friends to Foes
Editor: Istkhar Rao
ISBN: 979-8-89530-320-7

Keywords*:* predators, pest, pest suppression, biocontrol, behaviour

1. Introduction

Predators are the organisms that actively catch, kill, and prey upon other animals for their development, sustenance, and reproduction. In this kind of biological interaction, the predators benefit, and the prey animals are harmed. They occupy the third trophic level in a food chain or ecological pyramid. They must consume more than one individual to complete their life cycle. Predators are active organisms, and they adopt various strategies to obtain their prey, which include: hunting (adults of tiger beetles and ground beetles), stalking and ambush (nymphs of dragonflies and damselflies), trapping (grubs of antlions), provisioning (solitary wasps in the family Vespidae and Sphecidae), mass foraging (ants), deception and predation (fireflies), host feeding (female parasitoids feeding on oozes of ovipositional puncture), etc. Whether a predator is a generalist or specialist, they play a key role in manipulating the population of biocontrol agents for use in IPM. Different approaches are taken to increase the activity of predators in a crop ecosystem. Generalist predators are conserved, while specialist predators such as Vedalia beetles, *Novius cardinalis*, and *Rodolia cardinalis* are imported and augmented (Hagen et al. 1976). Prey specificity is generally much favoured in biological control; if the predator shows a high level of specificity, then it forms a key asset in determining its use in biological control. The first successful biocontrol agent, *R. cardinalis*, shows a high degree of specificity on *Icerya purchasi*. However, many biocontrol agents are not host-specific; they thrive on a wide range of hosts. For example, anthocorid bugs and green lacewing bugs. A specialist may take a longer time to arrive in a crop, and a generalist predator may already be present there. The predation by a generalist starts with the arrival of pests, even when the pest density is low. If the pest population has been checked since the beginning of the infestation, the pest may not be able to attain rapid growth. Thus, for a short duration, crop generalist predators may be more beneficial. Further, the spray of insecticide may upset the population of generalists, leading to a resurgence and outbreak of new secondary pests.

Predators are an important component in the agricultural ecosystem; they check the population of pests. Among the various insect predators, the members of Formicidae, Carabidae, Anthocoridae, Syrphidae, Cecidomyiidae, Reduviidae, Staphylinidae, Chrysopidae, and Coccinellidae are important in the agri-horticultural ecosystem. There are a few orders of insects that do not have any predatory insects, i.e., Phthiraptera, Phasmatodea, Strepsiptera, Isoptera, and Siphonaptera (Hagen, 1987). Based on their food habits, these predators can be classified as:

i. Generalist predators: They depend on a variety of prey consisting of different families under different orders of insects.
ii. Specialist predators: The food habits of these predators have narrowed down to some families under an order or even a specific pest.

Some of the important differences between generalist and specialist predators are mentioned in Table 1.

Table 1. Difference between generalist and specialist predators

Generalist predator	Specialist predator
A wide range of prey is available for predation; thus, a large area is inhabited by them.	Narrow range of prey species, they are restricted in their host environment.
They are generally bitrophic, they may feed on herbivores as well as carnivore animals. Many are omnivores and some are cannibalistic also.	They occupy a single trophic level, either third or fourth.
Classical biological control may pose a threat to the ecosystem.	They are preferred for the classical biological control of invasive species.
Example: Dragonflies, damselflies, carabid beetles, ants etc	Example: Vedalia beetle, giant bamboo beetle etc.

However, most of the organisms do not fit perfectly into either category. The terms 'generalist' and 'specialist' used in this chapter denote the range of food habits an organism may have. It is rare to find a specialist predator restricting its diet to a single species. Even the most successful example of a predatory species, *Rodolia cardinalis*, is not a true specialist predator, as it relies on other coccids besides predating on *Icerya purchasi.* Similarly, a lacewing grub that primarily predates on aphids may shift its diet to other insects and eggs, or if they are deprived of food, they may turn out to be cannibalistic. Compared to specialist predators, the diet of generalist predators may vary to a great extent. They may predate on herbivores and other carnivores, or they may even consume plants and fungi. Such generalists are true omnivores and predate on any small organism that comes their way. Niche breadth can vary according to the species. The dietary range of the Vedalia beetle is considered highly specific. Species like *Adalia bipunctata* and *Hippodamia convergens* are stenophagous and restrict feeding to a range of related taxa. Finally, other predators, e.g., the bug *Podisus maculiventris* and the coccinellid *Coleomegilla maculata*, are polyphagous (generalist) and are considered omnivores as they consume a wide variety of prey and non-prey items, including plant fluid and pollen. Further, biotypes of predatory insects can also exist, which can have different dietary requirements. For example, green lacewing *Chrysopa quadripuncata* (Tauber et al., 1995). Some of the important insect predators in agricultural systems are illustrated in Figure 1.

2. Strategy for Predation

Predation includes a complex mechanism for obtaining prey. This complex process has been broken down into four steps: search, recognition, capture, and handling (Hoback & Higley, 2000). Predators search for prey-rich environments. The search area for a polyphagous species is wider than that of a specialist predator (Zhang & Sanderson, 1993). Delimiting the range of food habits of a predator is difficult. A predator always selects prey that is nutritionally balanced and avoids prey that is distasteful and contains toxins. Once suitable prey is obtained, a predator searches for prey of the same kind in the same area. Once the suitable prey is depleted, a predator may shift its diet to less suitable prey, where the predator may have to pay a physiological cost for survival (Preisser & Bolnick, 2008). Hence, it can be said that at times when prey is abundant, a predator may act like a monophagus, and when the prey population is scarce, they may choose a wide range of prey species. The foraging efficiency of the predator is improved by learnt avoidance (Slansky & Rodriguez, 1987). Further, to capture prey, a predator should have specialised appendages and physiological specialisation. Specialised appendages are more prominent in parasitoids; for example, the ovipositor may be highly specialised to parasitise prey. Evolutionary processes may be operating at each step of predation. For example, a prey species may possess a chemical defence, which may harm the predator in a variety of ways. The ability to produce defensive chemicals comes at a physiological cost and may affect survival and reproductive rate (Preisser & Bolnick, 2008). At the same time, a predator may learn to avoid the defensive chemicals of the prey. This coevolutionary battle may be operating for a long time, and a predator may learn to specialise a prey.

Figure 1. Predators in agricultural ecosystems a. spider b. dragonfly c. praying mantis d. robber fly e. coccinellid (Micraspis yasumatsui), a stenophagous predator f. coccinellid (Pharoscymnus horni).

3. Ecological Interactions of Generalist and Specialist Predators

Predators necessarily occupy the third and fourth trophic levels in the food chain. A food web, which consists of many linear food chains interconnected with each other, shows the dependence of one organism on another for nutritional requirements. A more complex food web has a significant effect on species richness, ecosystem productivity, and stability (Ricklefs, 2008). A specialist predator has a lesser connection to a food web and is prone to disturbances in the ecosystem. He is also more likely to starve if the prey population gets too low, whereas a generalist has a wide range of options to choose from and will not be affected by minor disturbances. This spectrum of resource utilisation is called the ecological niche. Niche breadth (Figure 2) refers to the diversity of resources used by the species. When a particular niche is not evolving and the environment is relatively homogenous in space and time, to avoid competition, some predatory species evolve in a particular direction in such a manner as to avoid regions of niche overlap (Hardin, 1960). This strategy of natural selection is also called the 'competitive exclusion principle', also referred to as 'Gause's law.'

A specialist is essentially evolved from a generalist, and the three general pathways are followed (Tauber et al., 1993):

i) establishment of a population on a specific host
ii) the population evolutionary adaptations to specific hosts through natural selection and heritable traits that influence the survival and reproductive success of individuals on the new host and
iii) maintenance of host fidelity by evolving a reproductively isolated, newly evolved specialist.

In this course of evolution, specialists adapt to the seasonal fluctuations of their prey and the seasonal dynamics of the specialists are synchronised with those of their prey. On the other hand, the seasonal fluctuation of a single species in that ecosystem does not affect the population of generalists. Taniwaki et al. (1993) studied the population fluctuation of the beech sawfly about its specific predator, an ichneumonid parasitoid wasp, and a generalist predatory soldier beetle. The ichneumonid wasp showed a non-delayed synchronous response to fluctuations in beech sawfly, whereas the soldier beetle population showed a low-intensity delayed response. As compared to the generalist, specialist predators should arrive early after pest infestation and show a higher feeding efficiency (Sloggette et al., 2008). In comparison to a specialist predator of aphids and a generalist predator, the aphid population is present for a limited period. If the specialist predator arrives late in the season, the predator will not have a sufficient population of aphids for reproduction, and ultimately, the predator population will begin to decline (Hemptinne et al., 1992). The giant ladybird beetle *Isolemnia dilatata*, which specifically feeds on the woolly aphids of bamboo (Agarwala & Saha, 1984), such as *Ceratovacuna silvestrii*, *Ceratovacuna indica*, and *pseudodoregma bucktoni*, is chiefly considered for the management of woolly aphids in bamboo cultivation. The insecticides cannot reach the target pest, and the mealy secretion of the woolly aphid renders pesticide usage ineffective. On the other hand, giant ladybird beetles, being specific predators of woolly aphids, have very high feeding potential, where each adult female can consume as much as 1523.18±62.02 mg of aphid in its lifetime (Majumdar & Agarwala, 2013).

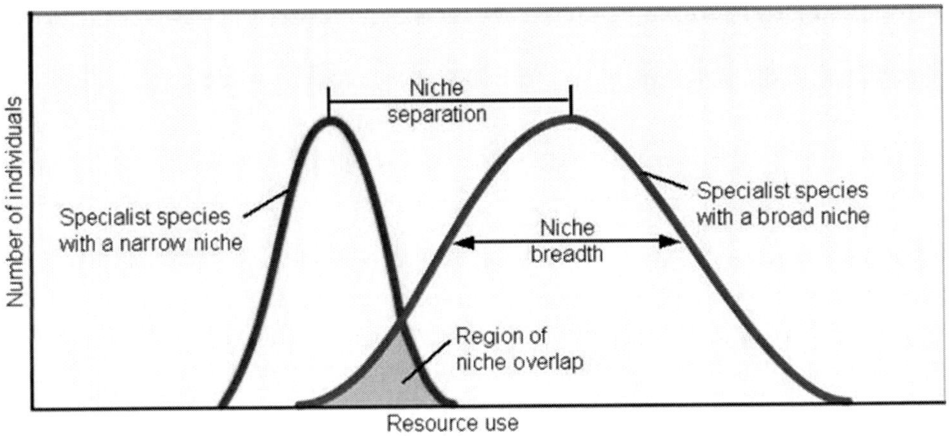

Figure 2. Specialist species have a narrow niche, and generalist species have a broad niche.

Generalist predators are in general bitrophic, i.e., they simultaneously occupy the third and fourth trophic levels. They may feed on herbivores, other predators, and parasitoids. This phenomenon is called intra-guilt predation (IGP). IGP is more intense when the prey species is common for the predators. A predator can prey on other predator species and members of the same species because cannibalism is common in generalist predators. There are several reasons for the cannibalistic behaviour of the predators, but the density-dependent mechanism is more often recognised as a prime cause by ecologists. Aphidivorous predator *Harmonia axyridis* Pallas (Coleoptera, Coccinellidae) shows cannibalism when the density of the prey is low; cannibalism is not observed when the aphid population is high (Burgio et al. 2002). Similar density-dependent cannibalism is observed in *Chrysoperla carnea* (Stephens) (Neuroptera, Chrysopidae). The grub of the green lacewing shows 100% cannibalism in the absence of its host, and with the restoration of the aphid population, negligible cannibalism is present (Rojht et al. 2009). In another example, sexual cannibalism, commonly seen in mantids, is not dependent on the amount of food or density of prey. The neurones in the heads of male mantids inhibit copulation; when the head is ingested by females during copulation, it stimulates copulation, and insemination is efficient (Alcock 2009). In many instances, IGP has resulted in failed or reduced biocontrol by specialist insects. Snyder and Ives (2001) studied the interaction between pea aphids (*Acyrthosiphon pisum*), a specialist parasitoid wasp (*Aphidius ervi*) that attacks the aphids, and a common generalist predator, carabid beetles (*Pterostichus melanarius*). When the carabid population was released, it led to a decrease in pest infestation by the aphids. However, this control was short-lived; the intraguild predation of the carabid beetle on *A. ervi* reduced parasitism.

4. Different Approaches to Biocontrol

Introduction, augmentation, and conservation are different approaches to the biocontrol of the pest. Introduction, also known as classical biological control, is largely dependent on the niche width of the organism. A narrow host range and the ability to thrive in broad environmental conditions are favoured (Taylor & Snyder, 2021). The successful biocontrol agent Vedalia beetle *R. cardinalis* saved the citrus industry in California and was introduced from Australia.

Similarly, in India, the beetle was released against cottony cushion scale *Icerya purchasi*. The beetle is known to be highly monophagous and thrives on the cottony cushion scale. Subsequently, *R. cardinalis* was released all over the world as a control measure against I. *purchasi*. This marked the birth of classical biological control.

A predator with a relatively broad range of hosts may have a significant impact on non-target organisms. An introduced polyphagous predator may suppress the local predator by competing with resources. The introduction of the seven-spotted coccinellid *Coccinella septumpunctata* in the South Dacota of the United States of America resulted in a reduction of native coccinellid species, including *Coccinella transversoguttata* and *Adalia bipunctata*, creating an adverse effect on the biocontrol of aphids (Elliott et al., 1996). Sometimes, a polyphagous predator may feed on several other predators, reducing the pest management efficiency of native natural enemies. The ground beetle *Pterostichus melanarius*, native to Europe but accidentally introduced to North America, fed on herbivores, predacious and detritivorous arthropods, and other prey species caused direct and indirect effects that impacted pest suppression (Hagley et al., 1982).

With the above examples, the utmost care and ecological assessment should be taken before the release of a natural enemy in an alien environment. Also, it is evident from the examples that predators with an increasing degree of specialisation are much safer than generalists for classical biological control. Once the natural enemy is imported, it is augmentatively released for the control of pests. Further, the effectiveness, cost-effectiveness, and ecological assessment of natural enemies are assessed before augmentative release. Conservation strategies in biological control often have many advantages over classical biological control or augmentation of natural enemies. Classical biological control is chiefly considered for invasive pests, and augmentation is not considered when the cost of augmentative release is high. Conservation of natural enemies plays a major role in suppressing native pests and in highly disturbed ecosystems such as agricultural lands, where mobile and opportunistic pests predominate. Here, natural enemies with similar ecological strategies are often required for effective biological control. These natural enemies have three attributes in common:

i) the predator should be able to adapt to the environment and keep a check on the spatial and temporal disruption of the habitat.
ii) the ability to survive when its target pest is not available, and
iii) a predator should have opportunistic feeding habits that allow for persistence and permit the enemy to rapidly exploit the pest population.

Generalist predators (including omnivorous insects) fulfil all three requirements. Under these circumstances, conservation of generalist natural enemies is favoured. Conservation biocontrol relies on locally available natural enemies that have high fitness in a given ecosystem. However, insect-reared natural enemies may not be fit for the environment where they are released. Also, genetic drift and inbreeding are other problems that will reduce the fitness of the natural enemy (Barbosa & Benrey, 1998).

5. Insect Predators of Agricultural Importance

Currently, many specialists and generalist predators have been employed for the control of insect pests in agriculture. Both categories are important for pest management. Some of these are given below in Table 2.

Table 2. Insect predators of agricultural importance

Taxonomic group	Scientific name	Host
Coleoptera		
Coccinellidae	*Coccinelia septumpunctata*	Soft-bodied insects
	Scymnus coccivora	Mealy bugs and scales
	Cheilomenes sexmaculatus	Soft-bodied insects
	Rodolia cardinalis	Cottony cushion scale
	Chilocorus nigritus	Sugarcane scale
	Cryptolaemus montrouzieri	Mealy bugs and scales
Carabidae (ground beetle)	*Parena lacticincta* *Ophionea sp.* *Paederus fuscipes*	Coconut black-headed caterpillar Rice BPH Rice leaf folder
Cicindellidae (Tiger beetle)		Small vertebrates and invertebrates
Hemiptera		
Reduviidae (cone nose bug)	*Rhynocoris fuscipes*	*Helicoverpa armigera*
	Platymeris laevicollis	Coconut rhinoceros beetle
Pentatomidae	*Eucanthecona furcellata*	*Helicoverpa armigera*
Miridae (Mirid bug)	*Cyrtohimus lividipennis*	Rice hoppers
Velidae (Riffle bug)	*Microvelia atrolineata*	Rice leaf and plant hoppers
Anthocoridae (Minute pirate bugs)		Small bodied insects
Giant water bug		Small insects floating on water
Lepidoptera		
Lycaenidae	*Spalgis epius*	Mealybugs
Epipyropidae	*Epiricania melanoleuca*	*Pyrilla perpusilla*
Pyralidae	*Dipha aphidivora*	Sugarcane woolly aphid *Ceratovacuna lanigera*
Noctuidae	*Eublemma amabilis*	Lac insect
Neuroptera		
Chrysopidae (Lacewingbug)	*Chrysoperla carnea*	Soft-bodied insects, lepidopteran larvae
Hemerobiidae (Brown lacewing bug)	*Micromus* sp.	Soft-bodied insects
Myrmeliontidae (Antlion)		Soft-bodied insects
Diptera		
Syrphids		Aphids and other soft-bodied insects
Robber fly		Generalist predator
Odonata		
Dragonflies		Generalist predator
Damselflies		Generalist predator
Mantodea		
Praying mantis		Generalist predator
Hymenopteran		
Ants		Generalist predator
Wasps		Caterpillars

6. Effect of Changing Climate on Generalist and Specialist Species

Climate change in the last few decades has wreaked havoc on the diversity of insects. There is an overall decline in insect abundance, biomass, species richness, and range sizes (Forister et al. 2019). In many organisms, a large shift in distribution and phenology has been observed (Bartomeus et al., 2011). The fitness of the organism may be significantly affected (Boggs, 2016), and because insects are ectothermic, their physiology will be affected. Thus, climate change will ultimately affect ecosystems by altering species interactions, including predator-prey interactions (Laws, 2017). Predators play an important role in the biocontrol of pest species (Bouvet et al., 2019), structuring community composition (Paine, 1969), and nutrient cycling in the ecosystem (Estes et al., 2011). Thus, it has become important to study the effect of changing climates on each group of predators. Among these, long-term research has shown that specialist insect species are being replaced by generalists (Clavel et al., 2011). There may be various disturbances that directly or indirectly influence the decline of specialist species, such as habitat destruction and the inability of insects to adapt to the changing climatic conditions. It is believed that specialisation has evolved from generalist insects in response to stable environments; on the other hand, generalist strategies are favoured by changing environments and a more heterogeneous group of organisms (Futuyma & Moreno, 1988). Habitat loss and fragmentation are affecting specialists more than generalists. This is reflected in the loss of diversity among specialist species, and as a result, biotic homogenisation is taking place (Rooney et al., 2004).

7. Negative and Positive Attributes of a Generalist Predator

A generalist predator may upset the efficiency of the inundated effort to augment the population of the biocontrol agents meant to suppress the population of the pest. *Perillus bioculatus* (F.) (Hemiptera: Pentatomidae), when released in the nymphal stage, provides excellent control over the Colorado potato beetle (*Leptinotarsa decemlineata*). If these predators are released in the egg or early nymphal stage, a satisfactory result is not obtained. On investigation, Hough-Goldstein et al. (1996) found that generalist predators such as spiders, harvestmen, carabids, ladybeetles, and gryllids preyed upon the eggs and early instar nymphs of *P. bioculatus*. Generalist predators may also upset the population of insects beneficial to humans. For example, *Acanthaspis siva* (Distant) (Hemiptera: Reduviidae) predates honeybees (Ambrose, 2000). Similarly, another reduviid bug, *Sycanus collarisis*, is known to be a pest of the Tasar silkworm (*Antheraea mylitta*) (David & Ramamurthy, 2011). The biological control potential of ants has been known since the Chinese used them to control citrus pests. They are generalist natural enemies of arthropods (Drummond & Choate, 2011). However, they are known to increase the honeydew-producing pest population by keeping the natural enemies of these pests away (Wielgoss et al., 2014). Also, predatory ants are sometimes keystone species in ecosystems where they are supposed to regulate the population of herbivores. Without these predatory species, it is supposed that the ecosystem will not function properly and lead to instability. However, some exceptions do exist; in a tropical ecosystem, the fire ant *Solenopsis geminat* reduced the overall species richness, indicating an unstable ecosystem (Risch & Carroll, 1982).

An invasive weed causes a nuisance in the country where it is introduced. This is mainly because of the absence of its natural enemies and the favourable environment for growth. In an attempt to control the weed, usually, a specialist herbivore is imported from the country of origin of the weed. This imported specialist herbivore is free from its natural enemy (usually a specialist), but the native generalist predator may upset the population of exotic herbivores. For example, *Altica carduorum* (Coleoptera: Chrysomelidae) is a promising biocontrol agent imported from Europe for the biocontrol of Canada thistle (*Cirsium arvense*) in Canada (Peschkenet et al., 1970). However, the establishment of this leaf-feeding chrysomelid was greatly reduced through predation by indigenous mites. The mites mainly caused the mortality of eggs and nymphs of chrysomelids (Peschkenet et al., 1971). Similarly, spiders and some hemipterans attacked all stages of *Amynothrips andersonii* meant for the biocontrol of alligator weed in the United States of America (Maddox et al., 1971). Also, the introduction of the generalist parasitoid *Bessa remota* (Diptera: Tachinidae) caused the extinction of *Levuana iridescens* (Lepidoptera: Zygaenidae) in the Fiji Islands (Hoddle, 2006).

Invasive pests pose a serious threat to agriculture; however, it is a very unlikely event. It is estimated that for every 1000 attempts made by the invasive pests, only one successful invasion is made (Williamson & Fitter, 1996). The fact that invasive pests do not arrive with their natural enemy (usually a specialist) is also responsible for their explosive spread (Gupta et al., 2019). However, generalist predators can eliminate the invasive species from the focal point of their invasion (Ehler, 1998); thus, greater diversity is always desired to maintain the balance of nature. The spread of the gypsy moth *Lymantria dispar* L. (Lepidoptera: Erebidae) in the USA was greatly checked by generalist predators such as ants and mice (Cook et al., 1995). Generalist predators can be a helpful tool for breaking the development of resistance in genetically modified crops. Riddicket et al. (1998) found generalist predators on transgenic as well as non-transgenic potatoes.

8. Recent Advances in Biological Control

Recent progress in the field of biotechnology and breeding methods has paved the way for the genetic improvement of natural enemies. Limitations of biological control, such as slow action, restricted host range, and limited persistence in the field, can be overcome using biotechnology (Arve et al., 2014). Attempts at artificial selection have been made in the past to increase the efficiency of natural enemies to control pests (Johnson & Tabashnik, 2018). However, artificial selection is time-consuming, and a limited gene pool is available for the incorporation of genes in an individual. Genetic engineering is an advanced way to incorporate genes of interest in an individual without much time, and genes of interest from outside the gene pool can also be thought to be incorporated. Maternal microinjections (Presnail & Hoy, 1992) containing the lacZ gene have been inserted in the western predatory mite *Metaseiulus occidentalis*, which persisted in more than 150 generations and was found equivalent to wild-type *M. occidentalis* in laboratory conditions, but these mites did not perform under field conditions (Li & Hoy, 1996). Further, molecular scissors for genome editing, such as CRISPER, a very powerful tool for the modification of genomes up to a single nucleotide level, can revolutionise the way we edit genomes. But we should be equally careful with such technologies. CRISPR gene drives can eradicate or suppress a globally invasive social wasp, *Vespula vulgaris* (Lester et al., 2020).

9. Discussion

Predators are very important in the ecosystem. An agricultural land presents a disturbed ecosystem where pest management is disrupted in many ways, such as pesticide usage, monocropping, etc. The choice of a specialist or generalist predator for pest management relies on many factors. The specialists have a narrow range of food habits and are thus dependent on a single or few prey species. During evolution, specialists have evolved to synchronise with the seasonal variability of their host population. They also arrive early when a pest infestation occurs and have higher feeding potential. Generalists, on the other hand, have a wider range of prey availability, which makes them stable in the ecosystem. The generalist predator can influence a single prey population but shows a delayed response to a pest attack. These attributes make specialist predators ideal for classical biological control. Also, there is a lesser chance for specialists to move to non-host prey.

Predators are sometimes released augmentatively for pest control when they cannot colonise the habitat naturally. In that case, they are mass multiplied in the insectaria and released for pest management. Augmentation of natural enemies may cause genetic drift and inbreeding of the population. Generalists are especially useful when opportunistic pests may cause disturbances in the habitat. Generalist predation can be promoted by the conservation of the habitat so that a continuous source of pollen and nectar, along with other non-pest prey populations, is present. A desired biological control will include a specialist predator along with generalist predators to maximise the level of control so that further interventions are not needed.

Climate change in the last few decades has affected insects in a variety of ways, which have influenced their phenology, distribution, and abundance. Anthropogenic disturbances, such as habitat destruction, play a detrimental role in manipulating populations in local habitats. The extinction of prey in an ecosystem may drive its specific predator away, but generalists will survive given the host range it feeds upon. Further, genetic improvement of predators can be done to improve the efficacy of the predators in controlling pests.

References

Agarwala, B. K., & Saha, S. (1984). Studies on *Ceratovacuna silvestrii* (Takahashi) (Homoptera: Aphididae) and its predator *Anisolemnia dilatata* (Fab.) on *Bambusa arundinacea*. *Records of the Zoological Survey of India*, *81*(3-4), 23-42.

Alcock, J. (2009). *Animal behavior: An evolutionary approach*. Sinauer associates.

Ambrose, D. P. (2000). Assassin bugs (Reduviidae excluding Triatominae). *Heteroptera of economic importance*, 695-712.

Arve, S. S., Chavan, S. M., Toke, N. R., & Patil, D. L. (2014). Scope of genetically engineered predator and parasitoid. *International Journal of Plant Protection*, *7*(1), 225-231.

Barbosa, P., & Benrey, B. (1998). The influence of plants on insect parasitoids: implications for conservation biological control. In *Conservation biological control* (pp. 55-82). Academic Press.

Bartomeus, I., Ascher, J. S., Wagner, D., Danforth, B. N., Colla, S., Kornbluth, S., & Winfree, R. (2011). Climate-associated phenological advances in bee pollinators and bee-pollinated plants. *Proceedings of the National Academy of Sciences*, *108*(51), 20645-20649.

Boggs, C. L. (2016). The fingerprints of global climate change on insect populations. *Current Opinion in Insect Science*, *17*, 69-73.

Bouvet, J. P. R., Urbaneja, A., Pérez-Hedo, M., & Monzó, C. (2019). Contribution of predation to the biological control of a key herbivorous pest in citrus agroecosystems. *Journal of Animal Ecology*, *88*(6), 915-926.

Burgio, G., Santi, F., & Maini, S. (2002). On intra-guild predation and cannibalism in *Harmonia axyridis* (Pallas) and *Adalia bipunctata* L. (Coleoptera: Coccinellidae). *Biological Control*, *24*(2), 110-116.

Clavel, J., Julliard, R., & Devictor, V. (2011). Worldwide decline of specialist species: toward a global functional homogenization? *Frontiers in Ecology and the Environment*, *9*(4), 222-228.

Cook, S. P., Smith, H. R., Hain, F. P., & Hastings, F. L. (1995). Predation of gypsy moth (Lepidoptera: Lymantriidae) pupae by invertebrates at low small mammal population densities. *Environmental Entomology*, *24*(5), 1234-1238.

David, B. V., Ramamurthy, V. V., 2011. Elements of Economic Entomology. Revised and Enlarged, sixth ed. Namrutha Publications, Porur, Chennai. p. 386.

Drummond, F., & Choate, B. (2011). Ants as biological control agents in agricultural cropping systems. *Terrestrial Arthropod Reviews*, *4*(2), 157-180.

Ehler, L. E. (1990). Introduction strategies in biological control of insects. *Critical issues in biological control/edited by Manfred Mackauer and Lester E. Ehler, Jens Roland.*

Ehler, L. E. (1998). Invasion Biology and Biological Controll. *Biological Control*, *13*(2), 127-133.

Elliott, N., Kieckhefer, R., & Kauffman, W. (1996). Effects of an invading coccinellid on native coccinellids in an agricultural landscape. *Oecologia*, *105*, 537-544.

Estes, J. A., Brashares, J. S., & Power, M. E. (2013). Predicting and detecting reciprocity between indirect ecological interactions and evolution. *The American Naturalist*, *181*(S1), S76-S99.

Forister, M. L., Pelton, E. M., & Black, S. H. (2019). Declines in insect abundance and diversity: We know enough to act now. *Conservation Science and Practice*, *1*(8), e80.

Futuyma, D. J., & Moreno, G. (1988). The evolution of ecological specialization. *Annual Review of Ecology and Systematics*, *19*(1), 207-233.

Hagen, K. S. (1987). Nutritional ecology of terrestrial insect predator. *Nutritional ecology of insects, mites, spiders, and related invertebrates.*

Hagen, K. S., Bombosch, S., & McMurtry, J. A. (1976). The biology and impact of predators. *Theory and practice of biological control*, 93-142.

Hagley, E. A. C., Holliday, N. J., & Barber, D. R. (1982). Laboratory studies of the food preferences of some orchard carabids (Coleoptera: Carabidae). *The Canadian Entomologist*, *114*(5), 431-437.

Hardin, G. (1960). The Competitive Exclusion Principle: An idea that took a century to be born has implications in ecology, economics, and genetics. *Science*, *131*(3409), 1292-1297.

Hemptinne, J. L., Dixon, A. F. G., & Coffin, J. (1992). Attack strategy of ladybird beetles (Coccinellidae): factors shaping their numerical response. *Oecologia*, *90*, 238-245.

Hoback, W. W., & Higley, L. G. (2000). Insect predation, prey defense, and community structure. In *1999 ABLE Workshop/Conference Proceedings: Tested Studies for Laboratory Teaching* (Vol. 21, pp. 293-304).

Hoddle, M. S. (2006). Historical review of control programs for *Levuana iridescens* (Lepidoptera: Zygaenidae) in Fiji and examination of possible extinction of this moth by *Bessa remota* (Diptera: Tachinidae). *Pacific Science*, *60*(4), 439-453.

Hough-Goldstein, J., Janis, J. A., & Ellers, C. D. (1996). Release Methods for *Perillus bioculatus* (F.), a Predator of the Colorado Potato Beetle. *Biological control*, *6*(1), 114-122.

Johnson, M. W., & Tabashnik, B. E. (2018). Laboratory selection for pesticide resistance in natural enemies. *Applications of genetics to arthropods of biological control significance*, 91-106.

Laws, A. N. (2017). Climate change effects on predator–prey interactions. *Current opinion in insect science*, *23*, 28-34.

Lester, P. J., Bulgarella, M., Baty, J. W., Dearden, P. K., Guhlin, J., & Kean, J. M. (2020). The potential for a CRISPR gene drive to eradicate or suppress globally invasive social wasps. *Scientific Reports*, *10*(1), 12398.

Li, J. and Hoy, M. A. (1996). Adaptability and efficacy of transgenic and wild type *Metaseiulus occidentalis* (Acari: Phytoseiidae) compared as part of a risk assessment *Experimental and Applied Acarology*, 20: 563.

Maddox, D. M., Andres, L. A., Hennessey, R. D., Blackburn, R. D., & Spencer, N. R. (1971). Insects to control alligator weed: an invader of aquatic ecosystems in the United States. *Bioscience*, *21*(19), 985-991.

Majumder, J., & Agarwala, B. K. (2013). Biology and population dynamics of giant ladybird predator An *isolemnia dilatata* (F.) (Coleoptera: Coccinellidae): a specialized predator of woolly aphids of bamboo plants in Northeast India. *World Journal of Zoology*, *8*(1), 55-61.

Paine, R. T. (1969). The Pisaster-Tegula interaction: Prey patches, predator food preference, and intertidal community structure. *Ecology*, *50*(6), 950-961.

Peschken, D. P. (1971). *Cirsium arvense* (L.) Scop., Canada thistle (Compositae). *Commonwealth Institute of Biological Control Technical Communication.*

Peschken, D., Friesen, H. A., Tonks, N. V., & Banham, F. L. (1970). Releases of *Altica carduorum* (Chrysomelidae: Coleoptera) against the weed Canada thistle (*Cirsium arvense*) in Canada. *The Canadian Entomologist*, *102*(3), 264-271.

Preisser, E. L., & Bolnick, D. I. (2008). The many faces of fear: comparing the pathways and impacts of nonconsumptive predator effects on prey populations. *PloS one*, *3*(6), e2465.

Presnail, J. K., & Hoy, M. A. (1996). Maternal microinjection of the endoparasitoid Cardiochiles diaphaniae (Hymenoptera: Braconidae). *Annals of the Entomological Society of America*, *89*(4), 576-580.

Ricklefs, R. E. (2008). *The economy of nature*. Macmillan.

Riddick, E. W., Dively, G., & Barbosa, P. (1998). Effect of a seed-mix deployment of Cry3A-transgenic and non-transgenic potato on the abundance of *Lebia grandis* (Coleoptera: Carabidae) and *Coleomegilla maculata* (Coleoptera: Coccinellidae). *Annals of the Entomological Society of America*, *91*(5), 647-653.

Risch, S. J., & Carroll, C. R. (1982). Effect of a keystone predaceous ant, *Solenopsis geminata*, on arthropods in a tropical agroecosystem. *Ecology*, 1979-1983.

Rojht, H., Budija, F., & Trdan, S. (2009). Effect of temperature on cannibalism rate between green lacewing larvae (*Chrysoperlacarnea* [Stephens], Neuroptera, Chrysopidae). *Acta agriculturae Slovenica*, *93*(1), 5.

Rooney, T. P., Wiegmann, S. M., Rogers, D. A., & Waller, D. M. (2004). Biotic impoverishment and homogenization in unfragmented forest understory communities. *Conservation Biology*, *18*(3), 787-798.

Scheiner, S. M. (1993). Genetics and evolution of phenotypic plasticity. *Annual review of ecology and systematics*, *24*(1), 35-68.

Slansky, F., & Rodriguez, J. G. (1987). *Nutritional ecology of insects, mites, spiders, and related invertebrates*. Wiley.

Sloggett, J. J., Zeilstra, I., & Obrycki, J. J. (2008). Patch residence by aphidophagous ladybird beetles: Do specialists stay longer? *Biological Control*, *47*(2), 199-206.

Snyder, W. E., & Ives, A. R. (2001). Generalist predators disrupt biological control by a specialist parasitoid. *Ecology*, *82*(3), 705-716.

Taniwaki, T., Watanabe, K., Komine, H., Tochigi, K., Yamane, M., & Koike, S. (2022). Response of specialist and generalist predators to nonprogressive annual fluctuations in herbivorous insect populations. *Biological Control*, *165*, 104810.

Tauber, C. A., Tauber, M. J., & Milbratif, L. R. (1995). Individual repeatability and geographical variation in the larval behaviour of the generalist predator, *Chrysopaquadri punctata*. *Animal Behaviour*, *50*(5), 1391-1403.

Tauber, M. J., Tauber, C. A., Ruberson, J. R., Milbrath, L. R., & Albuquerque, G. S. (1993). Evolution of prey specificity via three steps. *Experientia*, *49*, 1113-1117.

Taylor, J. M., & Snyder, W. E. (2021). Are specialists safer than generalists for classical biocontrol? *BioControl*, *66*(1), 9-22.

Wielgoss, A., Tscharntke, T., Rumede, A., Fiala, B., Seidel, H., Shahabuddin, S., & Clough, Y. (2014). Interaction complexity matters: disentangling services and disservices of ant communities driving yield in tropical agroecosystems. *Proceedings of the Royal Society B: Biological Sciences*, *281*(1775), 20132144.

Williamson, M., & Fitter, A. (1996). The varying success of invaders. *Ecology*, *77*(6), 1661-1666.

Zhang, Z. Q., & Sanderson, J. P. (1993). Spatial scale of aggregation in three acarine predator species with different degrees of polyphagy. *Oecologia*, *96*, 24-31.

Chapter 12

Predatory Insects Management for Sustainable Agriculture

Neha Bisht[1]
and Amar Prakash Garg[2,*]
[1]School of Biological Engineering & Life Sciences, Shobhit Institute of Engineering & Technology, Modipuram, Meerut, India
[2]Dean Academics and Director Research, Swami Vivekanand Subharti University, Subhartipuram, Meerut, India

Abstract

In the production of many crops, insect infestations are a significant constraint. Pesticide reliance incurs severe economic and environmental consequences due to off-target spread, chemical residues, and resistance. Since long ago, as an alternative, Integrated Pest Management (IPM) has been offered. IPM approaches use ecological knowledge about pests, beneficial species, their life cycles, cultural practices, nonchemical methods, pesticides, and their engagement with the crop, as well as the provision of a variety of strategies that the producer might use to achieve economic and environmental sustainability. The emerging era of insect-resistant transgenic cotton provides a platform for more environmentally friendly, economically feasible IPM by incorporating a variety of non-chemical approaches and relying less on pesticides. This review looks at new and advanced approaches in synthetic biology and biotechnology used for controlling pests. We cover emerging technology for genetically engineering tolerant crops and populations of insects, as well as improvements in bio-manufacturing that enable the production of new pest control solutions.

Keywords: agricultural crops, IPM, pesticides, economical, technology, pest control

1. Introduction

The term "predator" refers to a species that relies on prey (i.e., other animals) for growth, nourishment, and reproduction. They are typically more active, greater in size than their prey, and require multiple prey to finish their life cycle. These scavengers hunt and kill their prey using a variety of strategies, and some of their body parts have been altered to facilitate

---

* Corresponding Authotr's Email: amarprakashgarg@yahoo.com.

In: Insects as Humans' Frenemies: From Friends to Foes
Editor: Istkhar Rao
ISBN: 979-8-89530-320-7

predatory movement (ladybird beetle). Generally, the mature and immature phases are scavengers, or occasionally the immature stage is (e.g., syrphid fly, green lacewing) (Van Driesche et al., 2008). According to their habits, these predators might either be migratory (ladybirds) or permanent (predatory mites). A vast range of ecological services provided by beneficial insects are crucial for human survival. The natural management of crop pests (through predatory or parasitic methods, "natural enemies") and the process of pollination are two of the most economically significant functions in agricultural systems. Most globally significant crop types are affected by pollination services (Lautenbach et al., 2012; Klein et al., 2007; Redhead et al., 2020; Seni & Halder, 2022), which have been estimated to be worth between US $235 and 577 billion (Schowalter et al., 2018), and crop pests, whose natural management is predicted to be approximately over US $400 billion, play an important role in many high-value crops (Colloff et al., 2013; Classen et al., 2014; Redhead et al., 2020; Seni & Halder, 2022). It is particularly beneficial to use these predatory or insectivorous arthropods as biological control agents when managing insect pests (Redhead et al., 2020). The members of the family Coccinellidae, Syrphidae, Chrysopidae, Staphylinidae, Anthocoridae, Staphylinidae, Carabidae, Cecidomyiidae, Formicidae, and Reduviidae are significant from a horticultural standpoint among the more than 30 families of predatory insects (Seni & Halder, 2022). When the Vadalia beetle, *Rodolia cardinalis* (Mulsant), was imported to California from Australia in 1888 to control the cottony cushion scale, *Icerya purchasi* Maskell of citrus, predators also ushered in the current paradigm of biological management (Van Driesche et al., 2008; Colloff et al.,2013). Insecticide, herbicide, and other pesticide resistance have resulted in higher rates of application, more crop losses, and a rising price for peasants who are running pesticide mills (Classen et al., 2014; Seni & Halder, 2022; Heap et al., 2014). Rising health concerns for vulnerable populations such as growers, agricultural labourers, and rural residents, as well as for customers, are increasingly related to increased pesticide use (Schowalter et al., 2018; Classen et al., 2014).

The word "sustainable" has a very broad meaning and has been defined and ill-defined in various forms by various scientists and countries that suit their business interests. To our mind, sustainable means "a process, technology, or stage that does not adversely affect the life, environment, earth, or globe and ensures their long-term co-existence of life forms without compromising their quality of life in future generations." The major issues of environmental degradation need to be addressed by reducing the use of fossil fuels to decrease carbon emissions, managing non-biodegradable e-waste, polythene, plastic, etc. by encouraging its re-use and recycling, rainwater harvesting to increase groundwater levels, and minimising the use of systemic antimicrobials to preserve and protect the microbial diversity and its populations to save soil fertility.

Pesticides harm the health of the soil, the quality of the water, and the habitat of wildlife. Certain commodity crop subsidies promote monocultures and the wasteful use of inputs in their production (Myers et al., 2016). These negative effects include non-market costs that can only be estimated, but they place a heavy weight on the entire world (Colloff et al., 2013; Seni & Halder, 2022; Kim et al., 2017). IPM and sustainable agriculture farming are two strategies that lessen the need for pesticides. IPM is a process of decision-making and setting scientific strategies utilising organic and conventional agricultural systems. It is a major component and offers an ecologically sound possibility for biological control (Muller et al., 2017; Baker et al., 2020). A type of "ecologically orientated pest management" known as "biological control" makes use of "natural enemies" to keep pesticide species under control (Myers et al., 2016;

Baker et al., 2020). Natural enemies include entomopathogenic nematodes, parasitoids, predators, illnesses, microorganisms that compete with or are hyperparasites of crop infection, herbivores feeding on weeds and weed seeds, resource competitors, and organisms producing toxins, which are referred to as antibiosis or allelopathy (Muller et al., 2017). Natural occurring biological control, traditionally imported and established foreign agents, native or overseas agents that have been freed, boosting the populations, and preserving or boosting the populations of agents that are either native or foreign are all examples of biological control (Seni & Halder, 2022; Muller et al., 2017; Baker et al., 2020). Inoculated augmentation releases create a group of people who are anticipated to take action across eras, whereas inundated augmentation releases release organisms that are anticipated to provide instantaneous decreases (Baker et al., 2020).

Early in the 1970s, the phrases "Integrated Pest Management and IPM" initially appeared in everyday speech (Stern et al., 1959). IPM is derived from the ideas of infestation management, which uses a range of tactics, including surveillance and response levels, to keep pest populations under control, and holistic management, which was developed as a remedy for pests in the early 1950s outbreaks caused by pesticide impacts on beneficial insects and pest communities that are pesticide resilient. (Schowalter et al., 2018; Baker et al., 2020; Stern et al., 1959). The United Nations' Food and Agriculture Organisation (FAO) describes IPM as "the careful consideration of all available pest control techniques and the subsequent integration of appropriate measures that discourage the development of pest populations and keep pesticides and other interventions at levels that are economically justifiable and reduce or minimise risks to human health and the environment" (FAO, 2020). IPM organic methods for controlling pests and growth in crops with a minimum of disruption to agricultural systems (Seni & Halder, 2022; Muller et al., 2017). Although IPM is primarily concerned with controlling pests, its strategies also aim to enhance soil health and manage nutrients for the best possible plant health. IPM techniques include those that must be avoided, monitored, and/or repressed in communities to maintain the integrity of the landscape and protect human health. These include "PAMS," or plant diseases, weeds, pests for managing insects, animal feed, nematodes, pests, and fibrous crops (Baker et al., 2020). Suppression techniques include cultural ones like mulching or cover crops, as well as physical ones like tillage, chemical insecticides, and biological ones (Seni & Halder, 2022; Baker et al., 2020; Bottrell & Bottrell, 1979). IPM, originally called Integrated Control, has helped to advance crop protection all over the world thanks to its guiding principles. The following are the fundamental tenets and concepts of IPM, as they were concisely stated in the 1960s (Baker et al., 2020; Bottrell & Bottrell, 1979):

- Extremely dangerous species will remain present in manageable numbers.
- The ecosystem serves as the competent authority.
- The usage of organic pesticides is increasing.
- Each management method could result in unintended consequences.
- It is crucial to use an interdisciplinary approach.

Moreover, the following are the main rules:

1) Assess the pest situation and set thresholds.

2) Come up with plans to move the equilibrium point lower.
3) In times of emergency, look for corrective actions that disturb the environment as little as possible.
4) Develop monitoring strategies.

The "IPM pyramid" notion serves as a reminder that IPM isn't just about "integrating pest control technology," although there ought to be prioritising or actions of hierarchy (Deguine et al., 2021). Without this prioritising, it is impossible to manage pests sustainably, leaving everything to chance and resulting in severe environmental externalities. Notwithstanding the concept's virtues and beyond its admirable ideas, there are reasons to doubt how well it has worked in practice, especially in light of recent and upcoming agricultural issues (Redhead et al., 2020). Certain crops with pest and disease resistance or tolerance types can benefit from host plant resistance, but not other crops. For some Lepidopteran and Coleopteran pests, pheromones are frequently utilised for mating disruption, mass capture, or monitoring, but not for some Hemipteran pests (Seni & Halder, 2022). For pests in greenhouses, biological management is frequently employed, but less frequently in the field. High-value crops like strawberries use mechanical instruments like bug vacuums, but because they consume fossil fuels, they don't offer a cost-effective option for non-crops of distinction and do not reduce greenhouse gas emissions (Baker et al., 2020). Although, in theory, chemical pesticides ought to be utilised as an emergency measure, they are nonetheless sometimes considered the line of defence first in reducing the regional circular of some indigenous or noxious pests and illnesses, or in seed protection and transplanting for recurrent and repeated insect difficulties. (Redhead et al., 2020; Baker et al., 2020). In recent years, artificial pesticide application to seeds, for example, has gained popularity as a preventative approach in several crops (Bottrell & Bottrell, 1979).

IPM is a method for an agricultural manufacturing technique that is impacted by a variety of aspects as well because crop production is a combination of art, science, and business (Deguine et al., 2021). Each grower uses a different approach to produce crops to reduce losses and, in some way, make a profit that is favoured by retailers, safe for customers, and less harmful to the natural world (Baker et al., 2020; AO, 2020). IPM is a strategy for managing pests in a novel manner that is economically acceptable in social circles and environmentally safe (Barzman et al., 2015). With the following direct or indirect goals in mind: current crop-growing and protecting breakthroughs, technology innovation in communication, and agricultural and commercial changes worldwide, a new IPM paradigm with business, management, and sustainability dimensions is offered. With the following direct or indirect goals in mind: current crop-growing and protecting breakthroughs, technology innovation in communication, and agricultural and commercial changes worldwide, a new IPM paradigm with business, management, and sustainability dimensions is offered (Deguine et al., 2021; Barzman et al., 2015).

1) As needed, modernise the IPM idea and urge reconsideration and regard it as sustainable.
2) Increase the confidence of consumers and knowledge through IPM-based agriculture, which is suitable for every agricultural product and condition and ensures the global shortage of food, as well as food waste-related social disparities.

3) Guarantee revenue of the production while letting informed clients, not special speciality societies or merchants, decide which foods to buy.
4) Reduce the possible harm that conventional methods that are not based on IPM or those that are viewed as environmentally friendly alternatives may do and the difficulties that come with controlling some pests.

2. Aspects of Management

The new IPM model has four key elements that cover various pest control alternatives: the grower's expertise and resources to solve the pest issue; planning and organising information to conduct suitable control measures; and sustaining efficient interactions to acquire knowledge about pests and manage them.

2.1. Pest Management

With the understanding that, save for recently imported invasive pests, an equitable method of insect population control that does not result in financial losses is preferable to destroying or eradicating them, the term "pest control" has evolved into "pest management" over time (Baker et al., 2020). Despite its frequent use in literature and conversation, the word "control" typically refers to managing (Klein et al., 2007). It is crucial to have a firm grasp of broad IPM ideas and the many pest management solutions available because some are preventive while others are therapeutic (Kim et al., 2017). The producer or the pest control specialist must select the option(s) that are most suitable for their circumstances because some of the advised techniques might not be feasible in all circumstances (Redhead et al., 2020). To achieve the necessary control, the film additionally serves as vital for comprehending the intrinsic as well as prospective interconnections among different options from management (Klein et al., 2007; Redhead et al., 2020). The following typical management methods are applied to prevent, lessen, and treat the presence of pests at different levels of plant growth (Deguine et al., 2021; Barzman et al., 2015). Although each of them offers a little degree of supervision, their combined impact may be quietly effective in lowering the loss of yield.

2.2. Resistance Between the Host and Plant

The methods that take advantage of cultivars that have been produced through conventional breeding or genetic engineering are both pest-resistant and pest-tolerant (Douglas, 2018). These cultivars possess physiological, morphological, or biological traits that render plants more resistant to pests or less conducive to their growth, development, or effective reproduction (Nelson et al., 2018; Nielsen et al., 2016). These cultivars are resistant to or tolerant of insect damage, which lowers yield losses (Douglas, 2018; Nelson et al., 2018; Nielsen et al., 2016). The IPM's first line of defence is this choice.

2.3. Biological Surveillance

Natural adversaries such as wasps that are parasites and microbes are quite sufficient in some situations, leading to considerable decreases in insect populations (Hajek et al., 2018). Frequent spraying is one of the most common ways to control endemic pests. Distribution of defenders of nature that are accessible through commerce, protection of communities of enemies in nature by providing shelter, and avoidance of actions that harm them are all examples of good stewardship. In greenhouses, biological control has been successfully used on speciality crops like strawberries that are cultivated outdoors (Van Lenteren, 1988; Zalom et al., 2018). The classical biological control strategy is commonly employed to handle invasive pest challenges. In this approach, naturally occurring enemies from the original area of the invading pest have been transported, replicated, and dispersed in the insect's new habitat (Kenis et al., 2017; Heimpel et al., 2018). Another biological management method that has proven to be effective in combating a wide range of pests is the discharge of infertile, zapped insects (Klassen & Curtis, 2005).

2.4. Control by Physical or Mechanical Means

According to several studies (Webb & Linda, 1992; Gamliel & Katan, 2012; Gogo et al., 2014; Dara, 2019), this strategy comprises the use of different biological or mechanical methods for pest marginalisation, including capturing (which is frequently equated with behavioural control), elimination, or eradication. Examples of pest control can be done physically or mechanically. Restriction by nets or row coverings, pest removal by picking or vacuuming it out, mechanised controlling weeds and rodent pests equipment and traps, and altering environment-related variables such as heat and humidity in greenhouses, sunlight, or steam disinfection, as well as visible or physical bird deterrents such as reflecting material or loud devices.

2.5. Microbial Management

Microbial control mainly refers to the use of entomopathogenic bacteria, plant parasitic nematodes, fungi, nematodes, microsporidia, and viruses against arthropod pests, and plant pathogens (Mankau, 1981; Paulitz & Bélanger, 2001; Dong & Zhang, 2006; Lacey et al., 2017).

2.6. Cultural Influence

The application of effective agronomist approaches to avoid or minimise pest infestations and damage is known as cultural control. To limit the possibility of pest introduction at the commencement of crop production, clean seedlings or plant matter must be used. By changing the planting dates, one can prevent the occurrence of pests by removing their vulnerable periods. In Uganda, early cowpea planting reduced thrips, aphid, and pod bug infestations (Karungi et al., 2000), *Clavigralla tomentosicollis*, the pod-sucking insect, the thrips of legume flowers (*Megalurothrips sjostedti*), and the legume pod borer (*Maruca vitrata*) in Nigeria (Asante et al., 2001). Modifications in the fertiliser time frame, irrigated practices, and other agricultural processes can reduce the pest's environmental popularity. In California, especially for organic strawberries, spider mites are controlled by micro-sprinklers planted on strawberry beds. Compared to strawberry plots with overhead aluminium sprinklers, micro-sprinkler-equipped plots of diseases with Mildew with powdery particles induced by *Podosphaera aphanis* and *Botrytis cinerea* decay of fruits seemed to be less severe. (Dara, 2016). Low potassium levels in plants trigger the production of jasmonic acid, which increases their resistance to some diseases and insect pests (Dara, 2018(a)). Arthropod infestations may get worse as a result of increased plant nitrogen. *Pistachio psyllids* (Agonoscenapistaciae) on Iranian orchard floors were less likely to overwinter as a result of winter ploughs (Kunjwal & Srivastava, 2018). Ploughing is a crucial control method that can be used to remove crop residue and reveal pests' soil-dwelling stages (Leach et al., 2017). Spotted-wing drosophila (*Drosophila suzukii*) crowd were minimised by sanitation measures including grabbing unsold berries and even switching the harvesting cycle from every three days to every 1-2 days (Leach et al., 2017). In various cropping systems, the method of crop rotation has been utilised to manage weeds, insects, and diseases (Curl, 1963; Wright, 1984; Liebman & Dyck, 1993; Mohler & Johnson, 2009). Trap crops are used to keep pests from harming the primary crop; intercropping non-host plants or pest-repelling plants are a few more cultural management techniques used in IPM (Pretty & Bharucha, 2015; Nielsen et al., 2016).

2.7. Chemical Regulation

Chemical pesticides are frequently used while discussing chemical control (Pimental, 2009). Even though they are classified as biological (Lasota & Dybas, 1991; Sarfraz et al., 2005; Dodia et al., 2010), microbe-derived toxic metabolites like spinosad and avermectin and botanical extracts like azadirachtin and pyrethrins are still chemical molecules and carry many of the same risks for human and environmental safety as chemical pesticides. According to their modes of action, chemical pesticides are divided into various groups (IRAC, 2018), and switching between chemicals from various groups is advised to lower the likelihood of resistance emergence (Sparks & Nauen, 2015). According to several studies (Larkin, 2008; Vleesschauwer & Höfte, 2009; Sharma et al., 2014; López-Bucio et al., 2015; Dara, 2018) certain biostimulants in holistic defence against pests, diseases, and environmental stresses are induced by minerals, microorganisms, extracts of plants, seaweed, or algae. They are, however, used as modifications with no claims of pest or disease control. These novel goods or innovations may fit into one or more of the pest management subcategories indicated above. To minimise dangers, each pest management option must be carefully chosen and implemented. For instance, several pests became resistant to transgenic crops containing toxic proteins from *Bacillus thuringiensis* (Tabashnik et al., 2013), as well as sowing non-transgenic crops alongside them. It has been discovered that resistant plants exist as one strategy to lessen the development of resistance (Tang et al., 2001; Huang et al., 2011). Biological and botanical pesticides might cause pests to become resistant if they are used too frequently (Dara, 2017). The effectiveness of this technique may be impacted by factors like mated female migration and alternate mate recruitment techniques of some species, even if mating disruption is an effective method for eliminating many pests. IPM implementation is especially tough for economically disadvantaged growers in developed or developing countries. According to Rezaei et al. (2019), ethical and intellectual concerns are examples of socio-psychological factors that were shown as the forces behind IPM implementation in Iran. In Sri Lanka, a survey of vegetable growers revealed that just 20% were familiar with IPM and that nearly 50% applied chemical pesticides according to a calendar before insect or disease onset (Jayasooriya & Aheeyar, 2016). IPM application in industrialised nations is hindered by a dearth of qualified IPM specialists and extension educators, according to a survey done by Parsa et al. (2014).

2.8. Pest Monitoring

Understanding the biology, seasonal population trends, life stage disturbances and habitats, damage to the environment and its economic impact, each life stage's susceptibility to one or more control options, the preferred alternative hosts and host, cropping trends, farming practices, environmental factors, and other factors that predict pest occurrence are critical for accurately identifying the pest (Table 1). Scanning the agricultural land regularly for the emergence and dissemination of pests is an essential part of protecting the crops. For instance, earlier detection handles the issue of pests through a low-cost location treatment for the elimination of pests. Continuous surveillance is required to evaluate the harm and select the most appropriate course of action. When outbreaks of pests continue to develop, it is the optimum time to begin farm-wide management. Monitoring is necessary to avoid applying pesticides based on the timetable, especially in areas with a smaller number of pests that don't need treatment. An affordable substitute for wheat fumigation in lifts on a calendar basis can be a fumigation-based sample combined with better cleanliness and other procedures (Adam et al., 2010). In several interactions with farmers and outreach partners, it seemed that the depletion of resources prevented completing research or reaching an option that was exclusively based on research. However, employing drones can improve surveillance reliability and accuracy of aerial images through the identification of areas that are totally or partially exposed to any stressors, including biotic or abiotic stressors (Vanegas et al., 2018; Yue et al., 2018). Some of these tools are currently offered for purchase by businesses.

Table 1. Important insect pests for various crops

Crop	Pest	Natural Enemy
Fruits, Vegetables, Oilseeds Pulses	Aphids; *Myzus persicae, Aphis craccivora, Lipaphis erysimi and Leafhoppers; Empoasca kerri, E. facialis, E. fabae* etc.	The ladybird beetles *Menochilus sexmaculatus, Brumoides suturalis, Harmonia dimidiate Coccinella septempunctata* and green lacewings, *Chrysoperla zastro wisillemi*
Brinjal, Sweet Peppers, Tomato, Cucumber	Spider mite; *Tetranychus spp. (Acari: Tetranychidae)*	*Phytoseilus persimilis* and several species of *Amblyseius spp.*
Okra, Brinjal, Papaya, Tomato, Lemon, Guava, Sapota	*Paracoccus marginatus* and *Phenacoccus solenopsis, green shield scale: Pulvinaria psidii*	*Cryptolaemus montrouzieri, Scymnus* spp.
Pepper, Tomato, Potato	*Tuta absoluta (Lepidoptera: Gelechiidae), Tomato pinworm*, etc.	Predatory mirid bugs, *Nesidiocoris tenuis*
Sweet pepper, Brinjal, Cucumber	*Thrips palmi, Frankliniella occidentalis* etc.	Predatory anthocorid bugs, *Orius sauteri*

Source: Halder et al., 2018; Seni & Halder, 2022.

Conclusion

Systemic chemical pesticides and insecticides are highly toxic for humans as their residues remain in the crops that enter the food chain and cause serious diseases, including cancer. These chemicals also disturb the microbiological communities of soils, resulting in soil infertility and the accumulation of undecomposed biomass. Presently, scientists prefer an integrated pest management (IPM) system for the control of insect pests, as IPM provides numerous advantages over conventional chemical control measures. IPM facilitates the management of the diseases to such a level that it does not cause significant losses and also does not create conditions that compel the insect pest to mutate. Biological control, on the other hand, is neither long-lasting nor ensures the control of diseases. Using modern drone-assisted technology, IPM can manage diseases over a large field in an effective manner. IPM also reduces the chances of host-pathogen resistance.

Acknowledgements

The authors are thankful to Professor Jamal Ahemad Siddiqui, Librarian, C.C.S. University, Meerut for the plagiarism check.

Conflict of Interests

The authors declare that they have no conflict of interests.

References

Adam, B. D., Siaplay, M., Flinn, P. W., Brorsen, B. W., & Phillips, T. W. (2010). Factors influencing economic profitability of sampling-based integrated pest management of wheat in country elevators. *Journal of Stored Products Research*, *46*(3), 186-196.

Asante, S. K., Tamo, M., & Jackai, L. E. N. (2001). Integrated management of cowpea insect pests using elite cultivars, date of planting and minimum insecticide application. *African Crop Science Journal*, *9*(4), 655-665.

Baker, B. P., Green, T. A., & Loker, A. J. (2020). Biological control and integrated pest management in organic and conventional systems. *Biological Control*, *140*, 104095.

Barzman, M., Bàrberi, P., Birch, A. N. E., Boonekamp, P., Dachbrodt-Saaydeh, S., Graf, B., & Sattin, M. (2015). Eight principles of integrated pest management. *Agronomy for sustainable development*, *35*, 1199-1215.

Bottrell, D. G. (1979). *Integrated pest management*. Council on Environmental Quality.

Classen, A., Peters, M. K., Ferger, S. W., Helbig-Bonitz, M., Schmack, J. M., Maassen, G., & Steffan-Dewenter, I. (2014). Complementary ecosystem services provided by pest predators and pollinators increase quantity and quality of coffee yields. *Proceedings of the Royal Society B: Biological Sciences*, *281*(1779), 20133148.

Colloff, M. J., Lindsay, E. A., & Cook, D. C. (2013). Natural pest control in citrus as an ecosystem service: Integrating ecology, economics and management at the farm scale. *Biological Control*, *67*(2), 170-177.

Curl, E. A. (1963). Control of plant diseases by crop rotation. *The Botanical Review*, *29*, 413-479.

Dara, S. K. (2016). Android version of the IPMinfo app with new features just released. *University of California Agriculture and Natural Resources eJournal Strawberries and Vegetables.*

Dara, S. K. (2017). Insect resistance to biopesticides. *UCANR E-Journal of Entomology and Biologicals.*

Dara, S. K. (2019). The new integrated pest management paradigm for the modern age. *Journal of Integrated Pest Management, 10*(1), 12.

Dara, S. K., & Peck, D. (2018). Evaluation of additive, soil amendment, and biostimulant products in Santa Maria strawberry. *CAPCA Adviser, 21*(5), 44-50.

Dara, S. K., & Peck, D. (2018). Evaluation of additive, soil amendment, and biostimulant products in Santa Maria strawberry. *CAPCA Adviser, 21*(5), 44-50.

De Vleesschauwer, D., & Höfte, M. (2009). Rhizobacteria-induced systemic resistance. *Advances in botanical research, 51*, 223-281.

Deguine, J. P., Aubertot, J. N., Flor, R. J., Lescourret, F., Wyckhuys, K. A., & Ratnadass, A. (2021). Integrated pest management: good intentions, hard realities. A review. *Agronomy for Sustainable Development, 41*(3), 38.

Dodia, D. A., Patel, I. S., & Patel, G. M. (2010). *Botanical pesticides for pest management.* Scientific Publishers.

Dong, L. Q., & Zhang, K. Q. (2006). Microbial control of plant-parasitic nematodes: a five-party interaction. *Plant and soil, 288*, 31-45.

Douglas, A. E. (2018). Strategies for enhanced crop resistance to insect pests. *Annual review of plant biology, 69*, 637-660.

Elbert, A. (1995). IRAC-Insecticide Resistance Action Committee. *Nachrichtenblatt des DeutschenPflanzenschutzdienstes, 47*(6), 158-158.

FAO (2020) NSP – Integrated Pest Management, FAO definition. http://www.fao.org/agriculture/crops/thematic-sitemap/theme/pests/ipm/en/.

Gamliel, A., & Katan, J. (2012). Solarization: theory and practice. *American Phytopathological Society, St. Paul, MN.*

Gogo, E. O., Saidi, M., Ochieng, J. M., Martin, T., Baird, V., & Ngouajio, M. (2014). Microclimate modification and insect pest exclusion using agronet improve pod yield and quality of French bean. *HortScience, 49*(10), 1298-1304.

Hajek, A. E., & Eilenberg, J. (2018). *Natural enemies: an introduction to biological control.* Cambridge University Press.

Halder, J., Pandey, K. K., & Behera, T. K. (2022). Evaluation and economic analysis of ecofriendly biological approaches for the management of shoot and fruit borer (Eariasvittella F.) of okra. *EntomologiaHellenica, 31*(2), 61-70.

Halder, J., Rai, A. B., Dey, D., & Singh, B. (2018). Abundance of important parasitoids in the vegetable ecosystem and their prospects in integrated pest management. *Journal of Entomology and Zoology Studies, 6*(4), 762-769.

Heap, I. (2014). Global perspective of herbicide-resistant weeds. *Pest management science, 70*(9), 1306-1315.

Heimpel, G. E., & Cock, M. J. (2018). Shifting paradigms in the history of classical biological control. *BioControl, 63*, 27-37.

Huang, F., Andow, D. A., & Buschman, L. L. (2011). Success of the high-dose/refuge resistance management strategy after 15 years of Bt crop use in North America. *EntomologiaExperimentalisetApplicata, 140*(1), 1-16.

Jayasooriya, H. J. C., & Aheeyar, M. M. (2016). Adoption and factors affecting on adoption of integrated pest management among vegetable farmers in Sri Lanka. *Procedia food science, 6*, 208-212.

Karungi, J., Adipala, E., Ogenga-Latigo, M. W., Kyamanywa, S., & Oyobo, N. (2000). Pest management in cowpea. Part 1. Influence of planting time and plant density on cowpea field pests infestation in eastern Uganda. *Crop protection, 19*(4), 231-236.

Kenis, M., Hurley, B. P., Hajek, A. E., & Cock, M. J. (2017). Classical biological control of insect pests of trees: facts and figures. *Biological Invasions, 19*, 3401-3417.

Kim, K. H., Kabir, E., & Jahan, S. A. (2017). Exposure to pesticides and the associated human health effects. *Science of the total environment, 575*, 525-535.

Klassen, W., Curtis, C. F., & Hendrichs, J. (2021). History of the sterile insect technique. In *Sterile insect technique* (pp. 1-44). CRC Press.

Klein, A. M., Vaissière, B. E., Cane, J. H., Steffan-Dewenter, I., Cunningham, S. A., Kremen, C., & Tscharntke, T. (2007). Importance of pollinators in changing landscapes for world crops. *Proceedings of the royal society B: biological sciences*, *274*(1608), 303-313.

Kunjwal, N., & Srivastava, R. M. (2018). Insect pests of vegetables. *Pests and their Management*, 163-221.

Lacey, L. (2017). Microbial control of insect and mite pests. *FNP: Academic Press, 446p.*

Larkin, R. P. (2008). Relative effects of biological amendments and crop rotations on soil microbial communities and soilborne diseases of potato. *Soil Biology and Biochemistry*, *40*(6), 1341-1351.

Lasota, J. A., & Dybas, R. A. (1991). Avermectins, a novel class of compounds: implications for use in arthropod pest control. *Annual review of entomology*, *36*(1), 91-117.

Lautenbach, S., Seppelt, R., Liebscher, J., & Dormann, C. F. (2012). Spatial and temporal trends of global pollination benefit. *PloS one*, *7*(4), e35954.

Leach, H., Moses, J., Hanson, E., Fanning, P., & Isaacs, R. (2018). Rapid harvest schedules and fruit removal as non-chemical approaches for managing spotted wing Drosophila. *Journal of Pest Science*, *91*, 219-226.

Liebman, M., & Dyck, E. (1993). Crop rotation and intercropping strategies for weed management. *Ecological Applications*, *3*(1), 92-122.

López-Bucio, J., Pelagio-Flores, R., & Herrera-Estrella, A. (2015). Trichoderma as biostimulant: exploiting the multilevel properties of a plant beneficial fungus. *Scientiahorticulturae*, *196*, 109-123.

Mankau, R. (1981). Microbial control of nematodes. *Plant parasitic nematodes*, *3*, 475-494.

Mehrnejad, M. R. (2018). Investigation into the overwintering and winter-management of the common pistachio psyllid, Agonoscenapistaciae (Hemiptera: Aphalaridae), a major pest in pistachio plantations. *Zoology and Ecology*, *28*(4), 384-388.

Mohler, C. L. (2009). *Crop rotation on organic farms: a planning manual*. Natural Resource, Agriculture, and Engineering Service (NRAES).

Muller, A., Schader, C., El-HageScialabba, N., Brüggemann, J., Isensee, A., Erb, K. H., & Niggli, U. (2017). Strategies for feeding the world more sustainably with organic agriculture. *Nature Communications*, *8*(1), 1-13.

Myers, J. P., Antoniou, M. N., Blumberg, B., Carroll, L., Colborn, T., Everett, L. G., & Benbrook, C. M. (2016). Concerns over use of glyphosate-based herbicides and risks associated with exposures: a consensus statement. *Environmental Health*, *15*(1), 19.

Nelson, R., Wiesner-Hanks, T., Wisser, R., & Balint-Kurti, P. (2018). Navigating complexity to breed disease-resistant crops. *Nature Reviews Genetics*, *19*(1), 21-33.

Nielsen, A. L., Dively, G., Pote, J. M., Zinati, G., & Mathews, C. (2016). Identifying a potential trap crop for a novel insect pest, Halyomorphahalys (Hemiptera: Pentatomidae), in organic farms. *Environmental Entomology*, *45*(2), 472-478.

Parsa, S., Morse, S., Bonifacio, A., Chancellor, T. C., Condori, B., Crespo-Pérez, V., & Dangles, O. (2014). Obstacles to integrated pest management adoption in developing countries. *Proceedings of the National Academy of Sciences*, *111*(10), 3889-3894.

Paulitz, T. C., & Bélanger, R. R. (2001). Biological control in greenhouse systems. *Annual review of phytopathology*, *39*(1), 103-133.

Pimentel, D. (2009). Pesticides and pest control. *Integrated pest management: innovation-development process: volume 1*, 83-87.

Pretty, J., & Pervez Bharucha, Z. (2015). Integrated pest management for sustainable intensification of agriculture in Asia and Africa. *Insects*, *6*(1), 152-182.

Pretty, J., & Pervez Bharucha, Z. (2015). Integrated pest management for sustainable intensification of agriculture in Asia and Africa. *Insects*, *6*(1), 152-182.

Redhead, J. W., Powney, G. D., Woodcock, B. A., & Pywell, R. F. (2020). Effects of future agricultural change scenarios on beneficial insects. *Journal of environmental management*, *265*, 110550.

Rezaei, R., Safa, L., Damalas, C. A., & Ganjkhanloo, M. M. (2019). Drivers of farmers' intention to use integrated pest management: integrating theory of planned behavior and norm activation model. *Journal of Environmental Management*, *236*, 328-339.

Sarfraz, M., Dosdall, L. M., & Keddie, B. A. (2005). Spinosad: a promising tool for integrated pest management. *Outlooks on Pest Management*, *16*(2), 78-84.

Schowalter, T. D., Noriega, J. A., & Tscharntke, T. (2018). Insect effects on ecosystem services—Introduction. *Basic and Applied Ecology*, *26*, 1-7.

Seni, A., & Halder, J. (2022). Role of Predators In Insect Pests Management For Sustainable Agriculture. *Agriculture Letters.*2(02).

Sharma, H. S., Fleming, C., Selby, C., Rao, J. R., & Martin, T. (2014). Plant biostimulants: a review on the processing of macroalgae and use of extracts for crop management to reduce abiotic and biotic stresses. *Journal of applied phycology*, *26*, 465-490.

Sparks, T. C., & Nauen, R. (2015). IRAC: Mode of action classification and insecticide resistance management. *Pesticide biochemistry and physiology*, *121*, 122-128.

Stern, V. M., Smith, R. F., van den Bosch, R., & Hagen, K. S. (1959). The integrated control concept. Hilgardia 29: 81–101.

Tabashnik, B. E., Brévault, T., & Carrière, Y. (2013). Insect resistance to Bt crops: lessons from the first billion acres. *Nature Biotechnology*, *31*(6), 510-521.

Tang, J. D., Collins, H. L., Metz, T. D., Earle, E. D., Zhao, J. Z., Roush, R. T., & Shelton, A. M. (2001). Greenhouse tests on resistance management of Bt transgenic plants using refuge strategies. *Journal of Economic Entomology*, *94*(1), 240-247.

Van Driesche, R., & Hoddle, M. (2009). *Control of pests and weeds by natural enemies: an introduction to biological control.* John Wiley & Sons.

Van Lenteren, J. E., & Woets, J. V. (1988). Biological and integrated pest control in greenhouses. *Annual Review of Entomology*, *33*(1), 239-269.

Vanegas, F., Bratanov, D., Powell, K., Weiss, J., & Gonzalez, F. (2018). A novel methodology for improving plant pest surveillance in vineyards and crops using UAV-based hyperspectral and spatial data. *Sensors*, *18*(1), 260.

Webb, S. E., & Linda, S. B. (1992). Evaluation of spunbonded polyethylene row covers as a method of excluding insects and viruses affecting fall-grown squash in Florida. *Journal of Economic Entomology*, *85*(6), 2344-2352.

Wright, R. J. (1984). Evaluation of crop rotation for control of Colorado potato beetles (Coleoptera: Chrysomelidae) in commercial potato fields on Long Island. *Journal of Economic Entomology*, *77*(5), 1254-1259.

Yue, Y., Cheng, X., Zhang, D., Wu, Y., Zhao, Y., Chen, Y., & Zhang, Y. (2018). Deep recursive super resolution network with Laplacian Pyramid for better agricultural pest surveillance and detection. *Computers and electronics in agriculture*, *150*, 26-32.

Zalom, F. G., Bolda, M. P., Dara, S. K., & Joseph, S. (2002). UC IPM pest management guidelines: Strawberry, insects and mites. *Philips PA, UC IPM Program, UC Cooperative Extension Ventura Co. NC Toscano, Entomology, UC Riverside UC ANR, Publication*, *3468*.

Chapter 13

Creepy Crawlers to the Rescue: Insects as Bioindicators

Shalini Tripathi
Poornima Pandey*
Narendra Kumar Sharma
and Arindam Kuila
Department of Bioscience and Biotechnology, Banasthali Vidyapith, Tonk, Rajasthan, India

Abstract

Insects, one of the oldest classes in the animal kingdom, belong to the phylum Arthropoda and are ubiquitous and diverse. They also demonstrate an intriguing multitude of behavioural patterns and traits indicating the changes in the environment they are in. This ability to suggest variation in the environment is termed bioindication. Bioindicators are the species that help analyse the ecological performance and sensitivity against pollution in various environmental domains like the lithosphere, hydrosphere, and atmosphere. As insects are close to harmful soil, water, and air elements, they are especially useful in studying environmental pollution. Species like mayflies, ants, honeybees, and butterflies are some examples of bioindicator insects that can signify a shift in the quality of their surroundings. Moreover, these insects can be beneficial in not only inspecting environmental pollution but also in the assessment of existing toxic pollutants.

Keywords: bioindication, ecological performance, environment, pollution, beneficial

1. Introduction

Insects have many impacts on the livelihoods of humans, both by beneficial and harmful means. These animals play many roles, such as pests, defoliators, decomposers, recyclers, pollinators, silk producers, biocontrollers, and pollutants monitoring animals. Insects are used as bioindicators to assess the health and quality of the environment (Chowdhury et al., 2023). This is because they are highly sensitive to changes in their surroundings and are often the first group of species to be affected by environmental disturbances. By monitoring the population of certain insect species in a given area, scientists can gain valuable insights into the overall health of the ecosystem (Holt & Miller, 2011).

---

* Corresponding Author's Email: purnimapandey@banasthali.in.

In: Insects as Humans' Frenemies: From Friends to Foes
Editor: Istkhar Rao
ISBN: 979-8-89530-320-7

In addition to providing information about environmental health, insects can be used as climate change indicators (Parmar et al., 2016). As temperatures rise and weather patterns shift, the ranges and behaviours of many insect species are likely to change as well. By monitoring these changes, scientists can gain a better understanding of the impacts of climate change on ecosystems and biodiversity. A good bioindicator provides information about pollutants and the intensity of exposure, and it has been reported that there is no better technique for monitoring than the biomonitoring agent. Overall, the use of insects as bioindicators can provide valuable information for environmental management and conservation efforts, helping to protect and preserve ecosystems for future generations (Asif et al., 2018).

2. Types of Bioindicators

Bioindicators are living organisms or groups of organisms that are used to monitor the health of an ecosystem and the environment's ability to support life. They may be classified based on types of organisms acting as bioindicators i.e., microbes, plants, and animals (Figure 1). Some of these organisms that act as bioindicators, including:

a. *Plant bioindicators:* These include species of plants that are sensitive to environmental changes such as changes in temperature, light, moisture, and soil quality. They are used to assess the impacts of environmental changes on terrestrial ecosystems (Manning et al., 2002).
b. *Animal bioindicators:* Animals, especially insects and birds, are useful bioindicators as they can reflect changes in the ecosystem, such as changes in food availability, habitat quality, and air quality (Gadzała-Kopciuch et al., 2004).
c. *Microbial bioindicators:* Microorganisms are important indicators of environmental quality as they can reflect changes in water quality, soil quality, and air quality (Moscatelli et al., 2005).
d. *Stream benthic macroinvertebrates:* These are small aquatic animals that live on the bottom of streams and rivers. They are sensitive to changes in water quality, such as increases in pollutants, and can be used as bioindicators to monitor the health of aquatic ecosystems.
e. *Mosses and liverworts:* These bryophytes are good indicators of air quality, especially for air pollution caused by heavy metals and acid rain (Kohler & Schneider, 2003).
f. *Algae:* Algae are sensitive to changes in water quality and can be used as bioindicators to monitor the impacts of human activities, such as nutrient pollution and changes in water temperature, on aquatic ecosystems (Salo & Salovius-Laurén, 2022).
g. *Fish:* Fish are good indicators of overall ecosystem health as they are sensitive to changes in water quality, food availability, and habitat quality (Chovanec et al., 2003).
h. *Insect bioindicators:* There are several insect species that can be used as bioindicators. In this chapter, we will discuss how insects can be used as bioindicators.
i. *Indicator species:* living organisms and the environmental conditions of their habitat, which can be examined without any difficulty, can be regarded as indicator species (Landres et al., 1988). Some insect species are highly sensitive to changes in their environment and will only thrive in specific conditions. For example, the presence of

certain aquatic insect species can indicate good water quality, while the absence of these species can indicate pollution or contamination.

j. *Biodiversity:* Insects are incredibly diverse, with over one million known species, and their presence can indicate high levels of biodiversity in an area. The higher the number of different insect species, the more likely it is that the ecosystem is healthy and functioning properly.

k. *Biomagnification:* Insects can also be used to monitor the biomagnification of toxins and pollutants in the environment. Biomagnification occurs when toxins or pollutants accumulate in the food chain, with each organism in the chain accumulating higher levels of the substance than the one before it. By monitoring the levels of toxins and pollutants in insect populations, one can gain insight into the overall health of the ecosystem. A very alarming example of biomagnification is the use of microplastics in the current ecosystem.

l. *Sensitivity to environmental changes:* Insects are highly sensitive to changes in their environment and can provide an early warning system for environmental problems. For example, changes in temperature or precipitation patterns can affect the timing of insect life cycles, such as the emergence of certain species of butterflies. By monitoring these changes, scientists can gain insight into the impacts of climate change on ecosystems. Even the reproductive cycles of many species indicate fluctuations in the environment. In conclusion, insects can provide valuable information for monitoring the health and quality of ecosystems, detecting environmental problems, and managing and conserving biodiversity.

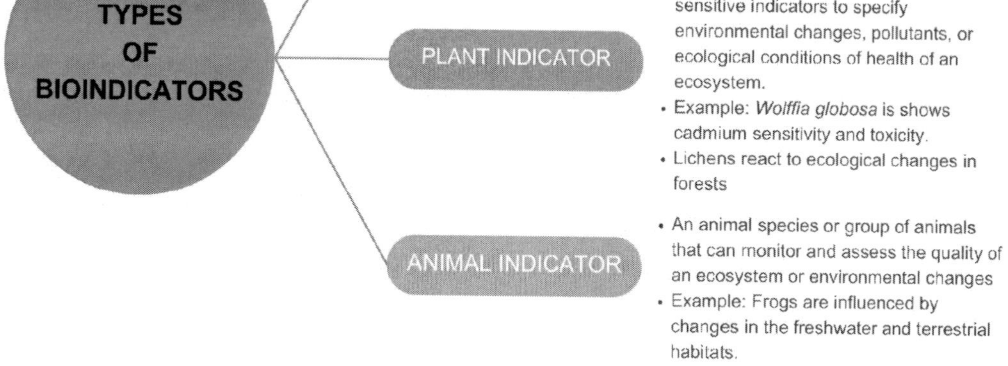

Figure 1. Various types of bioindicators.

These are just a few examples of the many types of bioindicators that are used to monitor the health of ecosystems. The choice of a particular bioindicator depends on the type of ecosystem being studied and the specific environmental changes being monitored. (Parikh et al., 2021).

3. Types of Insect Bioindicators

Insects are one of the most important groups of animal bioindicators because they are highly sensitive to changes in the environment, including changes in air and water quality, soil quality, and habitat quality, and are found in almost all ecosystems. Some examples of insect bioindicators include:

a. *Pollinators:* Pollinators, such as bees and butterflies, are important indicators of the health of terrestrial ecosystems. They are sensitive to changes in the quality and availability of flowers and other food sources, as well as to changes in air and soil quality.
b. *Dragonflies and damselflies:* These insects are sensitive to changes in water quality and are often used as bioindicators to monitor the health of aquatic ecosystems.
c. *Ladybugs:* Ladybugs are predators of other insects and are often used as bioindicators to monitor the health of terrestrial ecosystems. Changes in the populations of ladybugs can reflect changes in the abundance and diversity of other insect species, as well as changes in the quality of their food sources.
d. *Fireflies:* Fireflies are sensitive to changes in air and water quality, as well as changes in the quality of their food sources. They are often used as bioindicators to monitor the health of terrestrial and aquatic ecosystems.
e. *Mosquito larvae:* Mosquito larvae are sensitive to changes in water quality, including changes in temperature, pH, and the presence of pollutants. They are often used as bioindicators to monitor the health of aquatic ecosystems.
f. *Bark Beetles:* Bark beetles are important indicators of the health of forests, especially conifer forests. Changes in the populations of bark beetles can reflect changes in the quality of their food sources (tree bark) as well as changes in the overall health of the forest.

These are just a few examples of insect bioindicators. The choice of a particular insect bioindicator depends on the type of ecosystem being studied and the specific environmental changes being monitored. (McGEOCH, 1998). Insects can be classified based on principles and applications as mentioned in Table 1.

Table 1. Classification of Insects as bioindicator

Based on Principle				Based on the Application		
Diagnostic Indicator	Compliance Indicators	Early Warning Indicators	Accumulation Indicators	Environmental Indicator	Ecological Indicator	Biodiversity Indicator
Honeybees, Mosquitoes, Beetles, Dragonflies, Moths	Beetles, Butterflies, Honeybees	Butterflies, Caddis flies, Dragonflies	Honeybees, Carbide Flies, Dragonflies	Honeybees, Ground beetles, Moths, Butterflies, Mayflies	Beetles, Caddish flies, Sea skaters	Ants, Honeybees, Termites, Wasps

In the following sections, a brief description is given of important insects acting as bioindicators.

4. Ants as Bioindicators

Ants are fascinating and highly diverse insects that play crucial roles in ecosystems around the world. With over 12,000 known species, they are found in almost every terrestrial habitat, from tropical rainforests to deserts and even urban environments. Ants are social insects, living in colonies with an organised division of labour and communication systems. Ants are known for their remarkable collective behaviour and complex social structures. Within a colony, individuals fulfil specific roles such as workers, soldiers, and reproductive individuals (queens and males). They communicate through chemical signals called pheromones, which help them coordinate tasks, mark trails, and recognise colony members.

These tiny yet mighty creatures have proven themselves to be invaluable indicators of environmental health and ecological changes. As bioindicators, ants provide valuable insights into the condition and quality of ecosystems, acting as sentinels of environmental shifts. With their abundance, diversity, and sensitivity to changes in their surroundings, ants are well-suited for monitoring programmes aimed at assessing ecosystem health. Their responsiveness to variations in temperature, humidity, vegetation, and disturbance levels allows them to reflect the impacts of environmental disturbances with remarkable precision.

Ants' habitat specificity adds to their significance as bioindicators. Different ant species have specific habitat requirements, and their presence or absence can indicate the condition and quality of different habitats. By studying the composition and behaviour of ant communities, researchers can gain valuable information about the overall health of ecosystems.

Here are a few instances where ants have been used as bioindicators in various ecosystems:

a. *Habitat disturbance:* Ants have been used to assess the impacts of habitat disturbance, such as deforestation or land conversion. In a study conducted in the Amazon rainforest, researchers compared ant communities in undisturbed forest areas with those in areas subjected to different levels of logging. They found that the diversity and abundance of certain ant species decreased with increasing levels of disturbance, indicating the sensitivity of ants to habitat changes.
b. *Pollution monitoring:* Ants have been employed to monitor pollution levels in urban environments. Certain ant species are sensitive to heavy metal contamination, and their presence or absence can indicate the extent of pollution. In a study conducted in urban parks, researchers compared ant communities in polluted and non-polluted areas. They found that certain pollution-tolerant ant species were more abundant in polluted sites, while pollution-sensitive species were absent or scarce.
c. *Restoration assessment:* Ants can be used to evaluate the success of habitat restoration efforts. For example, in a study examining the restoration of abandoned agricultural lands, researchers monitored ant communities in restored sites over several years. They found that as the habitat quality improved, the diversity and abundance of ant species increased, eventually resembling those of nearby undisturbed habitats. This indicated the recovery of ant communities following restoration activities.
d. *Invasive species detection:* Ants can serve as indicators of invasive species and their impacts on native ecosystems. Invasive ants often outcompete native species, leading to changes in ant community composition. By monitoring ant populations, researchers can detect the presence and spread of invasive ants and assess their impact on local

biodiversity. For instance, the invasion of the Argentine ant (*Linepithema humile*) in various regions has resulted in the displacement of native ant species, and monitoring changes in ant communities has helped track the spread of this invasive species.

e. *Climate change effects:* Ants have also been used to study the impacts of climate change on ecosystems. Rising temperatures and altered precipitation patterns can affect the distribution and behaviour of ant species. Researchers have used ants to assess the effects of climate change by comparing ant communities in different temperature gradients or monitoring changes in species composition over time.

These examples illustrate how ants can provide valuable insights into ecosystem health, habitat quality, pollution levels, restoration progress, invasive species dynamics, and climate change impacts. By studying ants as bioindicators, scientists can better understand and manage ecosystems to ensure their long-term conservation and sustainability. (Andersen et al., 2002; Nithyatharani & Kavitha, 2018).

5. Butterflies as Bioindicators

Belonging to the order Lepidoptera, butterflies are a diverse group of insects, encompassing approximately 20,000 known species worldwide. They inhabit a wide range of ecosystems, from lush tropical rainforests to arid deserts and high mountain ranges. Their remarkable life cycle, from egg to caterpillar to chrysalis and finally to the resplendent adult form, is a testament to nature's ingenuity. Butterflies serve as bioindicators by providing valuable information about the health and condition of ecosystems. Bioindicators are species or groups of species that are sensitive to changes in their environment and can be used to assess the overall environmental quality. (Horn, 2003). Here's how butterflies fulfil this role:

a. *Sensitivity to habitat changes:* Butterflies are highly sensitive to habitat modifications and disruptions. They have specific requirements for food sources, host plants for larvae, and suitable breeding and wintering habitats. Therefore, changes in butterfly populations can indicate alterations in their habitat, such as habitat loss, fragmentation, pollution, or degradation.

b. *Rapid response to environmental changes:* Butterflies have relatively short lifespans and multiple generations within a year. This allows them to respond quickly to environmental changes. If a butterfly population declines or disappears from an area, it can indicate that the conditions in that area are deteriorating or becoming unsuitable for their survival.

c. *Species diversity and abundance:* The presence of a diverse and abundant butterfly community is indicative of a healthy ecosystem. Butterflies have different ecological requirements, and their presence reflects the availability of various resources like nectar plants, host plants, and suitable microhabitats. A decline in butterfly diversity or abundance can suggest ecological imbalances or disturbances.

d. *Sensitivity to climate change:* Butterflies are highly responsive to changes in temperature and climate patterns. Some species have specific temperature thresholds for their development and activity. Therefore, shifts in the distribution and phenology (timing of life cycle events) of butterflies can provide insights into the impacts of climate change on ecosystems.

e. *Indicator of pesticide use:* Butterflies can also be used to monitor pesticide contamination in ecosystems. Certain species are highly sensitive to pesticides, and their presence or absence in an area can indicate the level of pesticide use and its potential ecological impact.

Monitoring butterfly populations and their dynamics can provide valuable information about the health and functioning of ecosystems. It can help identify environmental changes, assess the effectiveness of conservation efforts, and inform decision-making processes regarding land use and habitat management (Horn, 2003).

6. Beetles as Bioindicators

Beetles, with their staggering diversity and abundance, offer a rich tapestry of bioindicators. With over 400,000 described species and likely many more awaiting discovery, beetles are a dominant and ecologically significant group within the insect world. Their ubiquitous presence in terrestrial and freshwater ecosystems, ranging from forests to wetlands, grasslands to urban environments, allows for comprehensive monitoring across diverse habitats and geographical regions. Beetles exhibit remarkable sensitivity to environmental changes, making them ideal bioindicators. Different beetle species have specific ecological requirements and sensitivities to factors such as temperature, humidity, air and water quality, and land-use changes. Their presence or absence, as well as shifts in their population dynamics, can provide critical insights into the overall health and condition of ecosystems, acting as ecological barometers (Ghannem et al., 2018).

a. *Species Diversity:* Beetles are one of the most diverse groups of insects, with an estimated 400,000 to 1 million species worldwide. Their diversity and abundance make them excellent indicators of ecosystem health. Changes in beetle species composition and diversity can indicate disturbances or changes in habitat conditions.
b. *Sensitivity to Environmental Changes:* Beetles are sensitive to various environmental factors, such as temperature, humidity, pollution, and land-use changes. Different species have specific habitat requirements, and their presence or absence can indicate the suitability of the environment for their survival. Some beetles are very specialised and can only survive in specific habitats, making them especially sensitive indicators.
c. *Feeding Habits:* Beetles exhibit a wide range of feeding habits, including herbivorous, carnivorous, detritivorous (feeding on decaying organic matter), and saprophagous (feeding on dead organisms). Changes in the abundance or distribution of beetles with specific feeding habits can indicate shifts in the availability of food resources or changes in the trophic structure of an ecosystem.
d. *Life Cycle and Habitat Associations:* Beetles often have specific requirements for their life cycles, such as breeding sites, host plants, or specific microhabitats. Monitoring the presence or absence of beetles with specific habitat associations can provide

insights into the health of those habitats. For example, the presence of certain beetles in aquatic ecosystems can indicate water quality and pollution levels.

e. *Response to Disturbances:* Beetles can respond quickly to changes in their environment, making them useful for assessing the impacts of disturbances or environmental changes. For instance, the presence of certain beetles known as pioneer species, which are among the first to colonise disturbed habitats, can indicate the recovery process following a disturbance.
f. *Longevity and Monitoring:* Beetles generally have relatively long lifespans compared to other insects, allowing for long-term monitoring of populations and environmental changes. Researchers can track changes in beetle populations over time to understand the impacts of various factors on ecosystems.

7. Honeybees as Bioindicators

Honeybees are small, highly social insects known for their important role as pollinators. They belong to the genus Apis and are commonly found in various parts of the world, especially in Europe, Asia, and North America. Honeybees are known for their ability to produce honey, hence their name. They have a complex social structure, with a queen bee, worker bees, and drones, each playing a specific role within the colony. Honeybees are vital for the pollination of numerous plants, including crops and wildflowers, making them essential for maintaining biodiversity and food production. They are also recognised as bioindicators, as their population health and behaviour can provide insights into environmental changes and ecosystem quality (Zhelyazkova, 2012).

Honeybees can indeed serve as bioindicators in various ways, as they are sensitive to environmental changes and can reflect the impacts of pollution, habitat degradation, and other ecological factors. Some ways honeybees indicate environmental fluctuations are:

a. *Pesticides and chemical pollution:* Honeybees are highly sensitive to pesticides and other chemical pollutants in their environment. They collect nectar and pollen from flowers, and if these sources are contaminated, the bees can accumulate harmful substances in their bodies. Monitoring honeybee colonies can provide valuable insights into the levels of pesticide contamination in an area, helping identify areas where chemical usage may pose a risk to both bees and other organisms.
b. *Habitat quality:* Honeybees require diverse and healthy habitats to thrive. They depend on a variety of flowering plants for their nutrition and on clean water sources for drinking. Changes in land use, such as urbanisation or intensive agriculture, can result in habitat loss and a decrease in floral resources. By monitoring honeybee populations, scientists can assess the overall health and diversity of the ecosystem and identify areas that need conservation efforts or habitat restoration.
c. *Climate change:* Honeybees are influenced by climatic conditions, particularly temperature and rainfall patterns. Climate changes can affect the timing of flowering and nectar availability, which can impact honeybee foraging and colony development. By studying honeybee behaviour and population dynamics, researchers can gain

insights into the impacts of climate change on ecosystems and potentially predict shifts in plant-pollinator interactions.

d. *Disease and parasites:* Honeybees are susceptible to various diseases and parasites, such as varroa mites and viral infections. Monitoring the prevalence and impact of these threats on honeybee colonies can indicate the overall health and stability of the ecosystem. Unusually high rates of disease or infestation can suggest environmental stressors or changes in the local ecology that need attention.

e. *Biodiversity assessment:* Honeybees are excellent pollinators and play a crucial role in maintaining biodiversity by facilitating plant reproduction. Monitoring honeybee populations can provide information on the availability of floral resources and the overall diversity of plant species. Changes in honeybee abundance or diversity can indicate shifts in the ecosystem and help identify potential conservation issues.

Overall, honeybees are valuable bioindicators due to their ecological sensitivity and importance in pollination. Studying honeybee populations and their interactions with the environment can provide valuable information for conservation efforts, land management practices, and the overall assessment of ecosystem health.

8. Mosquitoes as Bioindicators

Mosquitoes are small insects belonging to the order Diptera and the family Culicidae. They are found worldwide, with over 3,500 known species. They are found in almost every region on Earth, except in extreme polar regions. Mosquitoes have adapted to various habitats, ranging from tropical rainforests to arid deserts and urban areas. Mosquitoes play significant ecological roles, but they are often known for their nuisance and potential as disease vectors. These tiny creatures have captured human attention for centuries due to their irritating bites and the diseases they can transmit. While mosquitoes are remarkable creatures, with some species being harmless, many are known for their ability to transmit diseases to humans and animals. They are also associated with numerous health concerns. Certain species are capable of transmitting diseases such as malaria, dengue fever, the Zika virus, the West Nile virus, and others. These diseases pose significant global health challenges, particularly in tropical and subtropical regions (Dorvillé, 1996).

Mosquitoes, often regarded as bothersome pests, possess an unexpected ecological significance as bioindicators. These tiny insects can serve as valuable indicators of environmental conditions and ecosystem health. Their sensitivity to various ecological factors, such as water quality, habitat degradation, and climate change, makes them effective indicators of environmental changes. By monitoring mosquito populations, scientists and researchers can gain insights into the overall health of ecosystems, track the spread of diseases, and assess the impacts of human activities on natural environments. Understanding mosquitoes as bioindicators provides a unique perspective on these seemingly insignificant creatures and highlights their role in environmental monitoring and conservation efforts. Here's how mosquitoes can be used as bioindicators:

a. *Biodiversity and habitat quality:* Mosquito species diversity and abundance can reflect the overall biodiversity and habitat quality of an ecosystem. Higher diversity and abundance of mosquito species typically indicate a healthier and more diverse ecosystem, while a decline in mosquito populations may suggest environmental degradation or habitat loss.

b. *Water quality:* Mosquito larvae are aquatic and rely on standing water bodies like ponds, marshes, and wetlands for their development. Certain mosquito species have specific water quality requirements, and their presence or absence can indicate the water's quality. For example, some species are highly sensitive to pollution and contaminants, while others can tolerate more degraded conditions. Monitoring mosquito populations can provide insights into water pollution levels and the overall health of aquatic ecosystems. For instance, some species thrive in clean and unpolluted water, while others can tolerate more polluted environments. By studying the composition and abundance of mosquito populations in different water bodies, scientists can assess the water quality and identify potential pollution sources. A decline in mosquito populations or the dominance of pollution-tolerant species may indicate deteriorating water quality.

c. *Disease surveillance and outbreak prediction:* Mosquitoes are well-known vectors of various diseases, including malaria, dengue fever, and the West Nile virus. Monitoring mosquito populations allows scientists to track disease transmission patterns and predict potential outbreaks. Changes in the abundance and distribution of mosquito species can indicate shifts in disease dynamics. For example, an increase in the population of mosquitoes known to carry a particular disease may indicate an elevated risk of transmission in the area. This information is crucial for implementing timely public health interventions and vector control strategies.

d. *Climate change impact assessment:* Mosquito populations are sensitive to climate factors, such as temperature, rainfall patterns, and humidity. As the Earth's climate changes, mosquitoes respond to these alterations. By studying mosquito populations, scientists can gain insights into the impact of climate change on ecosystems. For instance, an expansion of mosquito habitats to higher altitudes or latitudes may indicate shifts in ecological boundaries due to warming temperatures. Additionally, changes in the phenology (timing) of mosquito life cycle stages, such as earlier breeding or prolonged activity periods, can provide evidence of climate-related influences on their biology.

9. Dragonflies as Bioindicators

Dragonflies, captivating creatures with intricate wings and agile flight, have fascinated humans for centuries. These ancient insects, belonging to the order Odonata, boast a complex life cycle that begins in water as nymphs before transforming into stunning winged adults. With their vibrant colours and exceptional hunting skills, dragonflies play vital roles as top predators in various ecosystems. Their presence in an environment serves as an indicator of environmental health and contributes to the overall balance of insect populations. Dragonflies are truly remarkable insects that continue to captivate and inspire observers of the natural world.

In addition to their captivating beauty and ecological significance, dragonflies have also emerged as valuable bioindicators in ecological studies. Due to their sensitivity to environmental changes and their position in the food chain, dragonflies can provide critical insights into the health and integrity of ecosystems. Researchers use dragonflies to monitor water quality, habitat conditions, pollution levels, and the impacts of climate change. By studying dragonfly populations, scientists can detect early warning signs of environmental degradation, inform conservation efforts, and make informed decisions for sustainable ecosystem management. The presence or absence of dragonflies in a particular area can serve as an important indicator of overall ecological well-being and the need for targeted conservation actions (Bulánková, 1997).

a. *Water pollution assessment:* Dragonflies have been utilised to assess the impact of water pollution on aquatic ecosystems. For example, researchers in a study conducted in China examined the diversity and abundance of dragonfly species in different sections of a river. They found that areas with higher pollution levels had lower dragonfly diversity and population sizes compared to less polluted areas. This indicated the negative impact of water pollution on dragonflies and suggested the overall degraded state of the river.
b. *Wetland conservation:* Dragonflies are commonly used as indicators of wetland health and conservation efforts. Wetlands provide crucial habitats for dragonfly species. By monitoring dragonfly populations in wetlands, scientists can assess the effectiveness of wetland restoration projects and evaluate the overall ecological condition of these vital ecosystems. Changes in dragonfly diversity and abundance can indicate the success or failure of wetland conservation initiatives.
c. *Climate change studies:* Dragonflies are highly sensitive to temperature changes, and their emergence patterns can be influenced by climate change. Researchers have used dragonflies to study the effects of climate change on ecosystems. For instance, a study in Europe examined the phenology (timing of life cycle events) of dragonflies over several years. The researchers found that warmer temperatures caused shifts in dragonfly emergence, affecting their interactions with other species in the ecosystem. This highlighted the potential consequences of climate change on the timing and synchrony of species interactions.
d. *Pesticide impact assessment:* Dragonflies are excellent indicators of pesticide contamination in aquatic environments. They can accumulate pesticides in their bodies through their prey or direct exposure to contaminated water. In a study conducted in agricultural areas, researchers analysed dragonfly larvae for pesticide residues and found that certain pesticides were present at levels that could have negative impacts on the dragonfly populations. These findings highlighted the potential ecological risks associated with pesticide use in agricultural landscapes.
e. *Habitat quality evaluation:* Dragonflies are particularly sensitive to changes in their habitat, making them useful indicators of habitat quality. By monitoring dragonfly populations in different habitats, researchers can assess the effects of habitat loss, degradation, and restoration efforts. For example, in a study conducted in a forested landscape, researchers compared dragonfly diversity and abundance in pristine forests, fragmented forests, and restored habitats. The results indicated that restored habitats

exhibited increased dragonfly diversity, indicating successful habitat restoration efforts.

These instances demonstrate the diverse applications of dragonflies as bioindicators in ecological research. Their responsiveness to various environmental factors makes them valuable tools for assessing the health of ecosystems, water quality, habitat conditions, and the impacts of human activities on natural environments.

10. Termites as Bioindicators

Termites, often regarded as pests due to their ability to damage wooden structures, are far more than mere nuisances. These social insects belong to the order Isoptera and play critical roles in ecosystems around the world. Termites are known for their remarkable architectural abilities, complex social structures, and vital ecological functions. They are highly adapted to various habitats, from tropical rainforests to arid savannahs, and their interactions with their environment have far-reaching consequences. Understanding the fascinating world of termites reveals the intricate balance they maintain within ecosystems and the crucial roles they play as decomposers, nutrient cyclers, soil engineers, and bioindicators (Nithyatharani & Kavitha, 2018).

In addition to their ecological significance, termites have emerged as valuable bioindicators in numerous ecosystems. These small but influential insects possess specific sensitivities and behaviours that make them excellent indicators of environmental health and quality. Termites' dependence on particular habitat conditions, such as temperature, humidity, and moisture, makes them highly responsive to changes in their surroundings. Alterations in termite populations, species composition, or foraging behaviours can signal shifts in habitat degradation, pollution levels, or the impacts of climate change. Their role as decomposers and nutrient cyclers further accentuates their value as bioindicators, as changes in termite activity and diversity can reflect alterations in ecosystem functioning and stability. By monitoring termite populations and studying their responses to environmental changes, researchers can gain valuable insights into the overall health and condition of ecosystems and guide conservation and management efforts accordingly.

Some specific instances that highlight the role of termites as bioindicators:

a. *Habitat degradation:* Termites are particularly sensitive to changes in habitat quality. For example, in tropical rainforests, certain termite species are highly abundant and contribute to soil formation and nutrient cycling. When logging or deforestation occurs, leading to habitat degradation, the populations of these termite species decline significantly. Monitoring termite abundance and diversity in such areas can provide insights into the extent of habitat degradation and help guide conservation efforts.
b. *Soil quality and restoration:* Termites are important contributors to soil health and structure. Some termite species construct complex underground galleries that improve soil porosity, water infiltration, and nutrient availability. In degraded or heavily compacted soils, termite populations may be significantly reduced or absent. By

studying termite presence and activity in such areas, researchers can assess soil quality and guide restoration efforts to enhance soil fertility and ecosystem resilience.

c. *Climate change impacts:* Climate change can affect termite populations due to alterations in temperature and moisture patterns. In certain regions, shifts in precipitation patterns and temperature regimes can disrupt termite behaviour and survival. For instance, studies have shown that increased temperatures can negatively impact termite colonies, reducing their abundance and altering their foraging patterns. Monitoring termite populations over time can help track the impacts of climate change on ecosystems.

d. *Pollution and toxic substances:* Termites can also serve as indicators of environmental pollution. Certain termite species are sensitive to pollutants such as heavy metals, pesticides, and industrial chemicals. Their populations can decline or exhibit abnormal behaviour in polluted environments. Monitoring termite communities near industrial sites or agricultural areas can provide early warning signs of contamination and prompt further investigation into the sources and extent of pollution.

e. *Land-use changes:* Termites are highly responsive to changes in land use, such as urbanisation and agriculture. When natural habitats are converted to urban areas or agricultural fields, termite populations can be significantly altered. Some species may disappear entirely, while others may thrive in disturbed environments. Monitoring termite diversity and abundance in areas undergoing land-use changes can help assess the impacts on local ecosystems and guide sustainable land management practices.

Conclusion

In conclusion, bioindicator insects play an indispensable role in our understanding of ecosystem health and environmental changes. Their sensitivity to specific conditions, pollutants, and alterations makes them valuable tools for detecting shifts in a variety of ecosystems. By observing the presence, absence, behaviour, or health of these insects, we can gain insights into the overall well-being of the environment and identify potential issues before they escalate. The need for bioindicator insects arises from the growing concern over human impacts on natural systems. These insects serve as early warning systems, helping us monitor and mitigate the effects of pollution, habitat loss, climate change, and other stressors. Their characteristics, such as being responsive to environmental changes, having relatively short life cycles, and occupying various niches, make them ideal candidates for assessing ecological conditions. As we navigate an era of environmental uncertainty, the utilisation of bioindicator insects not only aids in safeguarding biodiversity but also informs sustainable management practices and conservation efforts.

References

Andersen, A. N., Hoffmann, B. D., Müller, W. J., & Griffiths, A. D. (2002). Using ants as bioindicators in land management: simplifying the assessment of ant community responses. *Journal of Applied Ecology*, *39*(1), 8-17.

Asif, N., Malik, M., & Chaudhry, F. (2018). A review of on environmental pollution bioindicators. *Pollution*, *4*(1), 111-118.

Bulánková, E. (1997). Dragonflies (Odonata) as bioindicators. *Biologia, Bratislava*, *52*(2), 177-180.

Chovanec, A., Hofer, R., & Schiemer, F. (2003). Fish as bioindicators. In *Trace metals and other contaminants in the environment* (Vol. 6, pp. 639-676). Elsevier.

Chowdhury, S., Dubey, V. K., Choudhury, S., Das, A., Jeengar, D., Sujatha, B., Kumar, A., Kumar, N., Semwal, A., & Kumar, V. (2023). Insects as bioindicator: A hidden gem for environmental monitoring. *Frontiers in Environmental Science*, *11*, 273.

Dorvillé, L. F. (1996). Mosquitoes as bioindicators of forest degradation in southeastern Brazil, a statistical evaluation of published data in the literature. *Studies on Neotropical Fauna and Environment*, *31*(2), 68-78.

Gadzała-Kopciuch, R., Berecka, B., Bartoszewicz, J., & Buszewski, B. (2004). Some considerations about bioindicators in environmental monitoring. *Polish Journal of Environmental Studies*, *13*(5), 453-462.

Ghannem, S., Touaylia, S., & Boumaiza, M. (2018). Beetles (Insecta: Coleoptera) as bioindicators of the assessment of environmental pollution. *Human and Ecological Risk Assessment: An International Journal*, *24*(2), 456-464.

Holt, E., & Miller, S. (2011). Bioindicators: Using organisms to measure. *Nature*, *3*, 8-13.

Horn, H. S. (2003). Butterflies as bioindicators. *American Scientist*, *91*(6), 565-568.

Kohler, A., & Schneider, S. (2003). Macrophytes as bioindicators. *Large Rivers*, *14*(1-2), 17-31.

Landres, P. B., Verner, J., & Thomas, J. W. (1988). Ecological uses of vertebrate indicator species: a critique. *Conservation biology*, *2*(4), 316-328.

Manning, W., Godzik, B., & Musselman, R. (2002). Potential bioindicator plant species for ambient ozone in forested mountain areas of central Europe. *Environmental Pollution*, *119*(3), 283-290.

McGEOCH, M. A. (1998). The selection, testing and application of terrestrial insects as bioindicators. *Biological Reviews*, *73*(2), 181-201.

Moscatelli, M. C., Lagomarsino, A., Marinari, S., De Angelis, P., & Grego, S. (2005). Soil microbial indices as bioindicators of environmental changes in a poplar plantation. *Ecological Indicators*, *5*(3), 171-179.

Nithyatharani, R., & Kavitha, U. (2018). Termite soil as bio-indicator of soil fertility. *International Journal for Research in Applied Science and Engineering Technology*, *6*(1), 659-661.

Parikh, G., Rawtani, D., & Khatri, N. (2021). Insects as an indicator for environmental pollution. *Environmental Claims Journal*, *33*(2), 161-181.

Parmar, T. K., Rawtani, D., & Agrawal, Y. (2016). Bioindicators: the natural indicator of environmental pollution. *Frontiers in life science*, *9*(2), 110-118.

Salo, T., & Salovius-Laurén, S. (2022). Green algae as bioindicators for long-term nutrient pollution along a coastal eutrophication gradient. *Ecological Indicators*, *140*, 109034.

Zhelyazkova, I. (2012). Honeybees–bioindicators for environmental quality. *Bulgarian journal of agricultural science*, *18*(3), 435-442.

Chapter 14

The Role of Insects in Recycling Soil

Aasha Rana[1,*]
Sonu Shekhawat[1]
and Nisha Rana[2]
[1]Department of Zoology, Faculty of Basic and Applied Sciences, Madhav University, Pindwara (Sirohi), Rajasthan, India
[2]Central Research and Incubation Centre, Swami Vivekanand Subharti University, Meerut, India

Abstract

Nutrients are essential components not only for the health of human beings but also necessary for the growth and development of plants. When it comes to the cycling of nutrients into ecosystems, insects are essential. They are an essential component of biotic communities and can be both destructive and useful since they offer significant ecological services. The class Insecta is the most diverse group of organisms on Earth since it includes members of all continents and represents a variety of trophic niches and ecological roles in their native environments, including herbivores, predators, and detritus feeders. A wide variety of insects have an impact on human economics, both positively and negatively. They recycle nutrients on Earth and have an ecological role in pollination. The focus of the inquiry was insects' function in the cycle of nutrients in living things.

Keywords: ecosystem, ecological services, pollution, trophic niches, economics

1. Introduction

Since insects' processes produce vitamins and minerals, pollinate plants, spread seeds, maintain fertile soil and structure, manage pest populations, and serve as a significant source of food for other species, they provide the biological foundation for all ecosystems. Since many of them are relatively new human creations, their distinctive traits in the animal kingdom startle specialists in biomechanics and bioengineering. Since the beginning of soil agriculture, insects and humans have competed for the fruits of our labour. There are many kinds of scuttling insects on every continent, including Antarctica (Johnes, 2022; Parr & Bishop, 2022; Sridhar, 2023), which play a significant role in preserving the sustainability of the agricultural system and designing the environment. Insects are the movers and shakers of the ecosystem because

* Corresponding Author's Email: aasha.rana@madhavuniversity.edu.in; aasha.aasharana@ymail.com.

In: Insects as Humans' Frenemies: From Friends to Foes
Editor: Istkhar Rao
ISBN: 979-8-89530-320-7

they churn up the soil, disperse seeds, and are crucial to pollination and decomposition. Without insects, it would be impossible to carry out agricultural operations or replenish soil fertility. For instance, global variety may end in a few years if honeybees go extinct (Palmer, 2022). Insects are truly remarkable creators of life! They play an essential role in our ecosystems, breathing vitality into the world around us. In this light, it's clear that insects are not just beings; they are life-givers in countless ways! Common insects such as butterflies, dragonflies, ants, wild honeybees, etc. are often overlooked despite their importance to the environment. Like bees, butterflies serve as important pollinators. One of the most well-known insects, dragonflies, require clean aquatic systems and should not be allowed to use chemical pesticides or fertilisers since they are extremely sensitive to changes in their surroundings. When discussing the advantages of insects, it can be claimed that they help both the environment and humans in many different ways. Insects serve as specialists in cleanliness, controlling pest insects, and pollinating crops that humans depend on for a livelihood. They help clean up trash, preventing the earth from becoming overrun by waste material.

Understanding anything about hexapods aids in improving our understanding of the structure and function of ecosystems. The absence of ants and other insects would cause the collapse of most ecosystems. The biological base of all terrestrial ecosystems is created by insects. Insects assist in maintaining the fertility and structure of the soil, pollinating plants, distributing seeds, recycling nutrients, and managing the population of other species. The number of ants and their habitats are unknown to us. According to Kilgore's (2022) census, the population is estimated to be 20 quadrillion or 20 plus 15 zeros. However, very few people have ever seen an ant's dead body. It is a known truth that life on Earth would be very different for humans without insects. One cannot undervalue the importance of insects to ecology. In actuality, they influence both human health and agriculture, and they make up the base of the food pyramid. Insects are responsible for pollinating flowers, which in turn produce the fruits, vegetables, and clothes we consume. We may not have many growers of the delicious fruits and vegetables we eat; therefore, we must rely entirely on insect pollination to generate honey, silk, and other valuable goods. Similar to this, flies appear just after the monsoon and disappear the following day; their deadly remnants are nowhere to be discovered. Considering the countless billions of populations of different types of insects—some in the soil, others in the atmosphere, some in plant leaves, flowers, and fruits, etc.—as well as the numerous small and large animal species found on land, in the air, and deep within the ocean, one might wonder who maintains a record of their birth and death and which particular insect or animal should assume the form of which living being in light of their previous deeds. Undoubtedly, insects play a vital role in maintaining life on Earth. Experts believe that insects may be crucial in addressing several modern environmental challenges. Unfortunately, it is also true that insects have grown more and more vulnerable to extinction as a result of the growing climate crisis and the widespread and careless use of pesticides in agriculture, which is negatively affecting insects that are near farms and plantations.

2. Ecological and Economic Importance of Insects

Scavengers, such as beetles, in particular, recycle nutrients back into the soil by eating decaying organisms and fallen trees. Insects are decomposers and contribute to the formation of topsoil, the nutrient-rich soil layer that supports plant growth. Ants and beetles are examples of burrowing insects that create tunnels that act as water routes for plants. Blooming plants are pollinated by ants, wasps, butterflies, and bees. Because they regulate the size of some insect populations, such as aphids and caterpillars, which depend on new plant development, gardeners like the big-eyed bug and praying mantises. Lastly, the vitamins and minerals from their droppings are used by all insects to fertilise the soil. Insects are therefore very important economically. Certain insects yield valuable products, such as silk, honey, lacquer, and wax. For thousands of years, people have bred honeybees for honey. Silkworms have had a significant impact on human history. China's commerce with the rest of the world began when the Chinese discovered how to make silk from worms. Both adult insects—like crickets—and their larvae are frequently utilised as bait for fishing.

3. Insects as Food

Naturally, humans do not consume only insects. For many birds, mammals, amphibians, and reptiles, insects are the only source of food, which makes their functions in food chains and webs essential. If insect numbers drop, food webs can collapse. In certain regions of the world, people consume insects as food. Insects are considered delicacies in many third-world nations and are rich in protein, vitamins, and minerals. It is very hard to locate an insect that humans do not consume in one way or another. The most popular insects include wasps, cicadas, grubs, caterpillars, locusts, mantises, and ants. The idea of including a source of protein in the diet of humans is widely supported. Insects such as beetles and grasshoppers are roasted and consumed by people from South America to Japan.

4. Insects in Medicine

Medicine has also made use of insects. Fly larvae, or maggots, were sometimes used as a wound treatment to stop or prevent gangrene. The infection of decaying tissue is the cause of gangrene. Since maggots can only consume dead tissue, applying them to human corpses cleans up the wound and helps to keep infections from spreading.

5. Ecosystem Cycling

The quality, amount, and timing of plant detrital inputs are altered by insect herbivores, which may have a significant impact on the ecosystem cycle (Jankielsohn, 2018; Mattson & Addy, 1975; Belovsky & Slade, 2000). Because herbivorous grasshoppers transform living plant material into biomass, greenfall, and throughfall (Belovsky & Slade, 2000; Hunter, 2001), it is shown that they increase plant abundance. As a result, insect herbivores are essential players in ecological processes. They may also be responsible for a significant amount of the N and P fluxes that occur from above-ground to below-ground throughout whole ecosystems (Metcalfe et al., 2014). Approximately two-thirds of all plant species depend on insects for pollination, with over 75% of wild-blooming plant species in temperate climates requiring them. According to Schoonhoven et al. (2005), bees, beetles, butterflies, and flies are the most significant pollinators. As a result, insects indirectly increase animal variety through pollination and contribute to plant diversity.

6. Predation/Parasitism

Predators, parasites, and secondary or tertiary consumers that occupy higher trophic levels serve to regulate the population growth of primary consumers or phytophagous creatures below a certain point. Insect predators and parasitoids naturally regulate herbivorous insects that have the potential to become pests (Van Lenteren, 2012). While a significant portion of species in the orders Hemiptera (bugs), Coleoptera (beetles), Diptera (flies), and Hymenoptera (wasps, bees, and ants) are predators, either as larvae or in both larval and adult stages, all insect species in the orders Odonata (dragonflies) and Neuroptera (lacewings and ant lions) are predators. The order Hymenoptera has a variety of parasitic organisms that feed on the adults, larvae, or eggs of other insects. For instance, *Aphytis lingnanensis* parasitises scale insects, *Aphelinus asychis*, *Aphelinus varipes*, *Diaeretiella rapae*, *Aphidius colemani*, *Aphidius matricariae*, and *Aphidius ervi* parasitise cereal aphids, and *Trichogramma* parasitic wasps attack Lepidopteran eggs.

7. Decomposition

Approximately 4000 species of dung beetles have been identified, and they are crucial to the breakdown of manure. By burying dung, which has the benefit of eliminating surface waste and recycling nutrients that plants may consume, dung beetles primarily contribute to the upkeep of pasture health. Before the introduction of dung beetles that were adapted to cattle dung, Australia experienced negative environmental effects such as loss of grass cover, growth of unpalatable grass, leaching of nutrients in surface rainwater runoff, and the accumulation of large populations of dung-breeding flies. By raising the number of total proteins or nutrients such as calcium, magnesium, phosphorus, potassium, and nitrogen in the soil, dung beetles improve their overall health (Macfadyen et al., 2015). Additionally, dung beetles contribute to the carbon cycle by lowering greenhouse gas emissions by 7% to 12% (Nichols et al., 2008). Dead plant debris is cleared away by beetle larvae, flies, ants, and termites, who then break it down so that bacteria can continue to break it down. The macrofauna of dry, hot soil, namely termites and ants, is crucial to the growth of mineral nitrogen in the soil (Evans et al., 2011).

A dynamic succession of arthropod species, consisting of up to 100 species from tens of insect families, colonises cadavers in the beginning, with calliphorid flies being the first to take advantage of them (Farwig et al., 2014; Richards & Goff, 1997). However, the effectiveness of the breakdown is determined by the function that a particular species contributes. According to Farwig et al. (2014), the makeup of insect scavenger assemblages, rather than their abundance, determines the rate at which carrion decomposes. To effectively reduce livestock manure in Australian pastures, dung beetles were screened and brought to Australia from Europe and Africa (Davis, 1996; Tyndalebiscoe & Vogt, 1991).

For ecosystems to operate, one of the most crucial processes is decomposition. This is an intricate process, including interactions between biotic and abiotic elements in the ecosystem.

Mesofauna refers to organisms that fall between 100 and 200 mm in size, such as *Acari* mites, *Collembola* springtails, and small insects. On the other hand, macrofauna refers to larger organisms, such as some beetles (such as Scarabaeidae, Geotrupidae, or Silphidae), Diptera larvae (such as Muscidae, Sarcophagidae, Scatophagidae, or Calliphoridae), centipedes (Diplopoda), woodlice (Isopoda), and so on. Contributing to the phase of destruction, each of these categories is in charge of the fragmentation of plant or animal remnants. Both the production of soil components and the transfer of organic remnants are facilitated by them. Arthropod decomposers belong to these categories and are found in almost all terrestrial ecosystems, usually in large populations. Termites (Isoptera) and cockroaches (Blattodea) are two of the most well-known insect decomposers. Termites are obligatory carnivores that require symbiotic bacteria and protozoa to decompose wood. Diplopoda, or millipedes, have a unique role as decomposers. According to Fincher (1981), cow waste consistently covered about 300 hectares of grass each year. Cattle avoid feeding on the dense vegetation around dung pats, making their presence on pastures create an unpleasant environment. Cattle react negatively to waste because they find it irritating and will not feed on pasture grass growing next to it. Many authors have brought up the potential contribution of dung beetles to lessening worm parasite infection in cattle. Worm excrement left on the ground may serve as a breeding ground for parasitic worm larvae. Cattle parasitism can be reduced if dung beetles bury these excrements before the parasites become highly transmissible. Flies that reproduce in dung can also be effectively controlled by dung beetles.

8. Insects in Agriculture

Pests have affected crops for as long as people have engaged in agricultural cultivation, and insects have mostly been seen as rivals in the struggle for existence. The primary biotic interaction is between insects and plants (Samways, 1993). Of the insect species, over 50% are herbivorous, and the majority of these species consume plants that belong to one or a few related plant families (Schoonhoven, 2005). Chemical treatments are the primary means of controlling herbivorous insects, which cause harm to 18% of global agricultural productivity (Losey & Vaughan, 2006). Because they perform essential ecological tasks for agricultural production, insects are essential to human life. Approximately 72% of crops worldwide rely on insects for pollination (Dicke, 2017). Thirty per cent of the world's agricultural productivity is produced by insects that either increase or stabilise the yield of 75 per cent of all crop varieties (Schwägerl, 2016). Insect pollinators include hundreds of species of solitary bees, bumblebees, flies, beetles, and butterflies, and in several crops, wild bee species are more important for pollination than the honeybee, *Apis mellifera* (Garibaldi et al., 2013). It is estimated that insects provide 9.5% of the world's crop output yield through pollination services (Gallai et al., 2009). Dung beetles considerably enhance the number of total proteins, calcium, magnesium, phosphorous, potassium, and nitrogen in the soil through their activity, which is superior to artificial fertilisers in terms of increasing wheat plant output (De Groot, 2002).

8.1. The Role of Beetles

More species of beetles than any other insect group, including about a third of a million recognised species worldwide, are found in the biggest order, Coleoptera. Although most beetles are terrestrial, several have evolved to survive in aquatic habitats, including the real water beetle, whirligig, and water scavenger. Some beetles are very important to humans. Weevils, often called snout beetles, are a family of insects that are infamous agricultural pests that may ruin stored grains and developing crops. Certain kinds of beetles consume plant materials like wood, paper, and textiles. Because they eat harmful insect pests, certain beetles are important. For example, ladybirds, sometimes known as ladybird beetles, safeguard a wide range of flowers and crops by eating countless thousands of aphids every year. Numerous other beetles contribute to different ecosystems in more subtle but no less significant ways. In livestock regions, tumblebugs, sometimes known as dung beetles, eat large amounts of manure. The larvae of carrion beetles are scavengers that consume dead animals. Numerous beetles are crucial to pollination as well.

While the majority of beetles consume live or dead plant materials, some scavenge dead animal remains, while others hunt other insects, spiders, or other tiny creatures like snails, frogs, and small fish. Some species eat fungus as food. Many beetle species' adults and larvae consume a variety of plant roots, stems, fruits, seeds, and leaves. While some beetles are less picky about what they eat, others exclusively consume specific plant species and plant sections. In forest environments, the adults and larvae of many beetles consume rotting wood and aid in the decomposition of dead trees and other flora. Certain bugs are fierce predators, like tiger beetles. Tiger beetles in adulthood look for food that they can stifle with their strong jaws. The larvae wait in their little tunnels to catch insects and spiders that pass by. They are stationary.

Although most water-scavenger beetles consume algae and decomposing organisms as adults, they are predators as larvae. Certain beetle species exhibit very selective feeding behaviours, whereby they solely consume mites, ant larvae, aphids, or zooplankton. One of the most researched groups of decomposers is the dung beetle (Noriega et al., 2018). They support a variety of ecological processes and/or services, such as the decomposition of litter, nutrient cycling, ecosystem productivity, bioturbation, primary and secondary seed dispersal, plant growth, and the control of flies and parasites (Nichols et al., 2008; Tixier et al., 2015; Manning et al., 2016). Within the nest balls, where they grow in circumstances of moderate moisture and temperature, female dung beetles lay their eggs (Hanski & Cambefort, 1991). Dugout beetles can affect the physical and chemical characteristics of soil through their digging and dung displacement activities. Relocating dung and building nest balls aid in the activation of manure-based microorganisms engaged in the nitrogen cycle (Breymeyer et al., 1975; Kazuhira et al., 1991). Additionally, they link the above- and below-ground dung breakdown processes, which help to homogenise microbial populations (Slade et al., 2016). Nest ball production encourages manure moisture loss, promotes the proliferation of aerobic bacteria, inhibits ammonia volatilisation, and improves soil fertility by facilitating plant nitrogen absorption (Huerta et al., 2013). It also promotes nitrification in both the buried dung and the leftovers that remain at the soil surface (Kazuhira et al., 1991). Furthermore, dung beetles aid in nitrogen mineralisation by reducing soil compaction and increasing soil aeration through the construction of burrow tunnels (Manning et al., 2016). (Bertone et al., 2006). Termites and dung beetles both contribute significantly to the ecosystem by assimilating animal manure into the soil and encouraging pasture renewal (Schowalter, 2016).

8.2. The Role of Ants

Ants (Hymenoptera: Formicidae) are one of the most dominant terrestrial organisms worldwide and currently describe more than 17,000 ant species (www.antweb.org). Except for Antarctica, they are incredibly plentiful both in terms of raw numbers and biomass, and they have a complex relationship with a wide range of biological networks and processes (Tuma et al., 2020). We must comprehend the responses of several significant taxonomic groups, including ants, to the climatic changes that our planet is experiencing. Ants contribute to many ecosystem processes, which include seed dispersal, pest control, and soil bioturbation (Del Taro et al., 2012) and play the most important role and function in terrestrial ecosystems. Higher temperatures have a favourable impact on the growth and feeding of ants in temperate zones. In addition, the majority of ants in these regions are soil-nesting and can therefore take advantage of the buffering soil provides. An important ecological function is bioturbation, which is the reworking of soil by ants via movement, tunnelling, and nest-building activities. This process can have a significant impact on soil physical characteristics such as soil porosity as well as nutrient distribution (e.g., increasing P content in ant nests) (Dostál et al., 2005; Frouz et al., 2003; Nkem et al., 2000) and promote plant biodiversity (Wilkinson et al., 2009) and growth (Reichman & Seabloom, 2000). By consuming plant matter, debris, and seeds along with bioturbation, ants also change the availability of nutrients in ecosystems and their distribution. Ants are involved in the clearance of garbage in urban settings as well since they can recycle and eat human food sources. Invasive ants are particularly significant in this regard (Penick et al., 2015; Youngsteadt et al., 2015).

As widespread seed harvesters, ants have a significant impact on the variety and composition of plants in their habitats through seed predation (De Almeida et al., 2020; MacMahon et al., 2000). Ants are considered to significantly influence plant and animal populations since they are both competitors and predators. The presence of ants modifies the organisation of invertebrate communities and food webs, as clearly illustrated by manipulative experiments (Parr et al., 2016). There is some evidence to suggest that herbivores and decomposers thrive in environments with low ant abundances (Sanders and van Veen, 2011). According to Parr et al. (2016), reduced ant populations can also favour termites, the primary decomposer invertebrate in savannas, which would improve decomposition. This matters because we know dominant ants exert considerable pressure on local communities and can regulate community structure via competition and predation (Andersen and Patel, 1994; Parr, 2008).

Since ants are abundant plant-herbivore predators and can decrease herbivory, their value for pest control has also long been acknowledged (Hölldobler & Wilson, 1990). However, studies comparing the benefits of ants to other taxa have only recently been conducted; Piñol et al. (2010), for instance, showed that ants were more effective than birds at controlling arthropod populations on citrus. Studies have shown that ants may considerably increase crop yields, which makes their predation bio-control service particularly valuable in agricultural settings (e.g., in shade cocoa, Gras et al., 2016). Furthermore, ant pest control services also include weed management: throughout a single agricultural system, and presence nearly cut down on tumbleweed plants (Evans & Gleeson, 2016).

9. Factors Affecting Insect Population and Distribution

The population and distribution of insects are influenced by several biotic and abiotic variables. In comparison with biotic variables, which include biotic interaction, demography, and life features, among others, abiotic factors include temperature, moisture, and nutrients in the soil (Guisan & Thuiller, 2005; Chong et al., 2015). The biodiversity of insects is also influenced by several variables, including climate change and human activity. The quantity of insects is regulated by a variety of environmental elements, such as chemical, physical, and biological factors (Hughes et al., 1997). According to studies, they need wet, nutrient-rich soil and are responsible for the ecosystem's bottom-up nitrogen cycle in Antarctica (Convey & Block, 1996; Rico & Quesada, 2013; Chown & Convey, 2016; Scudder, 2017; Bokhorst et al., 2019).

Conclusion

We are running out of choices to guarantee food security for a growing population because the majority of the world's fertile land is used for agriculture and arable land regions cannot be further extended. Utilising the current land more sustainably is the only remaining choice. Insects can offer the required remedies in this context. Insects are essential to the functioning of every ecosystem because they contribute significantly to its overall health. Insects that provide important ecological services must be a basic component of the agricultural systems that we manage. Communities can be defined in a variety of ways, and species diversity is simply one of them. Other factors to take into consideration when presenting a community include its size and ecological role, the relationships between various groups, and its overall structure. With knowledge of the ecosystem functions provided by insects, we can then accommodate these insects in agricultural systems by changing management practices to increase the functional diversity in these systems. Because of the large numbers of insects and their great intra- and interspecific variety, the functional significance of insects is enormous, and the ecosystem services provided by insects are vital. Insects are the key components in diverse ecosystems and play a major role in the functioning of ecosystem processes. Since insects are mostly perceived as pests or potential pests, the ecological importance of insects often goes unnoticed. Insects have an important role in pollination, decomposition, predation, parasitism, and the ecosystem cycle. Insect herbivores can significantly alter ecosystem cycling by altering the kind, amount, and duration of plant detrital inputs. Lastly, several dung beetle species exhibit unique feeding and nesting habits.

To speak about the benefits of insects, it can be said that insects provide useful services to mankind and the environment in several ways. In addition to pollinating the crops that provide us with food and controlling pest insects, insects also serve as specialists in cleanliness, clearing away waste to prevent an overabundance of excrement from the planet. Everything we know about them helps us understand how ecosystems function and are put together. The absence of ants and other insects would cause the collapse of most ecosystems. The biological base of all terrestrial ecosystems is created by insects. Insects assist in maintaining the fertility and structure of the soil, pollinating plants, distributing seeds, recycling nutrients, and managing the population of other species. One cannot undervalue the importance of insects to ecology. They influence both human health and agriculture, and they make up the base of the food pyramid. Insects are responsible for pollinating flowers, which in turn produce the fruits, vegetables, and clothes we consume. We may not have many growers of the delicious fruits and vegetables we eat; therefore, we must rely entirely on insect pollination to generate honey, silk, and other valuable goods. Similar to this, flies appear just after the monsoon and disappear the following day; their life-threatening remnants are unable to be discovered. One is sometimes forced to consider whether or not rebirth phenomena occur. Taking into account the countless billions of populations of diverse insect species, some of which are found in the earth, others in the air, some on plant leaves, flowers, fruits, etc. Aside from the many different species of large and tiny animals that live on land, in the air, and in the depths of the ocean, one wonders who records when they reincarnate, and which bug or animal should assume the appearance of which living creature depending on their previous actions. Given that it appears that heaven and hell exist only during a person's lifetime, it is unclear why certain living things are born malformed, which raises the question of whether good actions and sin happen. A number of the environmental problems that the world faces today can be solved by insects, according to experts. Unfortunately, it is also true that insects are becoming more and more vulnerable to extinction as a result of the growing climate crisis and the widespread and careless use of pesticides in agriculture, which is negatively affecting insects that are near farms and plantations. Many insects are viewed by people as pests. But insects are also highly significant for a variety of reasons.

References

Ahmad, M. & Dar, N. (2020). Rashtriya Krishi. Hind Agricultural Research and Training Institute. 15(1): 15-16.

Belén Maldonado, M., Aranibar, J. N., Serrano, A. M., Chacoff, N. P., & Vázquez, D. P. (2019). Dung beetles and nutrient cycling in a dryland environment. *CATENA. 179,* 66-73. https://doi.org/10.1016/j.catena.2019.03.035.

Belovsky, G. E. & Slade, J. B. (2000) Insect Herbivory Accelerates Nutrient Cycling and Increases Plant Production. *Proceedings of the National Academy of Sciences of the United States of America, 97*, 14412-14417. https://doi.org/10.1073/pnas.250483797.

Bornemissza, G. F. (1979). The Australian Dung Beetle Research Unit of Pretoria. *South African Journal of Sciences 75*, 257-260.

Byrd, J. H., & Castner, J. L. (2001). Forensic entomology. The utility of arthropods in legal investigations. CRC Press, New York, New York.

Davis, A. L. V. (1996) Seasonal Dung Beetle Activity and Dung Dispersal in Selected South African Habitats: Implications for Pasture Improvement in Australia. *Agriculture, Ecosystems and Environment, 58*, 157-169. https://doi.org/10.1016/0167-8809(96)01030-4.

De Groot, R. S., Wilson. M. A. & Boumans, R. M. J. (2002) A Typology for the Classification, Description and Valuation of Ecosystem Functions, Goods and Services. *Ecological Economics, 41*, 393-408. https://doi.org/10.1016/S0921-8009(02)00089-7.

Dicke, M. (2017). Ecosystem Services of Insects. In: Van Huis, A. and Tomberlin, J. K., Eds., Insects as Food and Feed: From Production to Consumption, Wageningen Academic Publishers, Wageningen, The Netherlands, 61-76.

Evans, T. A., Dawes, T. Z., Ward, P. R. & Lo, N. (2011). Ants and Termites Increase Crop Yield in Dry Climate. *Nature Communications*, 2, Article No. 262. https://doi.org/10.1038/ncomms1257.

Farwig, N., Brandl, R., Siemann, S., Wiener, F. & Muller, J. (2014). Decomposition rate of carrion is dependent on composition not abundance of the assemblages of insect scavengers. *Oecologia, 175,*1291–1300. https://doi.org/10.1002/9781118945568.ch2).

Farwig, N., Brandl, R., Siemann, S., Wiener, F. & Muller, J. (2014). Decomposition Rate of Carrion Is Dependent on Composition Not Abundance of the Assemblages of Insect Scavengers. *Oecologia, 175*, 1291-1300. https://doi.org/10.1007/s00442-014-2974-y.

Fincher, G. T. (1981). The potential value of dung beetles in pasture ecosystems. *Journal of the Georgia Entomological Society 16*,316-333.

Gallai, N., Salles, J. M., Settele, J. & Vaissiere, B. E. (2009) Economic Valuation of the Vulnerability of World Agriculture Confronted with Pollinator Decline. *Ecological Economics, 68*, 810-821. https://doi.org/10.1016/j.ecolecon.2008.06.014.

Garibaldi, L. A., Steffan-Dewenter, I., Winfree, R., Aizen, M. A., et al. (2013) Wild Pollinators Enhance Fruit Set of Crops Regardless of Honey Bee Abundance. *Science, 339*, 1608-1611. https://doi.org/10.1126/science.1230200.

Halffter, G., & Edmonds, W. D. (1982). The nesting behaviour of dung beetles (Coleoptera: Scarabaeidae). An ecological and evolutive approach. Publication of the Institute of Ecology, Mexico.

Halffter, G., & Matthews, E. G. (1966). The natural history of dung beetles of the subfamily Scarabaeinae. Folia Entomologica Mexicana. México 12-14. Reprint for Medical Books di C. Cafaro (1999). Palermo. Italy.

Hanski, I., and Cambefort, Y. (eds.). 1991. Dung beetle ecology. Princeton University Press, Princeton, New Jersey.

Hunter, M. D. (2001) Insect Population Dynamics Meets Ecosystem Ecology: Effects of Herbivory on Soil Nutrient Dynamics. *Agricultural and Forest Entomology, 3*, 77-84. https://doi.org/10.1046/j.1461-9563.2001.00100.x.

Jankielsohn, A. (2018). The Importance of Insects in Agricultural Ecosystems. Advances in *Entomology, 6*, 62-73. doi: 10.4236/ae.2018.62006.

Johnes, R. (2022). Does Antarctica have any insects? Discover Wildlife. BBC Wildlife Magazine. In: The Importance of Insects (Scudder, G. G. E., Foottit, R. G., and Adler, P. H. Eds.).

Losey, J. E. & Vaughan, M. (2006). The Economic Value of Ecological Services Provided by Insects. *BioScience, 56*, 311-323. https://doi.org/10.1641/0006-3568(2006)56[311:TEVOES]2.0.CO;2.

Macfadyen, S., Kramer, E. A., Parry, H. R. & Schellhorn, N. A. (2015) Temporal Change in Vegetation Productivity in Grain Production Landscapes: Linking Landscape Complexity with Pest and Natural Enemy Communities. *Ecological Entomology, 40*, 56-69. https://doi.org/10.1111/een.12213.

Mattson, W. J. & Addy, N. D. (1975). Phytophagous Insects as Regulators of Forest Primary Production. *Science, 190*, 515-522. https://doi.org/10.1126/science.190.4214.515.

Metcalfe, D. B., Asner, G. P., Martin, R. E., Silva Espejo, J. E., Huaraca Huasco, W., Farfan Amezquita, F. F., Carranza-Jimenez, L., Galiano Cabrera, D. F., Durand Baca, L., Sinca, F., et al. (2014) Herbivory Makes Major Contributions to Ecosystem Carbon and Nutrient Cycling in Tropical Forests. *Ecology Letters, 17*, 324-332. https://doi.org/10.1111/ele.12233.

Nichols, E., Spector, S., Louzada, J., Larsen, T., Amezquita, S. & Favila, M. E. (2008) Ecological Functions and Ecosystem Services Provided by Scarabaeinae Dung Beetles. *Biological Conservation, 141*, 1461-1474. https://doi.org/10.1016/j.biocon.2008.04.011.

Putman, R. J. (1983). Carrion and dung. The decomposition of animal wastes. Edward Arnold, London, United Kingdom.

Richards, E. N. & Goff, M. L. (1997) Arthropod Succession on Exposed Carrion in Three Contrasting Tropical Habitats on Hawaii Island, Hawaii. *Journal of Medical Entomology, 34*, 328-339. https://doi.org/10.1093/jmedent/34.3.328.

Samways, M. J. (1993) Insects in Biodiversity Conservation: Some Perspectives and Directives. *Biodiversity and Conservation, 2*, 258-282. https://doi.org/10.1007/BF00056672.

Scholtz, C. H. & Mansell, M. W. (2009) Insect Biodiversity in the Afrotropical Region. In: Foottit, R. and Adler, P., Eds., Insect Biodiversity: Science and Society, Blackwell Publishing, Hoboken, NJ, 69-82. https://doi.org/10.1002/9781444308211.ch5.

Schoonhoven, L. M., Van Loon, J. J. A. & Dicke, M. (2005). Insect-Plant Biology. Oxford University Press, Oxford, UK, 400 p.

Schwägerl, C. (2016) What's Causing the Sharp Decline in Insects, and Why It Matters. Yale Environment 360, 6 July 2016.

Tyndalebiscoe, M. & Vogt, W. G. (1991) Effects of Adding Exotic Dung Beetles to Native Fauna on Bush Fly Breeding in the Field. *Entomophaga, 36*, 395-401. https://doi.org/10.1007/BF02377944.

Van Lenteren, J. C. (2012) Internet Book of Biological Control. International Organization for Biological Control, Zürich, Switzerland. http://tinyurl.com/zk3rdrr.

Chapter 15

Bugs for Benefits: Harnessing the Nutritional and Environmental Potential of Edible Insects for Human Consumption

Ramkumar[1,*]
Sunaullah Bhat[2]
Vinay Kumar[3]
Manisha Chaudhary[4]
Krishna Kant Prajapati[5]
Varun Arya[1]
Ram Kewal[1]
and Ashish Kumar Singh[6]

[1]Department of Entomology and Agricultural Zoology, Institute of Agricultural Sciences, Banaras Hindu University, Varanasi, Uttar Pradesh, India
[2]Department of Zoology, Kumaun University-SSJ Campus, Almora, Uttarakhand, India
[3]Division of Fruits and Horticultural Technology, ICAR- Indian Agricultural Research Institute, New Delhi, India
[4]Department of Entomology, Chaudhary Sarwan Kumar Himachal Pradesh Krishi Vishwavidyalaya, Palampur, Kangra, Himachal Pradesh, India
[5]Department of Entomology, Chandra Shekhar Azad University of Agriculture and Technology, Kanpur, Uttar Pradesh, India
[6]Division of Crop Protection, ICAR-Vivekananda Parvatiya Krishi Anusandhan Sansthan, Almora, Uttarakhand, India

Abstract

The importance of finding hygienic food became a topic of discussion after COVID-19 was classified as a worldwide pandemic by the WHO. Human beings have consumed insects as a source of food, especially protein, since ancient times, and now they're getting more attention as a possible food source. An edible insect could become an environmentally friendly, sustainable source of protein in the future. Chitosan, a polysaccharide derivative of insect chitin and silk fibroin (produced by *Bombyx mori*), a high-quality protein, is used in various industries, viz., agriculture and biomedicine. The natural lustrous properties of silk fibres have made it a highly valuable product in the textile

* Corresponding Author's Email: rkm1997@bhu.ac.in.

In: Insects as Humans' Frenemies: From Friends to Foes
Editor: Istkhar Rao
ISBN: 979-8-89530-320-7

industry. Honeybee workers are considered master chemists, and their products have been utilised by humans for various purposes. Honey is the best natural sweetener as well as an excellent source of energy for humans, while beeswax is used for its moisture-proof properties. Pollen and bee venom also have therapeutic properties that can be used to treat various inflammatory conditions. Lac, a natural resin produced by a type of insect, is another material that has a wide range of applications in pharmaceuticals, cosmetics, and many industrial uses for surface coatings. Lac culture, the process of cultivating these insects, is a sustainable and environmentally friendly way of obtaining this valuable material. Overall, using insect by-products and insects themselves has the potential to provide sustainable solutions to global food shortages and various industries.

Keywords: food security, protein source, chitosan, chitin, silkworm fibres, gold ornaments

1. Introduction

Insects are an important food source for initial human societies as they are readily available and easy to collect (Kourimská & Adámková, 2016; Imathiu, 2020; Kim et al., 2022). Due to this, insects were primarily consumed as food at that time in tropical countries, viz., Africa, Latin America, and Asia, and insects play an essential role in their staple diet, gastronomies, and culture (Garofalo et al. 2019). Since ancient times (from the last 1000 years), mankind has been consuming insects, and currently, two billion people have consumed these in approx. 113 nations (Lesnik, 2018). Since ancient times (from the last 1000 years), mankind has been consuming insects, and currently, these insects are food for two billion people from approximately 113 nations (Lesnik, 2018). Countries with the greatest number of insects utilising them include Thailand, Mexico, India, and China (Baiano 2020). Historical documents show that the Aztecs ate a variety of insect species, and the emperor Moctezuma II had a fondness for ahuauhtli, which are fly eggs found on the surface of still water channels and were extensively consumed in the community market of Mexico, as observed by Spanish Conquistadors in 1557 (van et al., 2013).

In the US, 25–50% of native communities have been observed to consume insect species such as red ants and caterpillars (Anderson, 2006). In recent years, numerous startups emerged, resulting in the establishment of around 130 edible insect companies in nations like South Asia, Europe, and North America (Melgar-Lalanne et al., 2019). As a novel protein source, people are promoting edible insects due to the limitations of the current food supply system and the projected 9.7 billion population by 2050 (Post, 2012; Berners-Lee et al., 2018). Edible insects have a high content of fat, protein, and dietary fibre; therefore, they increase the nutritional content of a plant-based diet (Osimani et al., 2018). Adding cinereous cockroaches to bread increases its protein as well as unsaturated fatty acid content, improving its nutritional value (de Oliveira et al., 2017). Grasshopper defatted powder added to bread improves its dietary fibre, fat, and protein content (Haber et al., 2019), and supplementing bread with *T. monitor* enhances its free amino acid, protein, and EAA content (Roncolini et al., 2019). Adding grasshopper powder and mealworm supplements increases the protein content as well as the crude fibre of egg pasta (Çabuk & Yılmaz, 2020). Silkworm and cricket powder are utilised to boost the protein content of buckwheat as well as the calories of durum wheat pasta, respectively (Duda et al., 2019; Biró et al., 2019).

Insects have numerous uses, including the production of chitosan, which has antimicrobial and antioxidant properties that can be used in the food and medicine industries (El Ghaouth et al., 1992; Casimiro et al., 2017; Soon et al., 2018; Shirvan et al., 2019; Kurakula & Naveen, 2021). Silk fibroin, one of the insect products, has numerous uses in the restoration and reconstruction of nerve tissues, bone tissues, and dental tissue maintenance, as well as 3D printing (Craig, 2003; Marei et al., 2018; Biswal et al., 2022; Wei et al., 2021). Honey, bee wax, royal jelly, bee venom, propolis, and lac are all useful products of insects. Honey is a rich source of carbohydrates, vitamins, and minerals, and it has antibacterial properties that can enhance immunity and haematology and is beneficial for anaemia sufferers (Ali et al., 1991; Somal et al., 1994; Steinberg et al., 1996; Ajibola et al., 2007; Chepulis, 2007; Molan, 2001; Osato et al., 1999). Beeswax exhibits antibacterial properties against several bacterial strains and yeasts (such as *Candida albicans*) (Ghanem, 2011). Propolis, a viscous product used by bees as a defence substance, contains balsamic substances, resinous, pollen, gummy, essential oils, and waxes collected from various plants with anti-inflammatory, antibacterial, and antioxidant properties (Anjum et al., 2019). Royal jelly contains carbohydrates, vitamins, and minerals and has antibacterial properties (Xue et al., 2017; Matsuzaki et al., 1996; Xue et al., 2009; Zhou et al., 2012; Bărnuţiu et al., 2011). Bee venom has numerous biological properties and is used as a treatment for certain human diseases (Wehbe et al., 2019; Carpena et al., 2020). Lac is a sticky exudate that comes from a female scale insect's body and has been used in Vedic science.

2. Insects as a Source of Human Food

Insects have been a significant source of food for human societies since ancient times due to their easy accessibility compared to other animals (Kim et al., 2022). Currently, people from various countries and diverse cultural communities consume edible insects, with over 80% of people on the ground utilising insects directly or indirectly as food sources (Kim et al., 2019b). Entomophagy is more common in sub-torrid and torrid zones compared to temperate zones due to the respiratory system of the insects, which can be influenced by the environmental circumstances of a particular region (Kirkpatrick, 1957; van Huis et al., 2013).

Cooking techniques for edible insects include broiling, steaming, frying, boiling, drying, grilling, and other conventional techniques. Cooking can reduce the crude protein, zinc, and ash contents of edible insects and generally improve their flavour, shelf life, safety, and digestibility (Williams et al., 2016). Globally, the importance of edible insects has increased in recent years, and legislation must be set up to use insects as food sources for human beings globally (Liceaga et al., 2022). The European Union has updated several regulations regarding the feeding, searching, hygiene, and slaughtering processing of consumable animals, and several edible insects have been approved as food sources for consumption (Lähteenmäki-Uutela et al., 2021).

Nearly all the insect groups have edible species; for example, butterflies and moths (Lepidoptera), beetles (Coleoptera), bees, ants, and wasps (Hymenoptera), hemimetabolous insects such as bugs (Hemiptera), crickets, locusts, grasshoppers (Orthoptera), and termites (Isoptera) contribute to most species of edible insects (Van Huis et al., 2013). A list of commonly edible insects is given in Table 1. Ideal edible insects are larger, emerge in large quantities, and are non-toxic. Insects with toxins can be made safe by removing the inedible body parts and inactivating the toxins (thermolabile) by boiling and heating processes (Grabowski et al., 2022). Different developmental stages of edible insects can help identify the best stages for consumption with non-toxic compounds and high nutritional values, and inedible body parts, such as the head, guts, antenna, legs, and wings, may need to be removed when consumed fresh (Liceaga et al., 2022). Some species of insects have international importance because they can provide food for human consumption, such as the stingless bee (*Tribe meliponini*), Mexican grasshopper (*Sphenarium* spp.), African bush cricket (*Ruspolia differens*), and *Teleogryllus mitratus* (Neto & Ramos, 2006; Van-Huis, 2020; Magara et al., 2021). Some locust species, like *Locust migratoria* and *Schistocerca gregaria*, are traditionally consumed during invasion (Tchibozo & Lecoq, 2017).

2.1. Nutritional Value of Edible Insects

Edible insects have a high nutritive value in terms of fat and protein content; the promotion of their inclusion in traditional foods can support this (Kim et al., 2022). Edible insects have conspicuous nutritional values (e.g., unsaturated fats, high-quality protein, minerals, and vitamins). Complete information related to the nutritional profile of 236 insects has been provided by Rumpold and Schlüter (2013). There are low environmental costs associated with the production of protein from edible insects as well as a high nutritional value (Oonincx & de Boer, 2012; Kim et al., 2019b; Bosch et al., 2019). The nutritious quality and composition of edible insects contain mostly protein with all the essential amino acids, accounting for 20 to 76% of fresh dry matter; chitin (2.7 to 49.8 mg); 2 to 50% of dry matter are lipids; and some mineral nutrients are K, Mg, P, and Ca (Womeni et al., 2012). The protein proportion of an edible insect depends upon the life cycle of the insect, the type of species, and the processing or preparation (e.g., mechanical treatments and thermal) (Van Huis, 2013; Baiano, 2020). For example, Longvah et al. (2011) demonstrated that mealworms (*Tenebrio molitor*) and eri silkworms (*Samia ricinii*) have different protein profiles depending on their developing phases (i.e., pre-pupae and pupae).

Proteins from edible insects comprise all essential amino acids and are suitable for digestibility. Rat models fed with insect dietary protein indicate that faecal N-levels showed protein digestibility of 86% to 90%, which is more than the plant proteins and lower than the animal proteins (casein, 99%; beef, 98%; egg, 95%) (Finke, 2013). The World Health Organisation and FAO can establish the exceeded requirements of amino acids from cricket and silkworm larvae (Hall et al., 2017; Wu et al., 2011).

Additionally, in addition to their excellent nutritional quality, insects are also admirable sources of bioactive peptides, as confirmed by enzymatic proteolysis. Bioactive peptides may be included in therapeutic or functional food products, and by enhancing physiological processes, they may promote health (Liceaga et al., 2022). Pepsin and pancreatin were used to enzymatically digest proteins to reveal the antioxidant properties of a variety of edible insects (Zielinska et al., 2017) (such as in *Amphiacusta annulipes*, *Gromphadorhina portentosa*, *Zophobas morio*, *Locusta migratoria*, and *Blaptica dubia*). Identification of anti-inflammatory, antihypertensive, and anti-glycemic peptide properties from the cricket (*Gryllodes sigillatus*) peptide by the gastrointestinal digests (Hall et al., 2020). As per their study, the cationic fraction can convert alpha-amylase, alpha-glucosidase, and angiotensin-converting enzymes in an *in vitro* study, and it inhibits the appearance of macrophages and NF-κB. Tao and their co-worker (2017) identified and purified peptides from silkworm (*Bombyx mori*) like GNPWM, pupal hydrolysate protein, having angiotensin-converting enzyme (ACE) inhibition activity.

Table 1. List of edible insects

Sl. No.	Common name of insects	The scientific name of Insects	Family	Order	Reference
International and National Importance					
1	Migratory locust	*Locusta migratoria*	Acrididae	Orthoptera	Althwab et al., (2021); Kwak et al., (2021); Kim et al., (2022)
2	House cricket	*Acheta domesticus*	Gryllidae	Orthoptera	Osimani et al., (2018); Gupta et al., (2020); Kim et al., (2022)
3	Lesser mealworm	*Alphitobius diaperinus*	Tenebrionidae	Coleoptera	Grabowski et al., (2022); Rumbos et al., (2019)
4	House fly	*Musca domestica*	Muscidae	Diptera	Grabowski et al., (2022)
5	Black soldier fly	*Hermetia illucens*	Stratiomyidae	Diptera	Grabowski et al., (2022); Lawal et al., (2021)
6	Desert locust	*Schistocerca gregaria*	Acrididae	Orthoptera	Haber et al., (2019); Kim et al., (2022); Lecoq, (2022)
7	Tropical banded crickets	*Gryllodes sigillatus*	Gryllidae	Orthoptera	Grabowski et al., (2022); Hall, (2020)
8	Mealworms	*Tenebrio molitor*	Tenebrionidae	Coleoptera	Lawal et al., (2021); Roncolini et al., (2019); Azzollini et al., (2018)
9	Death's head hawk	*Acherontia atropos*	Sphingidae	Lepidoptera	Grabowski et al., (2022); Blackford & Dinan, (1997)
10	Field cricket	*Gryllus assimilis*	Gryllidae	Orthoptera	da Rosa Machado and Thys (2019); Weissman et al., (2009)
11	Pigmy emperor	*Rohaniella pygmaea*	Saturniidae	Lepidoptera	Grabowski et al., (2022)
12	Convolvulus Hawkmoth-moth	*Agrius convolvuli*	Sphingidae	Lepidoptera	Grabowski et al., (2022); Nirazawa et al., (2017)
13	Fulvous hawk	*Coelonia fulvinotata*	Sphingidae	Lepidoptera	Grabowski et al., (2022)
14	Arrow Sphinx	*Lophostethus dumolinii*	Sphingidae	Lepidoptera	Grabowski et al., (2022)
15	Comms nephele hawk moth	*Nephela comma*	Sphingidae	Lepidoptera	Grabowski et al., (2022)
16		*Platysphinx stigmatica*	Sphingidae	Lepidoptera	Grabowski et al., (2022)
17	Red-veined dropwing	*Trithermis arteriosa*	Sphingidae	Lepidoptera	Grabowski et al., (2022)
18	Dragonflies	*Anax imperator*	Aeshnidae	Odonata	Liceaga et al., (2022); Minot et al., (2021)
19	Silkworm	*Bombyx mori*	Bombycidae	Lepidoptera	Biró, (2019); Park et al., (2017); Tojo et al., (1980); Chen et al., (1998)
20	Grasshopper	*Sphenarium purpurascens*	Pyrgomorphidae	Orthoptera	Ramírez-Rivera et al., (2021); Rodríguez-Miranda et al., (2019); Cerritos, and Cano-Santana, (2008)

Table 1. (Continued)

Sl. No.	Common name of insects	The scientific name of Insects	Family	Order	Reference
21	Apis spp.	*Apis cerana indica, Apis andreniformis, Apis dorsata, Apis lorea*	Apidae	Hymenoptera	Chen et al., (1998); Liceaga et al., (2022)
22	Cockroaches	*Blattella germanica, Periplaneta americana, B. orientalis*	Blattodea	Dictyoptera	Qiu et al., (2020); Liceaga et al., (2022)
23	Cicadas	*Cosmopsatria sp. Dtmdula sp. Rihana sp.*	Cicadjdae	Homoptera	Chen et al., (1998); Liceaga et al., (2022)
24	Stink bugs	*Tesseratoma javanica*	Pentatomidae	Heteroptera	Chen et al., (1998); Liceaga et al., (2022)
25	Damselflies (Stream glory)	*Neurobasis chinensis*	Calopterygridae	Odonata	Kalita et al., (2014); Liceaga et al., (2022)
26	Ant	*Carebara lignata Oecophylla smaragdina*	Formicidae	Hymenoptera	Chen et al., (1998); Liceaga et al., (2022)
27	Wasps	*Vespa affinis, Vespa mandarinia, Vespa tropica*	Vespidae	Hymenoptera	Chen et al., (1998); Liceaga et al., (2022)
28	Scarab beetle	*Oryctes rhinocerus Psilophosis sp. Xylotrupes gideon*	Scarabaeidae	Coleoptera	Chen et al., (1998)
29	Bamboo borer	*Omphisa* sp.	Pyralidae	Lepidoptera	Chen et al., (1998)
30	Powder post Termite	*Kalotermes flavicollis*	Kalotermiridae	Isoptera	Chen et al., (1998); Liceaga et al., (2022)

3. Chitosan

Chitosan is obtained from chitinous insects and other organisms, like marine annelids and microorganisms. Some important sources of chitosan are shown in Table 2. The amount of deacetylation reaches approximately 50%, and chitin may become soluble in acidic aqueous solutions, depending on its origin. This substance is known as chitosan (Rinaudo, 2006). Chitosan is generated from chitin, a line-like copolymer made of N-acetyl-D-glucosamine and D-glucosamine units joined together by glycosidic bonds β-(1 → 4). Chitosan is a natural polysaccharide that is biodegradable, biocompatible, and non-toxic with anti-oxidative and antimicrobial characteristics (Soon et al., 2018). Chitin is a polysaccharide chain made of highly combined groups of hydroxyls, amino acids, and acetyls with hydrogen bonds. So, standard solvents cannot dissolve chitin, making it unusable (Philibert et al., 2017). After deacetylation, it may be converted into chitosan; it is excellent for usage in various industries and is easily soluble in common solvents. (Shirvan et al., 2019). Only chitosan is a pseudo-natural cationic polymer. Chitosan is solubilised by the protonation of NH_2 activities on the C-2 place of the D-glucosamine repeat unit. Chitosan solution properties depend not only on the average hydroxyl and amino group but also on the distribution of acetyl groups throughout the main chain as well as the molecular weight (Rinaudo & Domard, 1989; Aiba, 1991; Kubota et al., 1997). Due to this, the presence of acetic acid can examine the role of the protonation of chitosan (Rinaudo et al., 1999a) and HCl on solubility and reveal that the pK and pH of the acid affect the degree of ionisation (Rinaudo et al., 1999b).

3.1. Use of Chitosan

3.1.1. Food Industries

Chitosan has antimicrobial properties, so it is used for the packing of food materials for a long period. It is effective against strawberry pathogens like *Rhizopus stolonifera* and *Botrytis cinerea* (El Ghaouth et al., 1992). Presently, innovative nanoscience approaches are being developed for packaged food applications. Comparison between the chitosan film made from shrimp (SCSF) and *Periplaneta americana* (PaCSF). Both films are examined by microscopic and visual methods and look similar, but the PaCSF film has better anti-oxidative and antimicrobial activities, making it a potential replacement for the packaging of food (Chen et al., 2021). In addition, fresh, innovative chitosan-based products have been developed, and they exhibit outstanding physical properties, mainly their porous design and antimicrobial functions. Black soldier flies were also used by scientists to make an industrial chitosan film. They contrasted natural chitosan-only film with semi-synthetic chitosan film made of glutaraldehyde and glycerol (Figure 1).

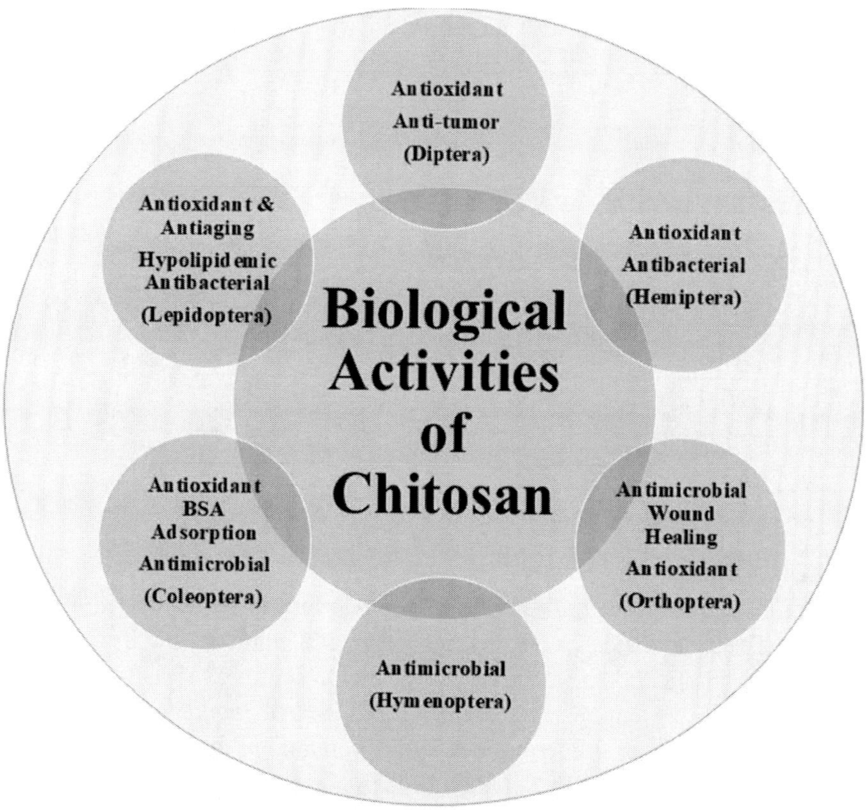

Figure 1. Biological activities of insect's chitosan on human.

Table 2. List of insects used as a source of chitosan

Insects (Common name)	Scientific name	Family	Order	References
Soybean hawkmoth	*Clanis bilineata*	Sphingidae	Lepidoptera	Wu, (2011); Gao et al., (2021); Ma et al., (2022)
Silkworm moth	*Bombyx mori*	Bombycidae	Lepidoptera	Luo et al., (2019); Bastiaens et al., (2019)
Mealworm Beetle	*Tenebrio molitor*	Tenebrionidae	Coleoptera	Shin et al., (2019)
Rhinoceros Beetle	*Allomyrina dichotoma*	Tenebrionidae	Coleoptera	Shin et al., (2019); Rinaudo, (2006)
Superworm	*Zophobas morio*	Tenebrionidae	Coleoptera	Soon et al., (2018); Rumbos & Athanassiou, (2021)
Colorado potato beetle	*Leptinotarsa decemlineata*	Chrysomelidae	Coleoptera	Kaya et al., (2014)
Dung beetle	*Catharsius molossus*	Scarabaeidae	Coleoptera	Ma et al., (2015); Liu et al., (2011)
House fly	*Musca domestica*	Muscidae	Diptera	Kim et al., (2016); Bastiaens et al., (2019)
Black soldier fly	*Hermetia illucens*	Stratiomyiidae	Diptera	Khayrova et al., (2020); Bastiaens et al., (2019)
Two-spotted field crickets	*Gryllus bimaculatus*	Gryllidae	Orthoptera	Kim et al., (2017)
Short-horned grasshopper	*Decticus verrucivorus*	Tettigoniidae	Orthoptera	Sani et al., (2017)
Moroccan locust	*Dociostaurus maroccanus*	Acrididae	Orthoptera	Latchininsky, (1998); Erdogan and Kaya, (2016)
Italian bee	*Apis mellifera*	Apidae	Hymenoptera	Nemtsev et al., (2004); Bastiaens et al., (2019)
Black ant	*Camponotus compressus*	Formicidae	Hymenoptera	Rinaudo & Domard, (1989); Ramakrishanan and Raja, (2020); Rinaudo, (2006)
Black cicada	*Cryptotympana atrata*	Cicadidae	Hemiptera	Wu et al., (2013)
European Cicadas	*Cicada mordoganensis*	Cicadidae	Hemiptera	Luo et al., (2019)
Water scorpion	*Ranatra linearis*	Nepidae	Hemiptera	Schaefer & Panizzi, (2000); Kaya et al., (2014)
American cockroach	*Periplaneta americana*	Blattidae	Dictyoptera	Rinaudo & Domard, (1989); Kaur and Dhillon, (2015)
Tebo worms (Chilean moth)	*Chilecomadia moorei*	Cossidae	Lepidoptera	Finke, (2013); Bastiaens et al., (2019)
Tobacco hornworm	*Manduca sexta*	Sphingidae	Lepidoptera	Holl et al., (1992); Bastiaens et al., (2019)
Butterfly	*Pieris spp.*	Pieridae	Lepidoptera	Kaur and Dhillon, (2015)

3.1.2. Medicinal Industries

Chitosan from insects has been utilised in the medical industry for wound healing since 1990 (Khor, 2014). The two major characteristics of insect chitosan are its antibacterial and biocompatibility activities. It can be used in biomedical research with medical tools, particularly bone replacement (Kim, 2010; Khor, 2014) and implants (Khor & Lim, 2003). Chitosan from insects was examined to identify its potential as an antihypertensive and anti-inflammatory agent (Nwe et al., 2011). Additionally, insect chitosan can be recognised as a stimulating molecule for drug transportation. (Blue). Nanomaterials of chitosan might be used for the encapsulation of several functional molecules for the treatment of cancer (Felt et al.,

1998) and nucleic acid-based and antiviral drugs (Vivek et al., 2013). Chitosan nanoparticles could be used for cancer treatment (Shanmuganathan et al., 2019).

3.1.2.1. Using for Wound Healing

It is one of the essential biological processes that includes the development of tissues and tissue regeneration. It depends on a series of connected processes in which various cellular components work together to rebuild or repair damaged tissues (blue). Liu et al. (2018) found that chitin and chitosan in hydrogels increase wound healing. Alternative methods for treating chronic and acute skin wounds include hydrogels. However, several reaction pathways, like aggregation, haemostasis, cell redevelopment, and propagation, are essential for wound healing (Figure 2) (Marei et al., 2017; Shirvan et al., 2019).

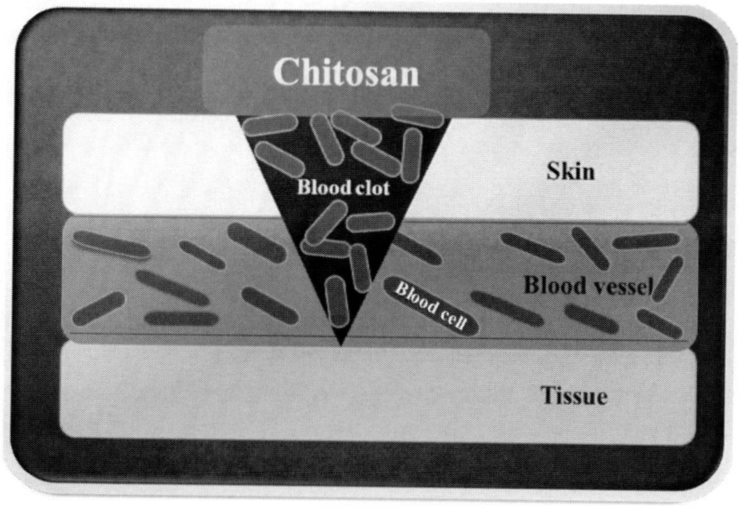

Figure 2. Chitosan dressing for wounds can stop blood loss.

3.1.2.2. Bone Regeneration

Biomaterials are used with bioactive and biomimetic molecules; it is a novel idea for bone regeneration that offers many advantages to conventional prosthetic implants (Guarino et al., 2020). Chitosan has been successful in bone grafting, improving surfaces, filming, and using electrospraying techniques to release bioactive medications one at a time or onto orthopaedic plates (Kurakula & Naveen, 2021).

3.1.2.3. Use of Chitosan in the Regeneration of Tissue

Insect chitosan as well as its biomimetic derivatives have been used in damaged and diseased tissue regeneration (tissue engineering) and organs as they maintain a biological microenvironment throughout plastic surgery (Casimiro et al., 2017). Chitosan has been blended with synthetic and natural composites, such as three-dimensional lyophilised scaffolding hydrogels and coatings, to enhance its capabilities. Besides, the procedure of tissue generation can be enhanced by inserting drugs that modify tissue generation and cellular behaviour (Sivashankari & Prabaharan, 2016).

4. Silk Fibroin

Silk is defined as the fibre spun by the silkworm moth *Bombyx mori* larvae and collected from the pupal cocoon of insects by unravelling. However, other fibres with distinct characteristics, including those made by spiders, mites, and caddisflies, are also included under the umbrella term "silk." These traditional animal fibres, which include those produced by Hymenoptera, Lepidoptera, Diptera, Acarids, Neuroptera, and Araneids, have several characteristics in terms of their biophysical and biochemical composition. (Craig, 2003). *B. mori* can produce huge quantities of silk worldwide through commercial production. The quality of silk varies from species to species; spider silk is stronger and more extensive than any silkworm silk. (Vollrath & Porter, 2006) Textile industries can utilise silkworm fibres as natural materials (Kaplan & Vepari, 2007; Craig, 2003).

Silk fibroins are composed of fibrous proteins and are usually rich in amino acids such as alanine, serine, and glycine (Shimura, 1983). There are two chains, a light chain of 25 kD and a heavy (H) chain of 350 kD, connected with disulphide bonds (Gamo et al., 1977; Shimura et al., 1982). As for silk fibroin's light chain, it is nonfibrous and contains an abundance of the acidic amino acids leucine, isoleucine, and valine (Shimura, 1983). The genes responsible for heavy and light chains are located on dissimilar chromosomes on the posterior side of silk gland cells (Hyodo et al., 1980; Doira, 1983; Hyodo et al., 1984; Kimura et al., 1985). The posterior cells of the silk glands synthesise large amounts of silk fibroin during the larval 5th instar, which is then released into the lumen of the gland and kept there in a liquid condition until the end of the fifth instar when it is spun out (Takei et al., 1987). Therefore, the efficient mechanism cannot be clarified for the intercellular secretion, maintenance, and transport of large fibrous proteins in a liquid state in both the cells and lumen of the posterior silk gland (Takei et al., 1987).

4.1. Utilisation of Silk Fibroin

Recently, silk-based materials have found use in an expanding range of fields, including tissue engineering products, biodegradable medical scaffolds, and implanted functional devices.

4.1.1. Medical Purposes

4.1.1.1. Nervous Tissue System

The nervous system in any organism is the most complex part of its body. A very changing and sensitive application of silk fibroin is the restoration and reconstruction of nerve tissues in the CNS (Marei et al., 2018). Hydrogel is a polymer that can evolve as an admirable transport medium to overcome CNS problems and the blood-brain barrier (BBB). For transporting micro-molecules like drug molecules, peptide fragments, siRNA, and stem cells into the brain, the medium of the hydrogel can help by encapsulating molecules and bypassing the blood-brain barrier (BBB). Some polymers, like chitosan, polypyrrole (PPy), polyaniline (PANI), and polyethylene glycol (PEG), are easily available on the market and have outstanding efficacy in treating central nervous system (CNS) dysfunctionality.

Biopolymers performed efficiently in two ways (Biswal et al., 2022). First, they can be provided with the supporting base to insert the micromolecule in the brain. Second, they have pro-resurgent characters among them. The spinning of non-mulberry silk fibroin by electrospinning with PEG and PVA incorporation exhibits outstanding results during brain tissue rebuilding. Due to the bioactive characteristics of non-mulberry silk fibroin, silk fibroin is capable of providing support for accelerating cell development, interaction, and proliferation; it also helps in the disintegration of the polymeric surface (Boni et al., 2018). Additionally, fibroin from non-mulberry silk has a triple peptide motif called RGD that binds to cells but is absent from mulberry silk. Triple-peptide RGD influences the improvement of cell compatibility and cell affinity in biomaterials. For the repair of injured nerve tissue, 3-dimensional nanoscaffolds (3DNSs), tissue constructed with polyethylene glycol (PEG), polyvinyl alcohol (PVA), and SF are being utilised. The features of the 3-DNSs are significantly similar to the mixture of polymeric concentrations and their formation by the electro-spinning process (Biswal et al., 2022).

4.1.1.2. Muscle Tissue System

Muscle tissue injury is a common issue caused by external and internal factors. Muscle tissues are divided into three classes: smooth muscle, skeletal muscle, and cardiac muscle (myocardium). These muscle tissues are located in different body parts, like skeletal muscle cells on the skeleton, smooth muscle tissues on the surface of the hollow internal parts of the body like the intestines, pancreas, and liver, and cardiac muscle tissue on the heart surface. For all types of injured muscle tissues, silk fibroin is a suitable biological material. Silk fibroin-based scaffolds exhibit optimal proficiency in cardiac-muscle tissue injury (Patra et al., 2012). In cardiac muscle tissues, silk fibroin-based scaffolds can help to increase cell cycle regulation, cellular metabolism, and cell signalling (Patra et al., 2012). Electrospinning is used to process the silk fibroin nanofibres; it effectively suppresses myocardial infarction with an interruption of numerous manners (Biswal et al., 2022).

4.1.1.3. Bone Tissue System

During bone tissue engineering, the main advantage of using HFIP with adipose-derived stem cells (ASCs) on silk fibroin-based scaffolds for bone tissue creation is that the combination enhances alkaline phosphatase (ALP) activity and calcium absorption. When preparing scaffolds and arranging bone lamellae, the freezing-drying manufacturing method displays a distinct result. Methanol (CH_3OH) inclusion on the scaffolds has an impact on how the beta-sheet develops. Silk fibroin scaffolds (SFS) made of aqueous solution show better osteogenic breakdown results than HFIP-based silk fibroin scaffolds when human mesenchymal stem cells (hMSCs) are placed on them (Biswal et al., 2022).

In terms of cellular-based tissue regeneration treatment for bone tissue remodelling, human adipose-derived stromal cells (hASCs) show potential alternative features to bone marrow-derived MSCs. The seamless separation of the hASCs from adipose-derived tissues results in a larger production of cells per unit of tissue volume. Additionally, hASCs have specific traits that make them more effective at promoting bone formation than MSCs found in bone marrow, such as their ability to multiply quickly. Tissue repairing techniques produce great results using hASCs and scaffolds based on silk fibroin. Additionally, it has structural potential in terms of size, porosity, and geometry of pores, which provides a great biomechanical result for the application of bone tissue remodelling (Biswal et al., 2022).

4.1.1.4. Dental Tissues System

The most difficult clinical problem is fixing tooth pulp. Presently, therapy for a root canal is the oldest method of dealing with dental problems like periapical periodontitis and irreversible pulpitis (Wei et al., 2021). The previous standard of care was to completely remove the diseased dental pulp; however, this caused additional contamination of the dental pulp, which not only caused a person to lose a tooth and their ability to chew but also hurt their overall appearance (Biswal et al., 2022). Studies revealed that the incorporation of silk fibroin scaffold with basic fibroblast growth factor is a very promising scaffold option for upcoming treatment approaches in regenerative endodontics to preserve teeth.

4.1.2. Printing Industry

Silk fibroin is used for making the bioink, which can be used for 3D (three-dimensional) bioprinting. For 3D printing of organlike construction and cell-laden materials, which can be used in 3D bioprinting, it is suitable and gives better results. At present, the 3D bioprinting method has steadily gained ground as an unparalleled production technique that can create cell-laden 3D structures by closely replicating tissue and organ function (Biswal et al., 2022).

5. Honeybee Products

Honeybees serve a critical role in the livelihood of humankind as they are one of the most important pollinators of flowering crops. During foraging, while collecting rewards from flowering plants, honeybees share a great contribution in horticulture and agriculture towards the enhancement of fruits and seed sets. These tiny insects, i.e., honeybees, have very sophisticated organs hidden inside their bodies that produce honey, royal jelly, wax, and propolis. Honeybees produce these products through a sophisticated machinery mechanism that involves regurgitation, enzymatic activity, and water evaporation. The sole purpose of preparing these products is to feed the colony (consisting of three castes; workers, drones, and a queen), hive construction, colony maintenance, etc. However, these honeybee products are highly beneficial to mankind due to their medicinal as well as high nutritional values. Here are some of the important bee products:

5.1. Honey

Honey, being a sticky and viscous substance, is a delicious food used by mankind since antiquity. Honey is prepared by honeybees to feed the colony, utilising pollen and nectar collected from flowers through a well-established intelligent mechanism. While foraging for food, bees collect nectar from flowers and store it in the honey sac (the glandular system in bees having enzyme invertase) (Crane, 1990; Saralaya et al., 2021). This disaccharide sugar (i.e., sucrose) present in nectar is broken down by enzyme invertase into two monosaccharides, i.e., fructose and glucose (Suarez et al., 1996). When the honey sac is filled with nectar, the bees return to the hive along with pollen in their pollen basket and release the nectar as well as pollen to the honeycomb cells. Then young honeybees concentrate on nectar by removing water from it using two intelligent and laborious methods. First, the young bees pass the nectar from one bee to another and drink out the water of the nectar by absorbing it through their stomach wall (Binkley, 2014). The other approaches honeybees use to concentrate nectar are by generating heat and airflow in the hive, vibrating their wings and flight muscles, and allowing the nectar in the cells to evaporate.

Fully matured honey, containing almost 17.8% water, is stored in the comb cells by sealing the cells with a thin layer of bee wax. Sealed honey does not spoil even after 1000 years, as most microorganisms do not grow in honey (Jones, 1990). The main content of honey is carbohydrates with traces of vitamins, minerals, and antioxidants. The carbohydrates include fructose (38%), glucose (31%), water (17%), maltose (7%), and traces of trisaccharide and sucrose. The traces of vitamins include A, B1, B2, B3, B5, B6, B12, C, D, E, and K. Minerals contain calcium, copper, iron, magnesium, manganese, phosphorus, potassium, sodium, and zinc (White & Doner, 1980).

Honey has good medicine properties in immunity and haematology and is also beneficial to anaemia-stricken people (Ajibola et al., 2007). It enhanced the haemoglobin content in blood and erythrocyte numbers and raised the hematocrit in people who ate honey (Chepulis, 2007). It has antibacterial properties and is used to prevent bacterial problems in teeth (Molan, 2001; Steinberg et al., 1996). The gastroprotective effect of honey has been established. *Helicobacter pylori*, which causes peptic and gastritis ulcers, is effectively inhibited by it (Ali et al., 1991; Somal et al., 1994; Osato et al., 1999). Honey can also be used to treat eye diseases. The earliest people used Attica and Indian locust honey for eye disease treatments (Beck & Smedley, 1944; Fotidar & Fotidar, 1945).

5.2. Beeswax

Beeswax, a complex product, is secreted in liquid form by honeybees from a special wax gland present in the abdomen of younger worker bees (nurse bees, which are 12–18 days old). When this wax comes into contact with air, it solidifies and is used by the nurse bees in the construction of honeycombs (Bogdanov, 2004). Bee wax resists the action of acids as well as gastric juices and is insoluble in cold alcohol and water, but it is partially soluble in boiling alcohol and completely soluble in chloroform, carbon disulphide, and hot turpentine (Bogdanov, 2016).

Beeswax is a complex mixture (containing more than 300 active components) of hydrocarbons, esters of fatty acids, free fatty acids, fatty alcohol, diesters, and exogenous substances (Tulloch, 1980). The pharmaceutical use of beeswax dates back to ancient Egypt. Beeswax was the key ingredient in many formulas used for the preparation of ointments, which were used to treat burnt skin, help pull plugs, heal wounds, and calm joint pain. The father of medicine, Hippocrates, recommended the use of beeswax in cases of purulent tonsillitis. In ancient Rome, many doctors of the time used to apply a cream known as cold cream, which contained olive oil, beeswax, and rose water, for the treatment of burns, wounds, cuts, bruises, and fractures. During the past decade, the antimicrobial activity of natural products, especially by-products of the honeybee, has gained importance, and unlike other bee products, beeswax has only recently been studied. Crude beeswax showed antibacterial activity against several bacterial strains and the *Candida albicans* yeast (Ghanem, 2011). The utilisation of a mixture of beeswax, olive oil, and honey is very effective for the treatment of diaper dermatitis, psoriasis, and atopic dermatitis (Al-Waili et al., 2003).

5.3. Propolis

Propolis is a viscous product used by bees as a defence substance composed of resinous, gummy, and waxes, essential oils, balsamic substances, and pollen of honeybees collected from various plants that contain plentiful compounds with biological functions like antibacterial, antioxidant, and anti-inflammatory, among others. These properties make propolis the perfect component to be used as a functional ingredient in foods, beyond its known use as an additive. The composition of propolis varies according to the plant and geography. However, its overall composition remains almost unaffected. Resins and balsams are the major components (45–55%), followed by waxes (8–35%), 5–10% essential oils, 5% aromatic acids, 5% pollen, and 5% organic and mineral substances (Anjum et al., 2019).

Propolis has antimicrobial activities against a large number of microorganisms, viz., bacteria, viruses, parasites, fungi, and yeasts (Saeed et al., 2016). Many human pathogenic microorganisms, like adenovirus, herpes simplex virus type 1, influenza virus, parainfluenza virus, and coronavirus (SARS-CoV-2) (Berretta et al., 2020), or bacteria causing ulcers, are present in skin injuries, and food products against them propolis may be effective. Most of the studies agree that propolis has strong activity against Gram-positive bacteria but only limited activity against Gram-negative bacteria (Anjum et al., 2019). The antibacterial activity of propolis, mainly attributed to the flavonoids, phenolic acids, aldehydes, ketones, and terpenes, can be explained by several mechanisms, such as the alteration of membrane permeability, inhibition of protein synthesis, and bacterial motility. (Irigoiti et al., 2021; Vasilaki et al., 2019).

5.4. Royal Jelly

Royal jelly, a white-yellowish gelatinous-viscous substance, is secreted by the hypopharyngeal gland of nurse honeybees aged between 5 and 15 days. It has a slight characteristic smell of phenol that gives it its flavour (Fujita et al., 2013). One of the strongest beliefs is that all bee larvae are fed with royal jelly. The duration of feeding a female larva with royal jelly determines the development of the larva into a reproductive queen or sterile worker. All the female larvae were fed with royal jelly for 3 days (hatching to the third day of life) during their larval development, followed by honey and beebread, causing the development of sterile workers. On the other hand, the female larvae that have been selected to become queens by worker bees are fed with only royal jelly until the 5th day of the larval stage, and the one that becomes queen is exclusively fed with royal jelly throughout her life (Gunes & Gungormus, 2022).

Water constitutes about 60–70% of royal jelly, and it is acidic with a pH range of 3.6–4.2. Carbohydrates contribute 7.5–15% of royal jelly, of which 90% is made up of fructose and glucose, whereas sucrose contributes 0.8–3.6%; the remaining carbohydrate portion comprises traces of maltose, trehalose, melibiose, ribose, and erlose. Lipids contribute 7–18% of the total and consist of a small chain of carbon atoms (8–12) called hydroxy fatty acids and dicarboxylic acids; however, 90% of the lipid components present in royal jelly are rare (Xue et al., 2017). Among the vitamins (A, B1, B2, B5, B6, B8, B12, C, and E), B5 is the most abundant, with 52.8 mg/100g (Xue et al. 2009; Zhou et al. 2012; Bărnuţiu et al. 2011). Mineral constituents are made up of 1.5% royal jelly, which includes various elements such as potassium, sodium, magnesium, calcium, phosphorous, sulphur, copper, iron, zinc, aluminium, barium, strontium, bismuth, cadmium, mercury, lead, tin, tellurium, thallium, tungsten, antimony, chromium, manganese, nickel, titanium, valium, cobalt, and molybdenum. Flavonoids in the royal jelly contain flavanones (isosakuranetin, hesperetin, and naringenin), flavones (apigenin, acacetin, chrysin, glucosides, and luteolin glucosides), flavonols (kaempferol glucosides and isorhamnetin), and soflavonoids (genistein, formononetin, and coumetarol), etc. (Kocot, 2018). The antibacterial function of royal jelly has been established as early as eight decades ago. *In-vitro* experiments demonstrated the antibacterial effect of royal jelly against four bacterial strains, such as *Staphylococcus aureus*, *Eberthella typkosa* (Rawlin's strain), *Escherichia coli*, and *Bacillus metiens* (Matsuzaki et al., 1996).

5.5. Bee Venom

Among the many insects, only very few of them have the defence tactics of having a sting, and venom injection is one of them. Sting insects belong to the Hymenoptera group. As sting is believed to have evolved or been modified from the egg-laying apparatus of the hymenopteran, only female bees can sting. The sting apparatus is always present at or near the end of the abdominal segment, rather than the head. It is not a bite from a honeybee that causes pain, but the sting of a female worker bee when the honeybee is defending its colony.

Bee venom (also called apitoxin) is secreted by the abdominal gland of the worker honeybee. It is a transparent, acidic liquid that is odourless, which honeybees use as a defence technique against predators. It is a combination of different compounds. As reported, bee venom comprises numerous active biological compounds, viz., peptides and enzymes, including melittin (one of the major ingredients of bee venom), mast cell degranulating peptide, adolapin, apamin, and enzymes (phospholipase A2 and hyaluronidase), as well as non-peptide components such as histamine, dopamine, and norepinephrine (Wehbe et al., 2019). Considering the wide range of biological properties of bee venom, it would not be surprising that its use could reach therapeutic purposes for human diseases. Venom therapy, like many other complementary and alternative medicine approaches, has been used to treat many diseases (Carpena et al., 2020).

6. Lac

Lac is a sticky resinous exudation secreted by the female scale insect, *Kerria lacca* (Hemiptera: Kerridae). The secretion protects these scale insects from adverse environments. Considering its benefits, lac is of great significance to mankind (Kadam, 2014). India is the largest producer of lac in the world. Lac is widely cultivated in different states of India on several host plants. Tribal communities and forest dwellers are involved in lac cultivation in India. The north-eastern region, particularly Assam, is one of the biodiversity hotspots, with approximately 4000 species of plants, but the exploitation of this natural flora for commercial production of lac is very limited. The suitable hosts for lac cultivation include palas (*Butea monosperma*), ber (*Ziziphus mauritiana*), kusum (*Schleichera oleosa*), *Flemingia semialata*, *Ficus* spp., etc. (Sharma, 2017). Tender twigs of the host plant are preferred by female-scale insects. These scale insects are beneficial in terms of the commercial cultivation of lac.

Lac insects are generally bivoltine and produce rangeeni and kusmi strains, which are classified based on the insect's preference for specific host plants. Raw lac is the source of three valuable products, i.e., resin, dye, and wax. The major component of lac is resin (65%), followed by two important by-products, viz. lac dye (1%) and lac wax (5–6%) (Sharma et al., 2020). Lac resin is thought to be a polyester complex of straight-chain hydroxy fatty acids of the C14-C18 carbon chain (e.g., aleuritic acid, butolic acid), mono- and di-hydroxy acids. Lac dye is a mixture of anthroquinoid derivatives and contains erythro-laccin and laccaic acid (Kadam, 2014). Among the eight native host plants of lac insects (*Flemingia semialata*, *Flemingia strobilifera*, *Indigofera teysmannii*, *Ficus religiosa*, *Zizyphus mauritiana*, *Litchi chinensis*, *Hibiscus rosa-sinensis*, and *Cajanus cajan*), *F. semialata* showed the highest productivity of brood lac (0.30 kg) and the longest life cycle duration (164.5 ± 2.5 days) in winter lac cultivation. The Phenol content of the bark and the production of lac were positively correlated, whereas both alkaloid and flavonoid content had no significant correlation with the production of lac (Rahman et al., 2021).

6.1. Properties of Lac

Lac possesses many desirable properties as a so-called multipurpose resin as compared to any other natural or synthetic resin. Lac is soluble in both organic and inorganic solvents, but water is insoluble (alcohols and weak alkalies). It can form a uniform, durable film. It is a good adhesive by nature. It acts as a good sealer and undercoat primer. It possesses high scratch hardness. It is a heat-soluble substance with a melting temperature of 80 °C. It is non-conductive and non-toxic. Lac can allow quick rubbing with sandpaper without slicking or gumming (Kadam, 2014).

6.2. Uses of Lac

Lac is one of the most useful natural resinous products of insects. It can be used in different industries, like plastics, adhesives, electrical, leather, hat manufacturing, wood finishing, and other industries. The resin is used to manufacture varnishes, glazing, and adhesive agents for food. The raw material for incense and perfume is also obtained from them. For protection and decoration, it is also used in polishes, lacquers, and finishes. Other applications include tubes and laminated paper boards, laminated sheets, and moulded mica, as well as the glazing of coffee beans, the application of undercoats and enamels, and the preservation of archaeological and zoological specimens (Perveen et al., 2013; Bashir et al., 2022).

Conclusion

The COVID-19 epidemic has highlighted how crucial it is to discover sustainable and eco-friendly sources of food and materials. Insects and their by-products have been used by humans for 1000 years and have gained increasing attention as a potential solution to food shortages. Edible insect proteins, chitosan, silk fibroin, honey, bee venom, pollen, and lac are just a few examples of the many valuable materials that can be obtained from insects. These materials' qualities may be enhanced in several ways, making them appropriate for a variety of applications in various industries. Moving forward, it may be possible to find long-term solutions to the world's food shortages and other problems in various industries by using insects and their by-products. However, there are still challenges to be addressed, such as regulatory issues and public perception. Supplementary research is required to study the importance and possible benefits of insects as food sources and to develop sustainable and ethical methods of insect cultivation. Additionally, initiatives should be required to inform and educate the rural and urban public about the advantages of insect-based products.

References

Aiba, S. I. (1991). Studies on chitosan: 3. Evidence for the presence of random and block copolymer structures in partially N-acetylated chitosans. *International journal of biological macromolecules*, 13(1), 40-44.

Ajibola, A., Idowu, G. O., Amballi, A. A., Oyefuga, O. H., & Iquot, I. S. (2007). Improvement of some haematological parameters in albino rats with pure natural honey. *J Biol Sci Res*, 2(1), 67-69.

Ali, A. T., Chowdhury, M. N., & Al Humayyd, M. S. (1991). Inhibitory effect of natural honey on *Helicobacter pylori. Tropical gastroenterology: Official Journal of the Digestive Diseases Foundation*, 12(3), 139-143.

Althwab, S. A., Alhomaid, R. M., Ali, R. F., Mohammed El-Anany, A., & Mousa, H. M. (2021). Effect of migratory locust (*Locusta migratoria*) powder incorporation on nutritional and sensorial properties of wheat flour bread. *British Food Journal*, 123(11), 3576-3591.

Al-Waili, N. S. (2003). Topical application of natural honey, beeswax and olive oil mixture for atopic dermatitis or psoriasis: partially controlled, single-blinded study. *Complementary therapies in medicine*, 11(4), 226-234.

Anderson, M. K. (2006). The use of fire by Native Americans in California. *Fire in California's ecosystems. University of California Press, Berkeley, California, USA*, 417-430.

Anjum, S. I., Ullah, A., Khan, K. A., Attaullah, M., Khan, H., Ali, H., ... & Dash, C. K. (2019). Composition and functional properties of propolis (bee glue): A review. *Saudi Journal of Biological Sciences*, 26(7), 1695-1703.

Azzollini, D., Derossi, A., Fogliano, V., Lakemond, C. M. M., & Severini, C. (2018). Effects of formulation and process conditions on microstructure, texture and digestibility of extruded insect-riched snacks. *Innovative Food Science & Emerging Technologies*, 45, 344-353.

Baiano, A. (2020). Edible insects: An overview on nutritional characteristics, safety, farming, production technologies, regulatory framework, and socio-economic and ethical implications. *Trends in Food Science & Technology*, 100, 35-50.

Bărnuţiu, L. I., Mărghitaş, L. A., Dezmirean, D. S., Mihai, C. M., & Bobiş, O. (2011). Chemical composition and antimicrobial activity of Royal Jelly-REVIEW. *Scientific Papers Animal Science and Biotechnologies*, 44(2), 67-72.

Bashir, N. H., Chen, H., Munir, S., Wang, W., Chen, H., Sima, Y. K., & An, J. (2022). Unraveling the Role of Lac Insects in Providing Natural Industrial Products. *Insects*, 13(12), 1117.

Bastiaens, L., Soetemans, L., D'Hondt, E., & Elst, K. (2019). Sources of chitin and chitosan and their isolation. *Chitin and Chitosan: Properties and Applications*, 1-34.

Beck, B. F., & Smedley, D. (1944). Honey and your health.

Berners-Lee, M., Kennelly, C., Watson, R., & Hewitt, C. N. (2018). Current global food production is sufficient to meet human nutritional needs in 2050 provided there is radical societal adaptation. *Elementa: Science of the Anthropocene*, 6, 1-14.

Bessa, L. W., Pieterse, E., Sigge, G., Hoffman, L. C., & Hoffman, L. C. (2019). An exploratory study into the use of black soldier fly (*Hermetia illucens*) larvae in the production of a Vienna-style sausage. *Meat and Muscle Biology*, 3(1).

Bharucha, Z., and Pretty, J. (2010). The roles and values of wild foods in agricultural systems. *Philosophical Transactions of the Royal Society*, 365, 2913–2926.

Binkley, D. (2014). How bees make honey is complex process. *The Columbus Dispatch, Columbus, Ohio, USA*.

Biró, B., Fodor, R., Szedljak, I., Pásztor-Huszár, K., & Gere, A. (2019). Buckwheat-pasta enriched with silkworm powder: Technological analysis and sensory evaluation. *LWT*, 116, 108542.

Biswal, B., Dan, A. K., Sengupta, A., Das, M., Bindhani, B. K., Das, D., & Parhi, P. K. (2022). Extraction of silk fibroin with several sericin removal processes and its importance in tissue engineering: A review. *Journal of Polymers and the Environment*, 30(6), 2222-2253.

Blackford, M., & Dinan, L. (1997). The effects of ingested ecdysteroid agonists (20-hydroxyecdysone, RH5849 and RH5992) and an ecdysteroid antagonist (cucurbitacin B) on larval development of two polyphagous lepidopterans (*Acherontia atropos* and *Lacanobia oleracea*). *Entomologia experimentalis et applicata*, 83(3), 263-276.

Bogdanov, S. (2004). Beeswax: quality issues today. *Bee World*, 85(3), 46-50.

Bogdanov, S. (2016). Beeswax: History, uses and trade. *Bee Product Science*, 1-18.

Boni, R., Ali, A., Shavandi, A., & Clarkson, A. N. (2018). Current and novel polymeric biomaterials for neural tissue engineering. *Journal of Biomedical Science*, 25, 1-21.

Bosch, G., Van Zanten, H. H. E., Zamprogna, A., Veenenbos, M., Meijer, N. P., Van der Fels-Klerx, H. J., & Van Loon, J. J. A. (2019). Conversion of organic resources by black soldier fly larvae: Legislation, efficiency and environmental impact. *Journal of Cleaner Production*, 222, 355-363.

Çabuk, B., & Yılmaz, B. (2020). Fortification of traditional egg pasta (erişte) with edible insects: Nutritional quality, cooking properties and sensory characteristics evaluation. *Journal of Food Science and Technology*, 57, 2750-2757.

Carpena, M., Nuñez-Estevez, B., Soria-Lopez, A., & Simal-Gandara, J. (2020). Bee venom: an updating review of its bioactive molecules and its health applications. *Nutrients*, 12(11), 3360.

Cartay, R. (2018). Between shock and disgust: the consumption of insects in the Amazon basin. the case of *Rhynchophorus palmarum* (Coleoptera Curculionidae). *Revista Colombiana de Antropología*, 54(2), 143-169.

Casimiro, M. H., Lancastre, J. J., Rodrigues, A. P., Gomes, S. R., Rodrigues, G., & Ferreira, L. M. (2017). Chitosan-Based matrices prepared by gamma irradiation for tissue regeneration: Structural properties vs. preparation method. *Applications of Radiation Chemistry in the Fields of Industry, Biotechnology and Environment*, 121-145.

Cerritos, R., & Cano-Santana, Z. (2008). Harvesting grasshoppers *Sphenarium purpurascens* in Mexico for human consumption: a comparison with insecticidal control for managing pest outbreaks. *Crop Protection*, 27(3-5), 473-480.

Chen, P. P., Wongsiri, S., Jamyanya, T., Rinderer, T. E., Vongsamanode, S., Matsuka, M., ... & Oldroyd, B. P. (1998). Honey bees and other edible insects used as human food in Thailand. *American Entomologist*, 44(1), 24-29.

Chen, S., Wei, X., Sui, Z., Guo, M., Geng, J., Xiao, J., & Huang, D. (2021). Preparation of antioxidant and antibacterial chitosan film from *Periplaneta americana*. *Insects*, 12(1), 53.

Chepulis, L. M. (2007). The effects of honey compared with sucrose and a sugar-free diet on neutrophil phagocytosis and lymphocyte numbers after long-term feeding in rats. *Journal of Complementary and Integrative Medicine*, 4(1).

Craig, C. L. (2003). *Spiderwebs and silk: tracing evolution from molecules to genes to phenotypes. Oxford University Press.*

Crane, E. (1991). Honey from honeybees and other insects. *Ethology Ecology & Evolution*, 3(1), 100-105.

da Rosa Machado, C., & Thys, R. C. S. (2019). Cricket powder (*Gryllus assimilis*) as a new alternative protein source for gluten-free breads. *Innovative Food Science & Emerging Technologies*, 56, 102180.

de Oliveira, L. M., da Silva Lucas, A. J., Cadaval, C. L., & Mellado, M. S. (2017). Bread enriched with flour from cinereous cockroach (*Nauphoeta cinerea*). *Innovative Food Science & Emerging Technologies*, 44, 30-35.

Doira, H. (1983). Linkage map of *Bombyx mori*: status quo in 1983. *Sericologia*, 23, 245-269.

Duda, A., Adamczak, J., Chełmińska, P., Juszkiewicz, J., & Kowalczewski, P. (2019). Quality and nutritional/textural properties of durum wheat pasta enriched with cricket powder. *Foods*, 8(2), 46.

Dzerefos, C. M., Witkowski, E. T. F., & Toms, R. (2013). Comparative ethnoentomology of edible stinkbugs in southern Africa and sustainable management considerations. *Journal of Ethnobiology and Ethnomedicine*, 9(1), 1-12.

El Ghaouth, A., Arul, J., Grenier, J., & Asselin, A. (1992). Antifungal activity of chitosan on two postharvest pathogens of strawberry fruits. *Phytopathology*, 82(4), 398–402.

Erdogan, S., & Kaya, M. (2016). High similarity in physicochemical properties of chitin and chitosan from nymphs and adults of a grasshopper. *International journal of biological macromolecules*, 89, 118-126.

Felt, O., Buri, P., & Gurny, R. (1998). Chitosan: A unique polysaccharide for drug delivery. *Drug Development and Industrial Pharmacy*, 24(11), 979–993.

Finke, M. D. (2013). Complete nutrient content of four species of feeder insects. *Zoo Biology*, 32(1), 27-36.

Fotidar, M. R., & Fotidar, S. N. (1945). Lotus' honey. *Indian Bee J*, 7, 102.

Fujita, T., Kozuka-Hata, H., Ao-Kondo, H., Kunieda, T., Oyama, M., & Kubo, T. (2013). Proteomic analysis of the royal jelly and characterization of the functions of its derivation glands in the honeybee. *Journal of proteome research*, 12(1), 404-411.

Gao, Y., Zhao, Y. J., Xu, M. L., & Shi, S. S. (2021). Soybean hawkmoth (*Clanis bilineata tsingtauica*) as food ingredients: a review. *CyTA-Journal of Food*, 19(1), 341-348.

Garofalo, C., Milanović, V., Cardinali, F., Aquilanti, L., Clementi, F., & Osimani, A. (2019). Current knowledge on the microbiota of edible insects intended for human consumption: A state-of-the-art review. *Food Research International*, 125, 108527.

Ghanem, N. B. (2011). The antimicrobial activity of some honey bee products and some Saudi folkloric plant extracts. *JKAU Sci*, 23, 47-62.

Grabowski, N. T., Abdulmawjood, A., Acheuk, F., Barragán Fonseca, K., Chhay, T., Costa Neto, E. M., & Plötz, M. (2022). Insects—A source of safe and sustainable food? "Jein" (Yes and No). *Frontiers in Sustainable Food Systems*, 5, 701797.

Guarino, V., Iafisco, M., & Spriano, S. (2020). Introducing biomaterials for tissue repair and regeneration. In *Nanostructured biomaterials for regenerative medicine* (pp. 1-27). Woodhead Publishing.

Gunes, H., & Gungormus, A. (2022). Identification of honey bee (*Apis mellifera*) larvae in the hive with faster R-CNN for royal jelly production. *Journal of Apicultural Research*, 61(3), 338-345.

Gupta, Y. M., Tanasarnpaiboon, S., Buddhachat, K., Peyachoknagul, S., Inthim, P., & Homchan, S. (2020). Development of microsatellite markers for the house cricket, *Acheta domesticus* (Orthoptera: Gryllidae). *Biodiversitas Journal of Biological Diversity*, 21(9), 4094-4099.

Haber, M., Mishyna, M., Martinez, J. I., & Benjamin, O. (2019). The influence of grasshopper (*Schistocerca gregaria*) powder enrichment on bread nutritional and sensorial properties. *LWT*, 115, 108395.

Hall, F. G. (2020). Bioactive and allergenic properties of edible cricket (*Gryllodes sigillatus*) peptides.

Hall, F. G., Jones, O. G., O'Haire, M. E., & Liceaga, A. M. (2017). Functional properties of tropical banded cricket (*Gryllodes sigillatus*) protein hydrolysates. *Food Chemistry*, 224, 414-422.

Hall, F., Reddivari, L., & Liceaga, A. M. (2020). Identification and characterization of edible cricket peptides on hypertensive and glycemic in vitro inhibition and their anti-inflammatory activity on RAW 264.7 macrophage cells. *Nutrients*, 12(11), 3588.

Holl, S. M., Schaefer, J., Goldberg, W. M., Kramer, K. J., Morgan, T. D., & Hopkins, T. L. (1992). Comparison of black coral skeleton and insect cuticle by a combination of carbon-13 NMR and chemical analyses. *Archives of biochemistry and biophysics*, 292(1), 107-111.

Hyodo, A., Gamo, T., and Shimura, K. (1980). Linkage analysis of the fibroin gene in the silkworm, *Bombyx mori. The Japanese Journal of Genetics*, 55(4), 297-300.

Hyodo, A., Yamamoto, T., Ueda, H., Takei, F., Kimura, K. I., & Shimura, K. (1984). Linkage analysis of the fibroin light chain gene in the silkworm, *Bombyx mori. The Japanese Journal of Genetics*, 59(3), 285-296.

Imathiu, S. (2020). Benefits and food safety concerns associated with consumption of edible insects. *NFS journal*, 18, 1-11.

Irigoiti, Y., Navarro, A., Yamul, D., Libonatti, C., Tabera, A., & Basualdo, M. (2021). The use of propolis as a functional food ingredient: A review. *Trends in Food Science & Technology*, 115, 297-306.

Jones, R. (1999). Beekeeping as a Business. Commonwealth Secretariat.

Kadam S. (2014). Lac: Distribution, Properties and Products. Article, Zoology Notes.

Kalita, G. J., Boruah, B., and Das, G. N. (2014). An observation on odonata (damselflies and dragonflies) fauna of Manchabandha reserve forest, Baripada, Odisha. *Pelagia Research Library*, 5(1), 77-83.

Kaplan, D. L., & Vepari, C. (2007). Silk as a biomaterial. *Prog Polym Sci*, 32, 991-1007.

Kaur, S., & Dhillon, G. S. (2015). Recent trends in biological extraction of chitin from marine shell wastes: a review. *Critical reviews in biotechnology*, 35(1), 44-61.

Kaya, M., Baran, T., Erdoğan, S., Menteş, A., Özüsağlam, M. A., & Çakmak, Y. S. (2014). Physicochemical comparison of chitin and chitosan obtained from larvae and adult Colorado potato beetle (*Leptinotarsa decemlineata*). *Materials Science and Engineering*, 45, 72–81.

Khayrova, A., Lopatin, S., & Varlamov, V. (2020, April). Obtaining chitin/chitosan-melanin complexes from Black Soldier Fly *Hermetia illucens*. In *IOP Conference Series: Materials Science and Engineering*, 809(1). 012020.

Khor, E. (2014). *Chitin: fulfilling biomaterials promise*. Elsevier.

Khor, E., & Lim, L. Y. (2003). Implantable applications of chitin and chitosan. *Biomaterials*, 24(13), 2339–2349.

Kim, H. W., Setyabrata, D., Lee, Y. J., Jones, O. G., & Kim, Y. H. B. (2016). Pre-treated mealworm larvae and silkworm pupae as a novel protein ingredient in emulsion sausages. *Innovative Food Science & Emerging Technologies*, 38, 116-123.

Kim, M. W., Han, Y. S., Jo, Y. H., Choi, M. H., Kang, S. H., Kim, S. A., & Jung, W. J. (2016). Extraction of chitin and chitosan from housefly, *Musca domestica*, pupa shells. *Entomological Research*, 46(5), 324-328.

Kim, M. W., Song, Y. S., Han, Y. S., Jo, Y. H., Choi, M. H., Park, Y. K., & Jung, W. J. (2017). Production of chitin and chitosan from the exoskeleton of adult two-spotted field crickets (*Gryllus bimaculatus*). *Entomological Research*, 47(5), 279-285.

Kim, S. K. (Ed.). (2010). *Chitin, chitosan, oligosaccharides and their derivatives: biological activities and applications*. CRC Press.

Kim, T. K., Cha, J. Y., Yong, H. I., Jang, H. W., Jung, S., & Choi, Y. S. (2022). Application of edible insects as novel protein sources and strategies for improving their processing. *Food Science of Animal Resources*, 42(3), 372.

Kim, T. K., Yong, H. I., Jeong, C. H., Han, S. G., Kim, Y. B., Paik, H. D., & Choi, Y. S. (2019a). Technical functional properties of water-and salt-soluble proteins extracted from edible insects. *Food science of animal resources*, 39(4), 643-654.

Kim, T. K., Yong, H. I., Kim, Y. B., Kim, H. W., & Choi, Y. S. (2019b). Edible insects as a protein source: A review of public perception, processing technology, and research trends. *Food science of animal resources*, 39(4), 521-540.

Kimura, K., Oyama, F., Ueda, H., Mizuno, S., & Shimura, K. (1985). Molecular cloning of the fibroin light chain complementary DNA and its use in the study of the expression of the light chain gene in the posterior silk gland of *Bombyx mori*. *Experientia*, 41, 1167-1171.

Kirkpatrick, T. W. (1957). Insect life in the tropics. *Insect Life in the Tropics*.

Kocot, J., Kiełczykowska, M., Luchowska-Kocot, D., Kurzepa, J., & Musik, I. (2018). Antioxidant potential of propolis, bee pollen, and royal jelly: possible medical application. *Oxidative medicine and cellular longevity*, 1-29.

Kourimská, L., & Adámková, A. (2016). Nutritional and sensory quality of edible insects. *NFS journal*, 4, 22-26.

Kubota, N., & Eguchi, Y. (1997). Facile preparation of water-soluble N-acetylated chitosan and molecular weight dependence of its water-solubility. *Polymer Journal*, 29(2), 123-127.

Kurakula, M., & Naveen, N. R. (2021). Electrospraying: A facile technology unfolding the chitosan-based drug delivery and biomedical applications. *European Polymer Journal*, 147, 110326.

Kwak, K. W., Ko, H. J., Kim, S. Y., Lee, K. Y., & Yoon, H. J. (2021). Developmental characteristics of ovary and egg of migratory locust, *Locusta migratoria* (Orthoptera: Acrididae). *Korean journal of applied entomology*, 60(2), 175-183.

Lähteenmäki-Uutela, A., Marimuthu, S. B., & Meijer, N. (2021). Regulations on insects as food and feed: a global comparison. *Journal of Insects as Food and Feed*, 7(5), 849-856.

Latchininsky, A. V. (1998). Moroccan locust *Dociostaurus maroccanus* (Thunberg, 1815): a faunistic rarity or an important economic pest. *Journal of Insect Conservation*, 2, 167-178.

Lawal, K. G., Kavle, R. R., Akanbi, T. O., Mirosa, M., & Agyei, D. (2021). Enrichment in specific fatty acids profile of *Tenebrio molitor* and *Hermetia illucens* larvae through feeding. *Future Foods*, 3, 100016.

Lecoq, M. (2022). *Schistocerca gregaria* (desert locust).

Lesnik, J. J. (2018). Edible insects and human evolution. *Archäologische Informationen*, 42, 319-321.

Liceaga, A. M., Aguilar-Toalá, J. E., Vallejo-Cordoba, B., González-Córdova, A. F., & Hernández-Mendoza, A. (2022). Insects as an alternative protein source. *Annual Review of Food Science and Technology*, 13, 19-34.

Liu, G. Q., Xiao, H. X., Wang, X. L., Zhao, Y., Zhang, Y. G., & Ren, G. P. (2011). Stimulated production of triterpenoids of *Ganoderma lucidum* by an ether extract from the medicinal insect, *Catharsius molossus*, and identification of the key stimulating active components. *Applied biochemistry and biotechnology*, 165, 87-97.

Liu, H., Wang, C., Li, C., Qin, Y., Wang, Z., Yang, F., Li, Z. (2018). A functional chitosan-based hydrogel as a wound dressing and drug delivery system in the treatment of wound healing. *RSC advances*, 8(14), 7533-7549.

Longvah, T., Mangthya, K., & Ramulu, P. J. F. C. (2011). Nutrient composition and protein quality evaluation of eri silkworm (*Samia ricinii*) prepupae and pupae. *Food Chemistry*, 128(2), 400-403.

Luo, Q., Wang, Y., Han, Q., Ji, L., Zhang, H., Fei, Z., & Wang, Y. (2019). Comparison of the physicochemical, rheological, and morphologic properties of chitosan from four insects. *Carbohydrate Polymers*, 209, 266-275.

Ma, J., Faqir, Y., Tan, C., & Khaliq, G. (2022). Terrestrial insects as a promising source of chitosan and recent developments in its application for various industries. *Food Chemistry*, 373, 131407.

Ma, J., Xin, C., & Tan, C. (2015). Preparation, physicochemical and pharmaceutical characterization of chitosan from *Catharsius molossus* residue. *International Journal of Biological Macromolecules*, 80, 547–556.

Magara, H. J., Niassy, S., Ayieko, M. A., Mukundamago, M., Egonyu, J. P., Tanga, C. M., & Ekesi, S. (2021). Edible crickets (Orthoptera) around the world: distribution, nutritional value, and other benefits—a review. *Frontiers in nutrition*, 7, 537915.

Marei, H. E., Hasan, A., Rizzi, R., Althani, A., Afifi, N., Cenciarelli, C., & Shuaib, A. (2018). Potential of stem cell-based therapy for ischemic stroke. *Frontiers in neurology*, 9, 34.

Marei, N. H., El-Mazny, W., El-Shaer, A., Zaki, K. D., Hussein, Z. S., & Abd-El-Samie, E. M. (2017). Enhanced wound healing activity of desert locust (*Schistocerca gregaria*) vs. *shrimp* (*Penaeus monodon*) chitosan-based scaffolds. *International Journal of Biological Macromolecules*, 97, 23–33.

Matsuzaki, K., Yoneyama, S., Murase, O., & Miyajima, K. (1996). Transbilayer transport of ions and lipids coupled with mastoparan X translocation. *Biochemistry*, 35(25), 8450-8456.

Melgar-Lalanne, G., Hernández-Álvarez, A. J., & Salinas-Castro, A. (2019). Edible insects processing: Traditional and innovative technologies. *Comprehensive Reviews in Food Science and Food Safety*, 18(4), 1166-1191.

Minot, M., Besnard, A., & Husté, A. (2021). Habitat use and movements of a large dragonfly (Odonata: *Anax imperator*) in a pond network. *Freshwater Biology*, 66(2), 241-255.

Mitsuhashi, J. (2016). *Edible insects of the world*. CRC press.

Molan, P. C. (2001). The potential of honey to promote oral wellness. *General dentistry*, 49(6), 584-590.

Nemtsev, S. V., Zueva, O. Y., Khismatullin, M. R., Albulov, A. I., & Varlamov, V. P. (2004). Isolation of chitin and chitosan from honeybees. *Applied Biochemistry and Microbiology*, 40, 39-43.

Neto, E. C., & Ramos-Elorduy, J. (2006). Los insectos comestibles de Brasil: etnicidad, diversidad e importancia en la alimentación. *Boletín Sociedad Entomológica Aragonesa*, 38, 423-442.

Nirazawa, T., Fujii, T., Seki, Y., Namiki, S., Kazawa, T., Kanzaki, R., & Ishikawa, Y. (2017). Morphology and physiology of antennal lobe projection neurons in the hawkmoth *Agrius convolvuli*. *Journal of insect physiology*, 98, 214-222.

Nwe, N., Furuike, T., & Tamura, H. (2011). Chitosan from aquatic and terrestrial organisms and microorganisms: production, properties and applications. *Biodegradable materials*, 28, 29-50.

Oonincx, D. G., & De Boer, I. J. (2012). Environmental impact of the production of mealworms as a protein source for humans–a life cycle assessment. *PloS one*, 7(12), e51145.

Osato, M. S., Reddy, S. G., & Graham, D. Y. (1999). Osmotic effect of honey on growth and viability of *Helicobacter pylori. Digestive diseases and sciences*, 44, 462-464.

Osimani, A., Milanović, V., Cardinali, F., Roncolini, A., Garofalo, C., Clementi, F., & Aquilanti, L. (2018). Bread enriched with cricket powder (*Acheta domesticus*): A technological, microbiological and nutritional evaluation. *Innovative food science & emerging technologies*, 48, 150-163.

Park, Y. S., Choi, Y. S., Hwang, K. E., Kim, T. K., Lee, C. W., Shin, D. M., & Han, S. G. (2017). Physicochemical properties of meat batter added with edible silkworm pupae (*Bombyx mori*) and transglutaminase. *Korean Journal for Food Science of Animal Resources*, 37(3), 351-359.

Patra, C., Talukdar, S., Novoyatleva, T., Velagala, S. R., Mühlfeld, C., Kundu, B., ... & Engel, F. B. (2012). Silk protein fibroin from Antheraea mylitta for cardiac tissue engineering. *Biomaterials*, 33(9), 2673-2680.

Perveen, A., Jahan, N., Wadud, A., & Alam, M. T. (2013). Medicinal benefits of lac described in Unani literature-an overview. *American Journal of Pharm Tech Research*, 3(5), 1-10.

Philibert, T., Lee, B. H., & Fabien, N. (2017). Current status and new perspectives on chitin and chitosan as functional biopolymers. *Applied Biochemistry and Biotechnology*, 181(4), 1314–1337.

Post, M. J. (2012). Cultured meat from stem cells: Challenges and prospects. *Meat Science*, 92(3), 297-301.

Qiu, L., Liu, Y. C., Wang, Z. Q., & Che, Y. L. (2020). The first blattid cockroach (Dictyoptera: Blattodea) in Cretaceous amber and the reconsideration of purported Blattidae. *Cretaceous Research*, 109, 104359.

Rahman, T., Das, P., Hazarika, L. K., Sharma, K. K., Mohanasundaram, A., & Roy, S. (2021). Evaluation of certain plant hosts for the lac insect. *International Journal of Tropical Insect Science*, 1-8.

Ramakrishanan, Y., & Raja, S. (2020). Chitosan extraction and its characterization from black ant *Camponotus compressus* (Fab,) (Hymenoptera: Formicidae). *International Journal of Creative Research Thoughts*, 8(7), 3011–3016.

Ramírez-Rivera, E. J., Hernández-Santos, B., Juárez-Barrientos, J. M., Torruco-Uco, J. G., Ramírez-Figueroa, E., & Rodríguez-Miranda, J. (2021). Effects of formulation and process conditions on chemical composition, color parameters, and acceptability of extruded insect-rich snack. *Journal of Food Processing and Preservation*, 45(5), e15499.

Rinaudc, M., Pavlov, G., & Desbrieres, J. (1999b). Solubilization of chitosan in strong acid medium. *International Journal of Polymer Analysis and Characterization*, 5(3), 267-276.

Rinaudo, M. (2006). Chitin and chitosan: Properties and applications. *Progress in polymer science*, 31(7), 603-632.

Rinaudo, M., & Domard, A. (1989). Chitin and chitosan. *Sources, Chemistry, Biochemistry, Physical Properties and Applications", eds. by Skjak-Braek G., Anthonsen T., Sandford P., Elsevier Science Publisher, London*, 71.

Rinaudo, M., Pavlov, G., & Desbrieres, J. (1999a). Influence of acetic acid concentration on the solubilization of chitosan. *Polymer*, 40(25), 7029-7032.

Rodríguez-Miranda, J., Alcántar-Vázquez, J. P., Zúñiga-Marroquín, T., & Juárez-Barrientos, J. M. (2019). Insects as an alternative source of protein: A review of the potential use of grasshopper (*Sphenarium purpurascens* Ch.) as a food ingredient. *European Food Research and Technology*, 245(12), 2613-2620.

Roncolini, A., Milanović, V., Cardinali, F., Osimani, A., Garofalo, C., Sabbatini, R., & Aquilanti, L. (2019). Protein fortification with mealworm (*Tenebrio molitor* L.) powder: Effect on textural, microbiological, nutritional and sensory features of bread. *PloS one*, 14(2), e0211747.

Rumbos, C. I., & Athanassiou, C. G. (2021). The superworm, *Zophobas morio* (Coleoptera: Tenebrionidae): a 'sleeping giant'in nutrient sources. *Journal of Insect Science*, 21(2), 13.

Rumbos, C. I., Karapanagiotidis, I. T., Mente, E., & Athanassiou, C. G. (2019). The lesser mealworm *Alphitobius diaperinus*: a noxious pest or a promising nutrient source. *Reviews in Aquaculture*, 11(4), 1418-1437.

Rumpold, B. A., & Schlüter, O. K. (2013). Nutritional composition and safety aspects of edible insects. *Molecular nutrition & food research*, 57(5), 802-823.

Saeed, F., Ahmad, R. S., Arshad, M. U., Niaz, B., Batool, R., Naz, R., & Ansar Rasul Suleria, H. (2016). Propolis to curb lifestyle related disorders: An overview. *International Journal of Food Properties*, 19(2), 420-437.

Sani, M. G., Bashir, U. G., & Bawa, A. A. (2017). Comparative study of percentage yield and solubility of chitosan extracted from male and female short-horned grasshopper. *Bayero Journal of Pure and Applied Sciences*, 10(1), 570-574.

Saralaya, S., B. S., J., Thomas, N. S., & S. M., S. (2021). Bee wax and honey—a primer for OMFS. *Oral and Maxillofacial Surgery*, 25(1), 1-6.

Schaefer, C. W., & Panizzi, A. R. (2000). Water scorpions (Nepidae). In *Heteroptera of Economic Importance*, 605-612.

Shanmuganathan, R., Edison, T. N. J. I., LewisOscar, F., Kumar, P., Shanmugam, S., & Pugazhendhi, A. (2019). Chitosan nanopolymers: An overview of drug delivery against cancer. *International Journal of Biological Macromolecules*, 130, 727–736.

Sharma, K. K. (2017). Lac insects and host plants. *Industrial Entomology*, 157-180.

Sharma, K. K., Chowdhury, A. R., & Srivastava, S. (2020). Chemistry and applications of lac and its by-product. *Natural Materials and Products from Insects: Chemistry and Applications*, 21-37.

Shin, C.-S., Kim, D.-Y., & Shin, W.-S. (2019). Characterization of chitosan extracted from Mealworm Beetle (*Tenebrio molitor*, *Zophobas morio*) and Rhinoceros Beetle (*Allomyrina dichotoma*) and their antibacterial activities. *International Journal of Biological Macromolecules*, 125, 72–77.

Shirvan, A. R., Shakeri, M., & Bashari, A. (2019). Recent advances in application of chitosan and its derivatives in functional finishing of textiles. *The impact and prospects of green chemistry for textile technology*, 107-133.

Singh, R. (2007). Lac culture.

Sivashankari, P. R., & Prabaharan, M. (2016). Prospects of chitosan-based scaffolds for growth factor release in tissue engineering. *International journal of biological macromolecules*, 93, 1382-1389.

Somal, N. A., Coley, K. E., Molan, P. C., & Hancock, B. M. (1994). Susceptibility of Helicobacter pylori to the antibacterial activity of manuka honey. *Journal of the Royal Society of Medicine*, 87(1), 9-12.

Soon, C. Y., Tee, Y. B., Tan, C. H., Rosnita, A. T., & Khalina, A. (2018). Extraction and physicochemical characterization of chitin and chitosan from *Zophobas morio* larvae in varying sodium hydroxide concentration. *International Journal of Biological Macromolecules*, 108, 135–142.

Steinberg, D., Kaine, G., & Gedalia, I. (1996). Antibacterial effect of propolis and honey on oral bacteria. *American journal of dentistry*, 9(6), 236-239.

Suarez, R. K., Lighton, J. R., Joos, B., Roberts, S. P., & Harrison, J. F. (1996). Energy metabolism, enzymatic flux capacities, and metabolic flux rates in flying honeybees. *Proceedings of the National Academy of Sciences*, 93(22), 12616-12620.

Takei, F., Kikuchi, Y., Kikuchi, A., Mizuno, S., & Shimura, K. (1987). Further evidence for importance of the subunit combination of silk fibroin in its efficient secretion from the posterior silk gland cells. *The Journal of Cell Biology*, 105(1), 175-180.

Tao, M., Wang, C., Liao, D., Liu, H., Zhao, Z., & Zhao, Z. (2017). Purification, modification and inhibition mechanism of angiotensin I-converting enzyme inhibitory peptide from silkworm pupa (*Bombyx mori*) protein hydrolysate. *Process Biochemistry*, 54, 172-179.

Tchibozo, S., and Lecoq, M. (2017). Edible Orthoptera from Africa: preservation and promotion of traditional knowledge. *Metaleptea*, 37, 24–29.

Tojo, S., Nagata, M., & Kobayashi, M. (1980). Storage proteins in the silkworm, *Bombyx mori*. *Insect Biochemistry*, 10(3), 289-303.

Tulloch, A. P. (1980). Beeswax—composition and analysis. *Bee World*, 61(2), 47-62.

Van Huis, A. (2013). Potential of insects as food and feed in assuring food security. *Annual review of entomology*, 58, 563-583.

Van Huis, A. (2020). Importance of insects as food in Africa. *African edible insects as alternative source of food, oil, protein and bioactive components*, 1-17.

Van Huis, A., Van Itterbeeck, J., Klunder, H., Mertens, E., Halloran, A., Muir, G., & Vantomme, P. (2013). *Edible insects: future prospects for food and feed security* (No. 171). Food and Agriculture Organization of the United Nations.

Vasilaki, A., Hatzikamari, M., Stagkos-Georgiadis, A., Goula, A. M., & Mourtzinos, I. (2019). A natural approach in food preservation: Propolis extract as sorbate alternative in non-carbonated beverage. *Food Chemistry*, 298, 125080.

Vivek, R., Babu, V. N., Thangam, R., Subramanian, K. S., & Kannan, S. (2013). pH-responsive drug delivery of chitosan nanoparticles as Tamoxifen carriers for effective anti-tumor activity in breast cancer cells. *Colloids and Surfaces B: Biointerfaces*, 111, 117-123.

Vollrath, F., & Porter, D. (2006). Spider silk as a model biomaterial. *Applied Physics A*, 82, 205-212.

Wehbe, R., Frangieh, J., Rima, M., El Obeid, D., Sabatier, J. M., & Fajloun, Z. (2019). Bee venom: Overview of main compounds and bioactivities for therapeutic interests. *Molecules*, 24(16), 2997.

Wei, J., Sun, X. Q., & Hou, B. X. (2021). Evaluation of silk fibroin-RGD-stem cell factor scaffold effect on adhesion, migration, and proliferation of stem cells of apical papilla. *Stem Cells International*, 2021, 1-10.

Weissman, D. B., Walker, T. J., & Gray, D. A. (2009). The field cricket *Gryllus assimilis* and two new sister species (Orthoptera: Gryllidae). *Annals of the Entomological Society of America*, 102(3), 367-380.

White, J. W., & Doner, L. W. (1980). Honey composition and properties. *Beekeeping in the United States Agriculture Handbook*, 335, 82-91.

Williams, J. P., Williams, J. R., Kirabo, A., Chester, D., & Peterson, M. (2016). Nutrient content and health benefits of insects. In *Insects as sustainable food ingredients* (pp. 61-84). Academic Press.

Womeni, H. M., Tiencheu, B., Linder, M., Nabayo, C., Martial, E., Tenyang, N., ... & Parmentier, M. (2012). Nutritional value and effect of cooking, drying and storage process on some functional properties of *Rhynchophorus phoenicis*. *International Journal of Life Science and Pharma Research*, 2(3), 203-219.

Wu, Q. Y., Jia, J. Q., Tan, G. X., Xu, J. L., & Gui, Z. Z. (2011). Physicochemical properties of silkworm larvae protein isolate and gastrointestinal hydrolysate bioactivities. *African Journal of Biotechnology*, 10(32), 6145-6153.

Wu, S. (2011). Preparation of chitosan from *Clanis bilineata* larvae skin using enzymatic methods. *Carbohydrate polymers*, 83(2), 1008-1010.

Wu, S. J., Pan, S. K., Wang, H. B., & Wu, J. H. (2013). Preparation of chitooligosaccharides from cicada slough and their antibacterial activity. *International journal of biological macromolecules*, 62, 348-351.

Xue, X., Wang, F., Zhou, J., Chen, F., Li, Y., & Zhao, J. (2009). Online cleanup of accelerated solvent extractions for determination of adenosine 5′-triphosphate (ATP), adenosine 5′-diphosphate (ADP), and adenosine 5′-monophosphate (AMP) in royal jelly using high-performance liquid chromatography. *Journal of agricultural and food chemistry*, 57(11), 4500-4505.

Xue, X., Wu, L., & Wang, K. (2017). Chemical composition of royal jelly. *Bee products-chemical and biological properties*, 181-190.

Zayas, J. F. (1997). *Functionality of proteins in food*. Springer science & business media.

Zhou, L., Xue, X., Zhou, J., Li, Y., Zhao, J., & Wu, L. (2012). Fast determination of adenosine 5′-triphosphate (ATP) and its catabolites in royal jelly using ultraperformance liquid chromatography. *Journal of agricultural and food chemistry*, 60(36), 8994-8999.

Zielińska, E., Karaś, M., & Jakubczyk, A. (2017). Antioxidant activity of predigested protein obtained from a range of farmed edible insects. *International Journal of Food Science & Technology*, 52(2), 306-312.

Chapter 16

Global Hunger and Insects: An Alternative Food Source to Fulfill the Requirements of Eight Billion People

Mansi Gupta
and Istkhar Rao*
Department of Bioscience and Biotechnology, Banasthali Vidyapith, Banasthali, Rajasthan, India

Abstract

Eight billion people around the globe have food resources, but still, ~17% of the population is undernourished, and approximately one billion people do not have enough food to feed themselves, belonging to 93 countries. If the nutrient richness of insects is well documented, then why not make them an option? Insect rearing is the main factor that is to be considered while stating that insects are an edible resource because, to feed a large population, there needs to be a wide range of varieties and populations of insects as well. As a low-tech, low-capital investment choice, harvesting and rearing insects even opens doors to the lowest segments of society, including women and the landless, and offers jobs for people. Globally, around 2 billion people have selected insects as a choice of food too, but this practice is still primarily limited to African countries. It's a matter of time before we put effort into understanding the biodiversity of insects and their mass production as food resources. With their very nutritional composition, insects might be a great substitute for common staples like chicken, pig, beef, and even fish. This would ease the burden of meeting the dietary needs of the world's eight billion rapidly growing people. Insects are the major contributors to the world's animal population, dominating all habitats. They have always proved themselves as beneficiary to humans by being providers of various valuable products like silk and honey. In this regard, utilising insects as food is not a new fact, as it's an old-age practice, but the efforts to utilise them as an alternate option are still limited. The practice of entomophagy is greatly impacted by cultural and religious traditions. Religious writings from the Jewish and Christian religions also mention the custom of eating insects. Though mostly found in African nations, insects are already a common component of many regional and national diets. Hymenopterans, orthopterans, hemipterans, isopterans, odonatans, and dipterans are the major orders of insects of food importance, and around 10–15 species of edible insects are consumed regularly. When it comes to greenhouse gas emissions, edible insects emit far less than other common farm animals. Depending on the amount of money invested, raising insects can be done in a very simple or complicated way. Insect rearing binds people to them not only for their food requirements but can also awaken them to the needs of the earth and sustainable development.

* Corresponding Author's Email: istkharrao@gmail.com; istkhar@banasthali.in.

In: Insects as Humans' Frenemies: From Friends to Foes
Editor: Istkhar Rao
ISBN: 979-8-89530-320-7

Keywords: food resources, nutrients, entomophagy, insect rearing, edible insects

1. Introduction

The pest of hunger crises and malnutrition has been eating up people around the world for many generations. The world today is characterised by the coexistence of agricultural bounty and widespread hunger and malnutrition. Despite record food output globally and several years of falling hunger trends, the problem is still with us. The problem is even worse because of the COVID-19 pandemic in the years 2019–20. According to the trends, about 25,000 people are dying every day because of hunger, and 2 billion people worldwide are suffering from malnutrition. The major reason behind this is the absence of food resources available to produce nutritious food. Common food sources like vegetables, cereals, and even livestock animals have failed to meet the demands of nutritious food for the people. Thus, there arose the need to look for an alternative to cure the issue to at least some extent. Insects or pests were selected as a fantastic and nutrient-dense substitute for food, both for the current population and future generations. The world is becoming more and more interested in mass-producing insects as a sustainable source of food and nutrition.

2. The World at Food Crisis

According to the research done at the United Nations in the last few decades, it has been reported that the world population is increasing alarmingly. Since the Second World War, there has been an increase of 1 billion people every 12–15 years. From 1950 to the present, the increase accounts for 250%, from 2.6 billion to 7 billion (Population, 2020). It is further estimated that the population across the world will reach around 50 billion in the next 30 years (https://www.fao.org/state-of-food-security-nutrition). Populations in India and China have crossed the mark of 1.4 billion in each country. Various policies and efforts were made by different organisations to opt for transformative approaches to accelerate past progress, but none seems to have worked effectively as the world is not moving towards eradicating malnutrition and even ensuring safe, healthier, and adequate food for the people. (https://www.fao.org/state-of-food-security-nutrition) Even with the world experiencing slowdowns and downturns, it is recognised that the COVID-19 epidemic has made matters worse. As per the statistics from the Food and Agriculture Organisation of the United Nations (FAO), the COVID-19 pandemic caused between 720 and 811 million people to go hungry in 2020. If we consider the middle stage and take 768 million people to be starving, there will be 118 million more people in 2020 than in 2019 who faced the hell of starvation.

However, the severe food insecurity graph has slowed down since 2014, but the increase in 2020 is calculated to be equal to that of the last 5 years together. On average, one person in three did not have access to adequate food last year, the reasons being high costs of healthy food, high-income inequality, and worsening the situation. COVID-19 came in first and has just created havoc in the entire system, whether it is food or health.

Unfortunately, the pandemic has exposed all the bruises in our food system that warn us about the food security and nutrition of millions of people around the world. The rates of hunger and malnutrition are also higher in developing nations like Afghanistan, Madagascar, India, Brazil, Argentina, and Madagascar. With over 50% of the population suffering from chronic malnutrition, Madagascar is among the ten nations with the highest rates of malnutrition (UNICEF, 2013). If malnutrition is not treated in early childhood, the child may suffer potentially fatal consequences. The most significant risk factor for disease-related morbidity and mortality is chronic malnutrition, which causes irreversible stunting (WFP, 2015). Almost 25% of children in Sub-Saharan Africa and over half (52%) of children in South Asia are among the 150 million children globally who do not get enough calories or nutrients for healthy growth and development (UNICEF, 2001).

With such a high crisis, the concerns are shifting towards natural resources. It is questionable that finite planet resources like agricultural land and freshwater meet the ever-growing demands of this rapidly growing population (FAO, 2021). The imbalance between food production and consumption has increased the burden on the environment to provide sufficient food (Jin et al., 2020). Food production needs to be boosted by 75% by the next 30 years to ensure a proper food supply for generations to come (Mozhui et al., 2020). Large-scale deforestation is occurring to provide settlement areas for people and to create agricultural lands for the production of food for them, but these increasing needs of people are pressuring the planet and pushing us towards an unsustainable future.

The greatest issue of the twenty-first century is producing enough food in the required quantities while causing the least amount of environmental damage and preserving a healthy ecosystem. To address the primary causes of food insecurity, malnutrition, and hunger that affect millions of people, bold measures must be taken (https://www.fao.org/state-of-food-security-nutrition) (Figure 1). Concerns about food security are driving research into a variety of sustainable, high-nutrient alternative food sources for humans (such as fungi, cultured meat, micro and macroalgae, and insects) as well as feed for animals (such as food side streams, byproducts of biofuel production, and insects). With mass production of insects becoming more and more popular worldwide, insects are emerging as a viable food and feed source.

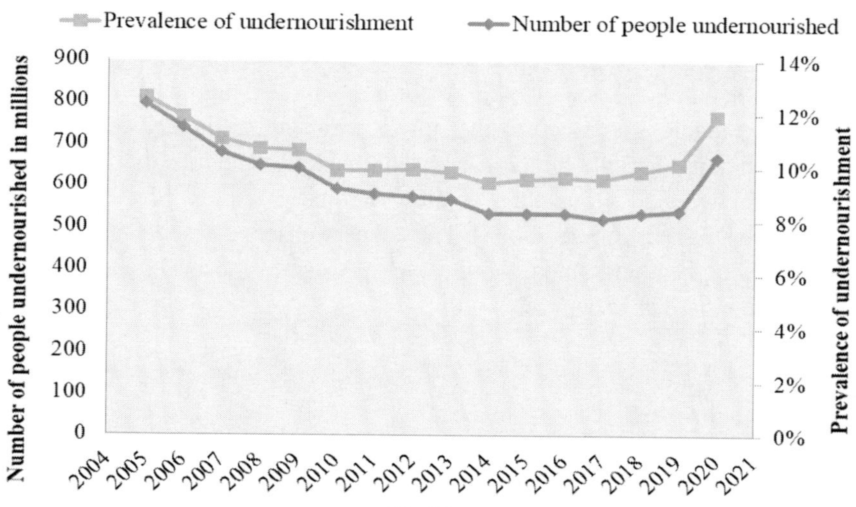

*Adopted from- https://www.fao.org/state-of-food-security-nutrition.

Figure 1. A graphical presentation of the rise in the number of people undernourished.

3. Insects as the Lead

Since insects have three parts to their bodies, the word "insect" comes from the Latin word "*insectum*," which means "with a notched or divided body," or literally "cut into sections." The most varied group of animals on Earth are insects, which belong to the phylum Arthropoda and are classified as class Insecta. The number of beetle species, insects, and terrestrial arthropod species estimated for the planet is 1.5 million, 5.5 million, and 7 million, respectively, based on averaging the new statistics. It now seems highly implausible that the 30 million species or more than previous estimates based on the host specificity of insects to plants were made (Stork, 2018).

3.1. What Makes Insects Extraordinary?

The following are some of the important features that make insects extraordinary:

- They are cold-blooded animals.
- They are the only winged invertebrates.
- They can eat a wide variety of foods of diverse types and qualities, increasing their chances of survival (Chakravorty, 2014).
- They have a short generation time and can reproduce very quickly (Chakravorty, 2014).
- They have an external skeleton (exoskeleton) made up of chitin, which provides support and protection for them.

Insects have always proved themselves as a beneficiary to humans by being providers of various valuable products like silk and honey. Bees deliver tonnes of commercial honey every year, and silkworms provide tonnes of silk every year. Engineering techniques and technology have also been influenced by insects. Spider silk proteins are robust and elastic and are employed as biomaterials (Lewis, 1992). According to van Huis et al. (2013), chitosan, a potentially biodegradable bio-based polymer for food packaging, is derived from chitin found in insects' exoskeletons. Termite hills can be used to create structures that effectively control temperature, humidity, and air quality because of their complex network of tunnels and ventilation systems (Turner & Soar, 2008). The concept of using insects as food is not new to the world; it has long been ingrained in our society. Around the world, many kinds of insects are consumed and regarded as delicacies.

4. What Is Entomophagy?

"Entomophagy" is the term for the practice of consuming insects, particularly by humans. The Greek words "*entomon*" (notched anima) and "*phagein*" (eating of a certain object) are the source of the name "entomophagy," as are the words "insectivorous" and "insectivore." Conversely, the term "insectivory," which refers to non-human animals consuming insects, is derived from the Latinate Linnaean class "Insecta" (Hooks, 1992; Leer & Kjaer, 2015).

Entomophagy is not a novel concept; rather, it is an ancient habit that has long been prevalent worldwide. For thousands of years, people have been gathering insects in all stages of development—eggs, larvae, pupae, and adults—from forests and other suitable places to consume. It has been estimated that at least 2 billion people regularly consume insects worldwide, ranging from ants to beetle larvae, which are consumed by tribes in Africa and Australia as part of their regular food habits, to the well-known crispy-fried locusts and beetles enjoyed in Thailand (van Huis et al., 2013). People are socialised to view food in ways other than its nutritional content. Because of this, certain dishes are seen as modern and universally acceptable, while others are seen as archaic and widely disapproved of. Therefore, until recently, anthropophagy was not a very common practice in the modern world.

Entomophagy is also heavily influenced by cultural and religious practices. The practice of eating insects can also be cited in religious literature in the Christian and Jewish faiths. The holy book of the Bible states that locusts are food (van Huis et al., 2013). There is a branch of entomology known as cultural entomology, which is the scientific study of insects to explore the influence of insects on culture, i.e., language, literature, art, and religion. Entomophagy is practised predominantly in tropical regions rather than temperate regions because many varieties of edible insects occur seasonally throughout the year. On the other hand, temperate regions experience a scarcity of insects in the winter. Additionally, it is quite easy to harvest insects in the tropics as compared to temperate zones (https://www.britannica.com/topic/entomophagy).

Indigenous people have a rich but limited traditional understanding of entomophagy because it is passed down orally through the generations, and part of it has been lost in translation over time. The younger generation knows very little or nothing about eating insects, particularly in urban settings. Authorities should encourage people to include edible insects in their meals to promote entomophagy in order to educate the youth about this. Edible insect flour ought to be added to foods like bread, energy bars, cereal, biscuits, and cookies to boost acceptance.

5. History of Entomophagy

The people of the past didn't have sophisticated tools or well-developed hunting abilities, so the consumption of insects as food was a common practice (Kim et al., 2019). The first reference to entomophagy was found to be in Europe (Greece), where cicadas were eaten as a cuisine (van Huis et al., 2013). In his book "*Historia Animalium*," Aristotle stated, "The larva of the cicada, on attaining full size in the ground, becomes a nymph; then it tastes best before the husk is broken [i.e., before the last moult]." Locust swarms were treated as a delicacy among the nomads of Arabia and Libya in 1550. In his dissertation, *De Animalibus Insectis Libri Septem*, published in 1602, Ulysse Aldovandi writes of German soldiers in Italy who would often and happily consume fried silkworms. Most Africans, certain Asians, and Arabs in particular eat locusts. They are widely available, roasted or grilled, on their market. They store well for a while when salted. They are used to supply ships and can be consumed as a dessert or with coffee. This cuisine is not disgusting to look at or associate with. It is similar to prawns but may have a more subtle flavour, especially in the females when they are stuffed with eggs. According to van Huis et al. (2013), Foucher d'Obsonville made observations about the laws and customs of different eastern nations in his 1783 philosophical studies on the manners of various foreign creatures. Insects are commonly consumed and used as food in many regions of the world. An elaborated global scenario of insect consumption is provided below.

6. Why Insects as a Dietary Component?

As stated earlier, the practice of eating insects has prevailed in the world since a long time ago. It is important to know about the facts that sustained the Entomophagy culture.

- *Health:* Various reports have been made to identify the nutrition profile of insects, and it has been made very clear by these that insects are a rich source of proteins, carbohydrates, fibres, and minerals. According to reports, proteins account for 4.9–77%, fats range from 0.7–77%, and fibres range from 0.9–29% (Xiaoming et al., 2010). They are therefore regarded as a superb substitute for common foods like chicken, pork, beef, and even fish. Numerous regional and national diets already include insects as a traditional component.
- *Environmental:* Insect rearing is the main factor that is to be considered while stating that insects are an edible resource because to feed a large population, there needs to be a wide range of varieties and populations of insects as well. Because it isn't always a land-based activity and doesn't require clearing land to increase output, it relieves a lot of the earth's burden and encourages the growth of vegetation. Compared to other typical livestock animals, edible insects create a significant reduction in greenhouse gas emissions (for example, only a few bug groups, such as termites and cockroaches, produce methane). Organic waste streams can provide food for insects. Because they have cold blood, they are adept at turning food into proteins (Van Huis et al., 2013).
- *Livelihood (economic and social factors):* Harvesting and rearing insects is a low-tech, low-capital investment option that even generates jobs for the populace by granting access to the most impoverished segments of society, including women and the landless. By selling extra produce as street food, these activities can help individuals lose weight and make money. Small cattle provide options for a living for both rural and urban residents. Depending on the amount of money invested, raising insects can be done in a very simple or complicated way. Raising insects links people to them not only for their dietary needs but also because it can make them aware of the importance of sustainable development and the needs of the planet.

7. Common Edible Insects

More than 1200 species of edible insects have been recorded in Asia, Africa, Oceania, and North and South America out of 1 million known insect species. The most commonly consumed species is Coleoptera (beetles), with 31 per cent of its species. The percentage consumption of Lepidoptera (caterpillars) in sub-Saharan Africa is estimated to be 18 per cent. Hymenoptera (bees, wasps, and ants) stand third with 14 per cent, followed by Orthoptera (grasshoppers, locusts, and crickets) with 13 per cent, Hemiptera (cicadas, leafhoppers, plant hoppers, scale insects, and true bugs) with 10 per cent, Isoptera (termites) with 3 per cent, Odonata (dragonflies) with 3 per cent, Diptera (flies) with 2 per cent, and other orders with 5 per cent (van Huis et al., 2013). Approximately 10–15 species of edible insects are consumed regularly (Feng et al., 2018). Below is Table 1 showing the insect's orders and their examples utilised by humans as food, and Figure 2 shows a percentage-wise comparison of the number of insects belonging to major insect groups of food importance.

Table 1. Orders of class Insecta and edible insects falling under them

Insect orders	Examples of edible members
Coleoptera	Pupae and larvae of beetles
Hemiptera	Large aquatic insects and cicadas
Hymenoptera	Bees and ants
Isoptera	Termites
Lepidoptera	larvae and pupae of moths and butterflies
Orthoptera	Grasses, locusts, and crickets

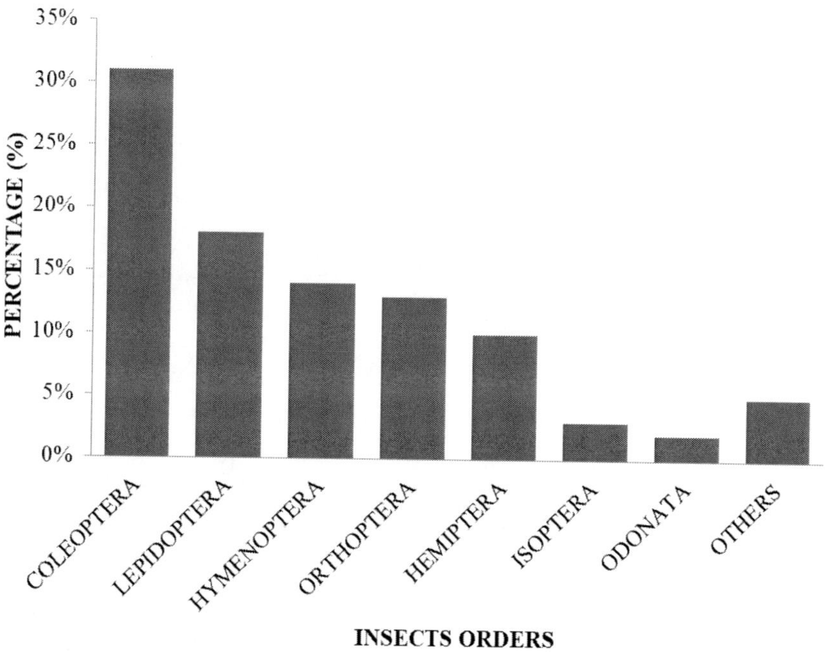

Figure 2. Percentage of insects' consumption according to the orders around the globe.

7.1. Order Coleoptera

The order Coleoptera generally comprises beetles, which form the largest population of edible insects around the world. According to recent studies, more than 460 species of beetles are being used as an edible food source (Elorduy, 2005). There are about 22 genera and 78 species of edible aquatic beetles in the world, which belong to the families Dytiscidae, Gyrinidae, Elmidae, Histeridae, Hydrophilidae, Haliplidae, and Noteridae. (Ramos-Elorduy et al., 2009) Normally, the larvae of these species are consumed worldwide. *Rhnchophorus*, or palm weevil, is found throughout Africa, South America, and southern Asia. In the Netherlands, the larvae of mealworm species that belong to the family Tenebrionideae, like *Tenebrio molitor* (yellow mealworm), *Alphitobius diaperinus* (lesser mealworm), and *Zophobas morio* (superworm), are reared as food for reptiles, fish, and aves. The nettles of the family Meloidae and Blaps are an integral part of folk Chinese medicines (Yang, 2015). Oak longhorn beetles are found in the forests of China and have immunomodulatory effects (Wang, 2011). *Holotrichia diomphalia* is the most commonly found species of white grubs in north-eastern China; they are not only edible but also have medicinal, anti-inflammatory, analgesic, and antioxidant activities (Oh et al., 2003). In countries like India, edible coleopterans are generally found from June to September and then get reduced during winter and early spring (Chakravorty, 2014). The Goliath Beetle is a kind of scarab that is pursued in parts of Africa (http://www.si.edu/spotlight/buginfo/insects-food). Coleopterans are found to be suitable for human consumption and are offered as human food in specialised shops.

7.2. Order Lepidoptera

Butterflies and moths form the major composition of the order Lepidoptera. Mostly, the larval stages (caterpillars) of butterflies and moths are eaten, but it has also been found that adult species are also consumed widely (van Huis et al., 2013). According to estimates made in 2015, about 255 species of lepidopterans are recorded as human food around the globe (Ramos-Elorduy, 2005). As reported by calculated data, 9.5 billion mopane caterpillars are harvested in southern Africa, but the consumption rate of other caterpillars is comparatively lower than theirs (Ghazoul, 2006). The bamboo caterpillar (*Omphisa fuscidentalis*) is a very popular food delicacy and is greatly promoted by the Thai Department of Forestry of the Ministry of Agriculture and Cooperatives in Asia (Yhoung & Viwatpanich, 2005).

Caterpillars of giant skippers in Mexico are taken as delicacies (https://www.si.edu/spotlight/buginfo/insects-food). Lepidopterans form the largest group of edible insects in China. The Chinese caterpillar fungus is well renowned for its excellent medicinal properties (Feng et al., 2018). Silkworm (*Bombyx mori*) is a highly beneficial insect, as it is not only used for silk production but also the pupae of two silk-producing moths, which are edible and are also part of health care food products. The pupae of the Chinese oak tassar moth (*Antheraea pernyi*) are nowadays consumed as food (Gao Yu et al., 2018). The eri-silkworm (*Philosamia ricini*) that is consumed widely in China and other countries originally originated in Assam, India (Li et al., 2017). The cultured *Cordyceps militarisi* (northern caterpillar fungus) is of great importance in medicine. Black rice worms (*Aglossa dimidiata*) and *Pyralis farinalis* are used in the production of insect tea, which is a traditional drink in China and is also exported (Gao Yu et al., 2018). In Mexico, the larvae of the moth *Comadia redtenbacheri* (red maguey worms) and the larvae of the butterfly *Aegiale hesperiaris* (white maguey worms) are a delicacy for Mexican farmers (van Huis et al., 2013).

7.3. Order Hymenoptera

Bees, wasps, and ants are the edible insect varieties belonging to the order Hymenoptera. These are consumed greatly in many parts of the world, especially in southern Asia, with about 355 species (Ramos-Elorduy, 2005). Honey and honey-based products are derived from bees, and they are well known for this, but on the other hand, in addition to harvesting bees for honey production, bee larvae and pupae are also used as eatables. Wasp larvae and pupae are also well-known edible insects (Feng et al., 2018). Ants are considered to be a delicacy all around the world, along with being an important part of ecological services. The larvae and pupae of bees and wasps are consumed as cooked or raw in various parts of the world (Feng et al., 2018). In Asia, the larvae and pupae of the reproductive form (queen brood) of weaver ants are a delicacy. The black weaver ant (*Polyrhachis dives*) is a nutritional ingredient used in various tonics or health foods. The black weaver ant is well-distributed in Bangladesh, India, Malaysia, and Sri Lanka (van Huis et al., 2013). In China, *Formica sanguinea* is used to treat rheumatic arthritis and other diseases (Gao Yu et al., 2018). In India, red ants, hornets, and termites are consumed by many tribes (Patel et al., 2019). In the annual Hebo festival held in Japan, the food products are made from yellow jacket wasps (*Vespula* and *Dolichovespula* spp.), commonly known as Hebo (Nonaka et al., 2008).

7.4. Order Orthoptera

The order Orthoptera comprises grasshoppers, locusts, katydids, and crickets. These are easy to collect because of their large size. Insects are eaten globally, particularly in Japan and Thailand. Around 80 species of grasshoppers are consumed worldwide (van Huis et al., 2013). In China, 43 species of orthopterans are consumed. Locusts occur in swarms, which makes them easy to harvest. Orthopterans are collected early in the morning because they are cold-blooded and immobile (van Huis et al., 2013). In China, fried nymphs and adult grasshoppers, locusts, and crickets are common dishes that are served. *Locusta migratoria manilensis* is a common variety of species that is harvested in plastic houses for the production of human food (Li et al., 2013). *Oxya chinensia*, a Chinese rice grasshopper, is a common edible grasshopper species in China (Osimani et al., 2017). In Latin America, chapuline (edible grasshoppers of the genus *Sphenarium*) has been a part of local diets for ages. The valley of Oaxaca is famous for the consumption of chapuline (Cohen et al., 2009). In Asia, *Gryllotalpa orientalis* (oriental mole cricket), *Gryllus bimaculatus*, *Teleogryllus occipitalis*, and *T. mitratus* species of crickets are harvested and consumed as food (van Huis et al., 2013). In Thailand, *Acheta domesticus* (house cricket) is reared and eaten commonly and is preferred over all other species because of its soft body. *Brachytrupes portentosus* (short-tail cricket), which has a large body and head, is also a common insect meal (van Huis et al., 2013).

7.5. Order Isoptera

Termites are the major isopterans that are eaten widely across the world. About 65 species of termites are being identified (Ramos-Elorduy, 2005). In China, about 30 species of termites are identified, out of which only 16 are named scientifically (Feng et al., 2018). The most commonly consumed termites are the large *Macrotermes* species. The adult and larval forms of this species are equally nutritive as its nest. In the Amazon, *Syntermes* are eaten largely (van Huis et al., 2013). Dried bodies of *Eupolyphaga sinensis* are used as medicine and as a novel food source (Gao Yu et al., 2018).

7.6. Order Hemiptera

The most commonly edible species of the order Hemiptera are cicadas, leafhoppers, aphids, and true bugs. More than 190 species are being identified across the world (Ramos-Elorduy, 2005). Cicada species like Iona, *Platypleura*, and *Pycna* are consumed as food in Malawi. *Dactylopius coccus* (cactus cochineal bug) is useful for the production of food products like carmine dye. In South Africa, pentatomid bugs are consumed, especially in Sub-Saharan Africa (van Huis et al., 2013). In China, 20 edible species of Hemiptera have been recorded (Feng et al., 2018). Cicadas, stinkbugs, and bugs are being described as edible items in the ancient literature of China (Liu, 1991). *Cryptotympana at*rata (black cicada) is a widely consumed insect and is also an essential component of Chinese medicines (Yu et al., 2018). In the Republic of Sudan, *Agonoscelis versicolor* is a pest that can be eaten roasted. Bugs can also be used to derive oils to prepare food and for the treatment of scab disease in camels (van Huis et al., 2013).

7.7. Order Diptera

The order generally comprises flies and cockroaches; these are not typical foods for people and are used as animal feed. Black soldier flies and houseflies are the different species of fly that feed on organic matter to convert organic waste into fertiliser. In this way, humans indirectly benefit from the order Diptera (Feng et al., 2018). The larva of a housefly (*Musca domestica*) is a source of high-quality proteins, and its powdered form is used for the treatment of pro-inflammatory responses (Yu et al., 2018). Cockroaches *Periplaneta americana* were used as medicines in ancient times and the present pharmaceutical industries (Hu et al., 2008; Luo et al., 2012). Apart from being used as medicine, cockroaches are also used as human food sources (Feng et al., 2018).

7.8. Order Odonata

Dragonflies are an edible species of the order Odonata throughout the world. About 29 species are recorded in the world (Ramos-Elorduy, 2005). Adult forms and nymphs are consumed by the people in China. Among various species of dragonflies, *Anax parthenope* is the most common edible insect (Feng et al., 2018).

8. Global Status of Entomophagy

8.1. Entomophagy in Oceania

Rhynchophorus ferrugineus, or sago grub, is a delicacy and a feature of an annual grub festival in Papua New Guinea (Kim et al., 2019). Locals in the area eat several species of locusts, crickets, mantises, and spiders (Meyer-Rochow & Changkija, 1997). Entomophagy is generally less known among the people of Australia. *Agrotis infusa* (Bogong moth) is reported to be a food of indigenous Australians (Flood, 2008). Among the variety of caterpillars, witchetty grubs are the most renowned in Australia (Meyer-Rochow, 2005). The insect market in Australia is increasing dramatically, and delicacies of insects are now being introduced to menus (Irvine, 1989). Wasps (*Vespula* spp. and *Vespa* spp.) are imported from New Zealand for consumption (van Huis et al., 2013).

8.2. Entomophagy in Asia

With over 9% of the Asia-Pacific edible insect and insect protein industry, Korea is one of the top markets for both edible insect consumption and edible insect protein. In Korea, rice-field grasshoppers are eaten as a snack, lunch, or side dish. In Korea, Thailand, and Japan, silkworm pupae are frequently consumed. It is also a delicacy of China and Vietnam as it is a rich source of proteins (Kim et al., 2019). In Japan, grasshoppers (*Oxya yezoensis*) are eaten fried or boiled as a side dish or snack (Nonaka, 2009). Rice is typically served with raw or boiled wasps or bee larvae and adults, which are considered delicacies (DeFoliart, 1999). China has long engaged in entomophagy, with an estimated 324 species from 11 orders being consumed (Feng et al., 2018). India is well known for using insects for food, medicine, and silk fertiliser, among other purposes. Northeast India's delicacy is the pupae of the eri silkworm (*Samia ricini*) (Peigler, 1993). In India, the locust *Schistocerca gregaria* is used as fertiliser and as food for humans (Das, 1945). Termites are used as a source of protein and carbs in Meghalaya, Orissa, and Karnataka. Thus far, India has identified over 225 different species of edible insects (Chakravorty, 2014). The Nyishi and Galo tribes in Arunachal Pradesh eat 102 different species of insects (Chakravorty et al., 2011). Due to its importance as a source of protein, fat, and other nutrients, the residents of the Thai village of Ubon alone consume 20–60 grammes of insects per day (Sungpuag & Puwastien, 1983). Figure 3 depicts a few of the regularly consumed insect cuisines.

8.3. Entomophagy in Africa

Insects emerge as an important food source on the African continent when staples are scarce. The availability of insects in Africa greatly depends on their seasonal availability. Caterpillars are found in abundance during the rainy season (van Huis et al., 2013). Termites are also commonly eaten edible insects in Africa. In Nigeria, various species of termites like *Macrotermes natalensos* and locusts are eaten in their adult form, whereas the larvae of beetles and weevils are consumed by the locals (DeFoliart, 1999). In Uganda, grasshoppers are considered a source of energy and proteins (Mbabazi, 2011). In the Democratic Congo, more than 65 species of edible insects are consumed, including beetles, silk moths, termites, and bees (DeFoliart, 1999). In total, about 470 species of insects are eaten across the African continent as local food or delicacies (Kim et al., 2019).

8.4. Entomophagy in America

It is thought that residents of Mexico devour 27 different species of caterpillars. There, two leafcutter ant species—Atta mexicana and A. cephalotus—are starting to be sold commercially (van Huis et al., 2013). The chapuline is arguably the most popular edible grasshopper in Latin America. This small grasshopper has long been used in traditional Mexican cuisines and is still eaten in some regions of the country. In addition to being a pest of lucerne, Sphenarium purpurascens is one of the most desirable culinary insects in Mexico. In Mexico, it is common to find people consuming both adults and nymphs of stink bugs (Hemiptera: Pentatomidae) (DeFoliart, 2002). Escamol, a fried bug meal with aromatic spices, is a popular delicacy in Mexican restaurants (DeFoliart, 1999). The Yukpa people of Colombia prefer insects to meat; however, they have been forced to reduce bug eating due to severe deforestation (Ruddle, 1973). Table 2 shows commonly consumed insects in different countries.

8.5. Western Phobia

In Western countries, eating insects is still considered taboo, despite the indisputable health benefits of doing so (Sidali et al., 2019). People in Western countries view them as a threat and a health risk, even though they provide an intriguing supply of high-quality nutrients (Looy et al., 2014). For a very long time, catching insects has been associated with the hunter-gatherer era and "primitive" food sources. Since the advent of agriculture and the rise of sedentary lifestyles, insects have been viewed as pests. This stands in sharp contrast to numerous tropical regions across the globe where insects are used for myth, folklore, dance, medicine, fun, and decoration (Meyer-Rochow, 1979; Yen et al., 2013). In Western countries, where the majority of protein is still derived from farmed animals, insects such as termites, flies, and mosquitoes are nearly synonymous with irritation since they are viewed as a nuisance to humans. Butterflies and ladybirds are two rare insects that do not elicit feelings of aversion, avoidance, distaste, or disgust. (van Huis et al., 2013). Furthermore, religious concerns are associated with this limited usage of edible insects (Messina et al., 2019).

Various academics collaborated to look into the causes behind the widespread rejection of edible insects in Western countries, and several factors were discovered. Many studies have recently found that food neophobia is one of the most important indicators of consumer readiness to eat insects as an individual trait (Megido et al., 2014). It is rejection based on dislike for the organoleptic features, fear of bodily harm, or disgust based on a preconceived notion of what the product is or where it came from (Rozin & Fallon, 1980). Food exposure boosts product familiarity and influences the desire to try new foods, lowering neophobic reactions (Mancini et al., 2019). In contrast to products that are advertised as having insects in their composition, western consumers are more accepting of insects when they are ground into flour or paste because they are unaware of their presence, despite the well-documented aversion to insects as food among people in Western societies (Meyer-Rochow & Hakko, 2018). A survey of consumers was carried out in some European countries to find out if consumer information affects their inclination to eat bug meals. Of those surveyed, 34% said they would eat and 38% said they would probably eat. Only 3% of respondents indicated they would never consume insects, compared to 22% who were unsure of success (Piha et al. 2018). The main conclusions of the study showed that knowledge influences a person's propensity to eat insect food; the effects of knowledge differed according to the nation or location, with central and northern Europe having different consequences. Finally, it was demonstrated that the possible marketing of insect feeds in northern European countries would be more successful (Piha et al. 2018).

Table 2. Most consumed insects as food resources in different countries

Country	The scientific name of consumed insects	Common name of consumed insects	Consumed type
Korea	*Bombyx mori*	Silkworm	Pupae
	Oxya velox	Grasshopper	Adult
Japan	*Oxya yezoensis*	Grasshopper	Adult
	Vespula lewisii	Bee	Larvae, Pupae
	Bombyx mori	Silkworm	Pupae, Female Adult
China	*Bombyx mori*	Silkworm	Pupae
	Antheraea pernyi	Tassar Silkworm	Pupae
	Tenebrio molitor	Mealworm	Larvae
	Polyrhachis dives	Weaver Ants	Adult
	Apis cerana	Honey Bee	Larvae, Pupae
	Locusta migratoria	Locust	Adult
	Odontotermes formosanus	Termites	Adult
India	*Samia cynthia*	Eri silkworm	Larvae, pupae, adult
	Schistocerca gregaria	Locust	Adult
	Apis cerana	Honey bee	Adult
Thailand	*Allonemobius fasciatus*	Ground cricket	Adult
	Anurogryllus arboreus	Short-tailed cricket	Adult
	Bombyx mori	Silkworm	Pupae
	Omphisa fuscidentalis	Bamboo caterpillar	Larvae
Papua New Guinea	*Rhynchophorus ferrugineus*	Sago grub	Larvae
		Locust	Adult
		Mantis	Adult
		Crickets	Adult
Nigeria	*Macrotermes natalensos*	Termits	Adult
	Microtermes bellicosus	Termites	Adult
	Rhynchophorus phoenicis	Palm weevil	Larvae
	Aphodius rufipes	Dung beetle	Larvae
	Apis mellifera	Honey bee	Egg, larvae, pupa, honey
	Gymnogryllus lucens	Cricket	Adult
	Zonocerus variegatus	Grasshopper	Adult
Uganda	*Macrotermes spp.*	Termites	Adult
	Ruspolia nitidula	Grasshopper	Adult
Congo	*Oryctes owariensis*	Rhinoceros beetle	Larvae
	Anaphe panda	Silkmoth	Larvae
		Termites	Adult
		Bees	Egg, larvae, pupa, honey
Mexico	*Rhynchophorus palmarum*	Palm weevil	Larvae
	Sphenarium	Grasshopper	Adult
	Schistocerca	Grasshopper	Adult
	Liometopum apiculatum	Ants	Larvae and Pupae
	Liometopum occidental	Ants	Larvae and Pupae
Colombia	*Aidemona azteca*	Grasshopper	Adult
	Podischnus agenor	Beetles	Larvae
	Atta	Ants	Adult
	Trigona trinidadensis	Bees	Honey
	Trigona clavipes	Bees	Honey

Grubs used as an insect cuisine

Chapuline made up of grasshoppers.

Fried Crickets

Locust Taco

Hebo-Wasp Cuisine

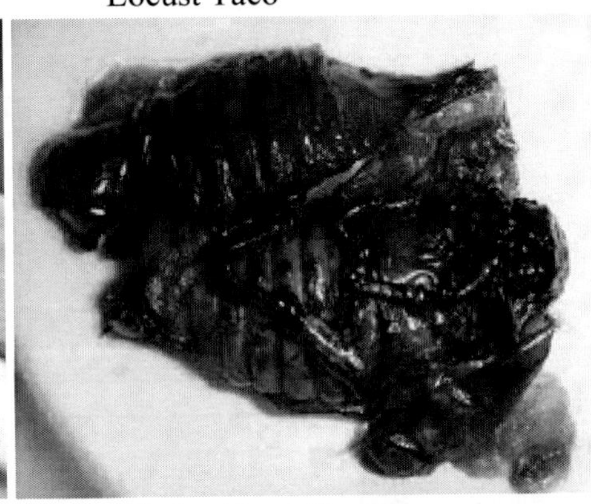

Choice Dung Beetle

Chocolate-covered insect snacks

Silkworm Toast

Figure 3. Some commonly eatable insects' cuisines.

According to DeFoliart (1999), "Westerners should become aware of the fact that their bias against insects as food has an adverse impact, resulting in a gradual reduction in the use of insects without replacement of lost nutrition and other benefits" (van Huis et al., 2013). Many studies are still being conducted to investigate the use of insects as food or food additives for human consumption, either in terms of consumer acceptance or product qualities (Guiné et al., 2020). According to some academics, scientists should acknowledge that insects have long been an important source of nutrition in the world's poorer regions and work to promote this fact rather than oppose or ignore it.

9. Edible Insects- Enriched Source of Nutrients

The issue of malnutrition has always been in our world. 825 million people suffer from malnutrition worldwide, with China and India having the highest rates (UN, 2014). This is because the food consumed in these countries is primarily plant-based, low in protein with a high biological value content, and has a low nutrient density. As a result, even with enough calories supplied, the diet may be lacking in important nutrients (WHO, 2008). Insects are an alternative to fighting against malnutrition. Many studies have been conducted across the globe, and it has been proven that insects are one of the richest sources of high-quality nutrients, and their consumption can also help to drive away the prevailing problem of malnutrition.

The nutritional value of insects is influenced by various factors such as species, growth conditions, sex, diet, developmental stage, and assessment techniques (van Huis & Oonincx, 2017). In terms of nutrients, insects are comparable to beef, pork, chicken, and fish. Fat, proteins, fatty acids, fibres, dietary minerals, and vitamins form the major composition of edible insects (https://www.fao.org/edible-insects/84625/en/). Below is a discussion of these elements:

9.1. Proteins and Amino Acids

Proteins are present in large proportions in insects. Protein content is generally higher than that of soybeans and comparable to that of poultry animals (Bosch et al., 2014). Raw protein concentration ranges from 20–70% in almost 100 edible insect species studied during the egg, larva, pupa, and adult stages (Xiaoming et al., 2010). Proteins are made up of more than 20 kinds of amino acids. Proteins contribute to the physical and sensory properties of the individual, along with providing food nutrition. Out of the 20 kinds of amino acids, eight are not synthesised in the human body and hence are to be taken from an outer source, which is food. Cereal proteins are a staple of many diets worldwide, although they often lack the amino acids tryptophan and threonine (found in, for example, maize) and are low in lysine. Certain bug species have high concentrations of these amino acids (Bukkens, 2005). For instance, the lysine amino acid score of numerous aquatic insects, palm weevil larvae, and caterpillars in the Saturniidae family is greater than 100 mg/100 g of pure protein.

When mopane caterpillars are dry-roasted (48%) as opposed to dried (57%), the protein content decreases (Bukkens, 1997). The protein composition of insects varies greatly across species. Meal variations also occur in the quantity of protein (e.g., vegetables, grains, or trash).

In Nigeria, grasshoppers given bran—which has a high concentration of important fatty acids—have almost twice as much protein as those fed maize. The protein content also varies depending on the stage of metamorphosis; it has been found that adults have a higher protein content than instars (Ademolu et al., 2010).

9.2. Fats and Fatty Acids

According to van Huis et al. (2013), fats make up the second-largest component of the edible insects' nutrient content. These are the food's most energy-dense macronutrients, which are necessary for the storage and production of energy as well as the maintenance and defence of many organs. Saturated and unsaturated fatty acids are the two additional divisions of fatty acids. In general, it is believed that unsaturated fatty acids are better for human health than saturated ones. The amount of fat varies between species and is also influenced by what insects eat. Most insects have a healthier ratio of saturated to non-saturated fatty acids—less than 40%—than fish or poultry. The fatty acid composition of grasshoppers and white ants was 60% and 75%, respectively. Silkworm pupae fat content ranges from 20–40%, with a high concentration of polyunsaturated fatty acids (Raheem et al., 2019).

Numerous analyses have shown that many edible insects have higher lipid contents in their larval and pupal stages than in their adult stages. For example, the fat content of adult *Oxya chinensis* is only 2.2%, whereas the average fat content of edible insects is estimated to be between 10 and 20 per cent. Lepidopteran larvae and pupa exhibit substantial fat contents, similar to those of *Ostrinia furnacalis* (46.08%) and *Pectinophora gossypeilla* (49.48%) (van Huis, 2016).

9.3. Carbohydrates

Insects have a comparatively lower amount of carbohydrates than fats and proteins, ranging from 1–10%, and that too varies from species to species. Insect tea is an unusual source because it has a great carbohydrate content (16.27%) (van Huis, 2016). Chitin, a macromolecular compound that forms the exoskeleton of insects, is known for its high nutritive and healthy food values. According to Raheem et al. (2019), they are said to have immunity-boosting properties by inhibiting the growth of harmful microbes and encouraging the growth of helpful bacteria. More than 90% of insect skin is composed of chitin; the percentage of chitin varies amongst edible insect species (5–15%), with *Bombyx mori* dried pupa having a chitin content of 3.73 per cent and de-fatted pupa having a chitin content of 5.55 per cent (van Huis, 2016).

9.4. Minerals

Micronutrients, including minerals, play a vital role in the nutritional value of food. In developing countries where micronutrient deficiency is common, edible insects are proven to be a boon to them in overcoming major health consequences. It has been determined that iron insufficiency is the most prevalent and pervasive nutritional problem worldwide. Nearly 40%

of preschoolers and one in two pregnant women in underdeveloped nations are thought to be anaemic, according to the WHO. Due to their undeniable high iron content, eating insects can help improve iron status and treat anaemia in developing nations. For instance, termites and caterpillars have respective iron contents of 35.5 mg and 27–29 mg per 100 grams of dry weight (Banjo et al., 2006; van Huis et al., 2013; 2016). According to Raheem et al. (2019), grasshoppers and weevils contain 7.54–8.21 mg of magnesium per 100g.

Zinc deficiency is another significant public health concern that has detrimental effects on the health of mothers and children. According to van Huis et al. (2013), deficiencies in zinc can cause skin lesions, growth retardation, delayed sexual and bone maturation, and an increased risk of infections. It is thought that most insects are a good source of zinc. The zinc content of the palm weevil, *Rhynchophorus phoenicis*, is 26.5 mg/100g (Bukkens, 2005).

9.5. Vitamins

A group of chemical compounds known as vitamins are necessary for human metabolism. The majority of consumable insects are rich in vitamins, which are essential for boosting the immune system and metabolic function. Although there hasn't been enough research done on vitamins in edible insects, it has been shown that they include carotene as well as vitamins A, B1, B2, B6, D, E, K, and C (DeFoliart 1991; Lu et al. 1992; Chen & Feng 1999).

It has been shown that certain caterpillars, including *Imbrasia oyemensis*, *I. truncata*, and I. *epimethea*, contain retinol and β-carotene (vitamin A). In general, insects don't provide the finest vitamin A. According to van Huis et al. (2013), the powdered and freeze-dried silkworm (*Bombyx mori*) has a comparatively high amount of vitamin E (9.65 mg/100 g). For every 100 grammes of dry weight, honey bee eggs contain 12.44 mg of vitamins A, B2, and C, whereas their larvae have 3.24 mg. 10.25 mg per 100 g dry weight are present in honey bee pupae (Bukkens, 1997).

10. Food Safety Characterisation of Insects

Insects are a great source of nutrition, but they can also contain physical risks, chemical and biological toxins, and other issues that scientists and consumers are concerned about (FAO, 2021). Only sometimes are edible insects consumed; they are gathered from their natural habitats and sold at public marketplaces. Contaminants can build up in an insect's food or housing materials. Eating edible insects raised in sanitary, regulated environments may provide different food safety concerns than eating insects collected in the wild (Li et al., 2016; Murefu et al., 2019). To develop insect-species-appropriate sanitary methods, it is necessary to characterise numerous food safety-associated dangers in edible insects.

It is commonly known that insects can carry a variety of microorganisms that are harmful to both human and animal health (Kooh et al., 2019). In addition to being pathogenic and opportunistic, several bacterial species from the genera *Staphylococcus*, *Streptococcus*, *Bacillus*, and *Pseudomonas* can shorten the shelf life of edible insects. To avoid this, strict biosafety protocols and the avoidance of contact with livestock animals should be followed (FAO, 2021). Despite the minimal risk of contracting foodborne viruses, including hepatitis A, hepatitis E, and nonvirus, from consuming edible insects, it's crucial to prevent the virus from entering the insect production process through substrates (Vandeweyer et al., 2020). Insects can potentially serve as replicative vectors for viruses that infect vertebrae. Foodborne fungus can cause food spoilage by deteriorating product quality and causing nutritional losses (FAO, 2021).

Insect species deemed suitable for mass production may be parasite vectors, and this risk must be taken into account. The parasitic dangers connected with edible insects that can impact humans, on the other hand, are understudied. Certain protozoan species have been found in edible insects, including *Entamoeba histolytica*, *Balantidium* spp., *Isospora* spp., *Giardia lamblia*, *Toxoplasma* spp., and *Sarcocystis* spp. (Gałęcki & Sokól, 2019). Humans who ate black army fly larvae grown on ripe, unwashed fruits that had been oviposited by flies developed intestinal myiases (Lee et al., 1995; Wang & Shelomi, 2017). Chicken dung contains coccidia parasites. If insects are raised on poultry dung, proper processing procedures must be created before the insects are fed to animals (FAO, 2021).

Insects grown on agricultural waste may contain dioxins, toxic metals, mycotoxins, and other chemical risks like pesticides used for crop protection. Manure from cattle and poultry used for insect rearing may contain antimicrobials and insecticides. The potential transfer and accumulation of these contaminants in insects used as food for humans or animal feed should be properly evaluated and treated by this developing sector (FAO, 2021). Aflatoxin levels of up to 50 g/kg were discovered in several commercial lots of "ready-to-eat" mopane worms in Botswana, highlighting the need for hygienic handling, processing, and storage of edible insects (Mpuchane et al., 1996, 2000). Pesticides employed on crops may be acquired by insects raised on treated plant-based side stream materials (Houbraken et al., 2016). Numerous factors, such as the kind of metal, the insect species, the stage of growth, the substrates used, and environmental contamination, have been connected to the accumulation of toxic metals by edible insects. The Mexican state of Oaxaca's chapulines, a term for dried grasshoppers, were found to contain high amounts of lead. FAO (2021). Before using insects as food or feed, the maximum concentrations of specific heavy metals such as arsenic, mercury, cadmium, and lead must be established based on the species of the insects.

11. FAO for Entomophagy

The director at FAO has stated that "forests contribute to the livelihoods of more than a billion people, including many of the world's neediest. Forests provide food, fuel for cooking, fodder for animals, and income to buy food. Wild animals and insects are often the main protein sources for people in forest areas, while leaves, seeds, mushrooms, honey, and fruits provide minerals and vitamins, thus ensuring a nutritious diet." The FAO is working on introducing edible insects in many countries worldwide. The contributions made by FAO towards this are:

- Spreading knowledge through publications, expert meetings, and a web portal on edible insects;
- Collaborating with media (e.g., newspapers, magazines, and TV) to raise awareness about edible insects;

- Supporting member countries to establish projects for the introduction of edible insects;
- Networking and interdisciplinary connections with different sectors both inside and outside FAO (e.g., stakeholders working with nutrition, feed, and legislation-related concerns).

12. Development of Edible Insect Programmes

12.1. Ray of Income for Poor

Harvesting insects can be done easily by farmers and villagers. In South Africa, mopane caterpillars are collected by village women from the wild; annually, 9.5 billion mopane larvae are harvested, worth about 85 million US dollars; out of these 85 million dollars, about 40% goes to producers, i.e., rural women. This way, many such programmes can be initiated to encourage villagers to opt for insect harvesting as their profession, which can subsequently address food security problems both by increasing incomes and by increasing the availability of nutritious food. Insect farming is emerging as a factor in fighting rural poverty in Cambodia. The edible insect market is expected to exceed 270 million dollars by 2024 (Guiné et al., 2021).

12.2. Technical Support for Rearing Edible Insects

When insects or other small creatures are generated or collected for commercial purposes, such as selling them for human consumption or animal feed, they might be referred to as "mini livestock" (Guiné et al., 2021). Production of insects is divided into two systems: 1) completely domesticated and raised in captivity, and 2) partially raised in captivity, keeping them connected to their wild populations but modifying the environment to boost yield (Tabassum-Abbasi et al., 2016; Feng et al., 2018; Ghosh et al., 2018). Edible scorpions are reared at an industrial scale in China. Some crickets and water bottles are reared on semi-industrial scales. Mass-rearing methods are available for the production of maggots and soldier flies as livestock feed.

12.3. Improved National Integrated Pest Management Programmes

According to FAO research, eating edible insects is a sensible and sustainable strategy to fight chronic hunger and open up new revenue-generating options. After malaria, malnutrition is the second-leading cause of death for children, affecting as many as 5 million of them. Improving nutritional practices is desperately needed, especially in urban slums. Protein and calorie deficits are the most prevalent forms of malnutrition, followed by micronutrient insufficiencies. High levels of protein, fat, and micronutrients can be found in insects.

The increase in demand for the production of food is going to be raised by mainly developing nations like India, Brazil, and Argentina due to the increase in the protein requirement for individuals for their survival, and this need is fulfilled by meat (produced by livestock animals). In the current situation, it is nearly impossible to meet the requirements of food production because of the limited availability of farmland. Currently, 30% of the global land area is consumed by farmland, and out of this, 70% is used for the production of animal feed (Yen, 2009). Livestock animals utilise 77 million metric tonnes of proteins and then produce 58 million metric tonnes of protein for human consumption (Steinfeld et al., 2006). This data makes the pressure on natural resources very clear and states the upcoming danger to the environment because of the rapidly growing population. We cannot ignore the demands for animal proteins as pet food in Western countries, which indicates a worst-case scenario.

Conclusion

Today, when the natural resources of the planet are ceasing to meet the growing need for nutritious food for the population, entomophagy is one of the best solutions to solve the need for nutrients globally. Along with serving our ecology, insects are now widely accepted as a part of people's diets because they provide high amounts of proteins, fats, vitamins, and mineral elements. The nutritional profile of insects is way better than that of livestock animals or staple foods that are generally consumed. Out of 1200 known species of edible insects, the highest number belongs to Coleoptera, making wasps and beetles widely consumed across the world. Insects are considered a delicacy in various countries like China, Japan, and Africa, but Western societies still do not accept insects as eatable. Westerners consider insects disgusting and a health hazard. In India, as well, insects are being consumed by many tribal groups. Various cuisines of silkworm larvae, beetles, wasps, crickets, and ants are part of the menus of numerous eateries.

People still need to be made aware of the need to include edible insects in their diets. The FAO is proposing several measures to promote the eating of edible insects since doing so will not only ensure a future population that is well-fed but also present chances to make a living by turning insect raising into a profitable enterprise. Countries must encourage social and scientific communities to collaborate in support of the use of insects as food and feed, in addition to global efforts. Policies and rules must be put in place to address social concerns and guarantee that the production and use of food and feed derived from insects do not damage the environment, hurt agriculture and forest plants, or hurt human health. Accurate information on edible insects must be provided through education and publicity to enlighten the public and boost consumption. People's perceptions about insects need to shift so that we can perceive them as a wholesome, nutrient-dense, and environmentally friendly food source for all people, not just a nuisance or a source of sustenance for the underprivileged and illiterate.

Acknowledgments

The authors are thankful to the Vice Chancellor, Banasthali Vidyapith and the Head of the Department of Bioscience and Biotechnology for providing the necessary facilities to facilitate the work. Facilities provided by DST-FIST to the University are highly acknowledged.

References

Abbasi, T., & Abbasi, S. A. (2016). Reducing the global environmental impact of livestock production: the mini livestock option. *Journal of Cleaner Production*, *112*, 1754-1766.

Ademolu, K. O., Idowu, A. B., & Olatunde, G. O. (2010). Nutritional value assessment of variegated grasshopper, *Zonocerus variegatus* (L.) (Acridoidea: Pygomorphidae), during post-embryonic development. *African Entomology*, *18*(2), 360-364.

Bukkens, S. G. (1997). The nutritional value of edible insects. *Ecology of Food and Nutrition*, *36*(2-4), 287-319.

Bukkens, S. G. (2005). Insects in the human diet: nutritional aspects. *Ecological implications of mini livestock: potential of insects, rodents, frogs and snails*, 545-577.

Caparros Megido, R., Sablon, L., Geuens, M., Brostaux, Y., Alabi, T., Blecker, C., & Francis, F. (2014). Edible insects acceptance by Belgian consumers: promising attitude for entomophagy development. *Journal of Sensory Studies*, *29*(1), 14-20.

Chakravorty, J. (2014). Diversity of edible insects and practices of entomophagy in India: an overview. *Journal of Biodiversity, Bioprospecting and Development*, *1*(3), 124.

Chakravorty, J., Ghosh, S., & Meyer-Rochow, V. B. (2011). Practices of entomophagy and entomotherapy by members of the Nyishi and Galo tribes, two ethnic groups of the state of Arunachal Pradesh (North-East India). *Journal of ethnobiology and ethnomedicine*, *7*, 1-14.

Cohen, J. H., sáánchez, N. D. M., & montiel-ishino, F. (2009). Chapulines and food choices in rural Oaxaca. *Gastronomica*, *9*(1), 61-65.

Das, S. (1945). Locust as food and manure. *Indian Farming*, *8*.

DeFoliart, G. R. (1999). Insects as food: why the Western attitude is important. *Annual review of entomology*, *44*(1), 21-50.

DeFoliart, G. R. (2002). The human use of insects as a food resource: a bibliographic account in progress. University of Wisconsin, Madison, WI, USA. *Edible insect value chains in Africa*.

FAO. 2021. Looking at edible insects from a food safety perspective. Challenges and opportunities for the sector. Rome.

Feng, Y., Chen, X. M., Zhao, M., He, Z., Sun, L., Wang, C. Y., & Ding, W. F. (2018). Edible insects in China: Utilization and prospects. *Insect Science*, *25*(2), 184-198.

Flood, J. (1980). The moth hunters: Aboriginal prehistory of the Australian Alps. *(No Title)*.

Gałęcki, R., & Sokół, R. (2019). A parasitological evaluation of edible insects and their role in the transmission of parasitic diseases to humans and animals. *PLoS One*, *14*(7), e0219303.

Gao, Y., Wang, D., Xu, M. L., Shi, S. S., & Xiong, J. F. (2018). Toxicological characteristics of edible insects in China: A historical review. *Food and Chemical Toxicology*, *119*, 237-251.

Ghazoul, J. (2006). Mopane Woodlands and the Mopane Worm: Enhancing rural livelihoods and resource sustainability. Final Technical Report.

Ghosh, S., Jung, C., & Meyer-Rochow, V. B. (2018). What governs selection and acceptance of edible insect species? *Edible insects in sustainable food systems*, 331-351.

Hooks, B. (2014). Black looks: Race and representation. South End Press, Boston, MA.

Houbraken, M., Spranghers, T., De Clercq, P., Cooreman-Algoed, M., Couchement, T., De Clercq, G., & Spanoghe, P. (2016). Pesticide contamination of *Tenebrio molitor* (Coleoptera: Tenebrionidae) for human consumption. *Food Chemistry*, *201*, 264-269.

http://www.si.edu/spotlight/buginfo/insects-food.

https://www.britannica.com/topic/entomophagy.

https://www.fao.org/edible-insects/84625/en/.

https://www.fao.org/state-of-food-security-nutrition.

https://www.fao.org/state-of-food-security-nutrition.

Irvine, G. (1989). Putting insects on the Australian menu.

Joint, F. A. O. (2019). Hazards associated with animal feed: Joint FAO/WHO expert meeting FAO headquarters, Rome, Italy 12–15 May 2015.

Kim, T. K., Yong, H. I., Kim, Y. B., Kim, H. W., & Choi, Y. S. (2019). Edible insects as a protein source: A review of public perception, processing technology, and research trends. *Food science of animal resources*, *39*(4), 521.

Kooh, P., Ververis, E., Tesson, V., Boué, G., & Federighi, M. (2019). Entomophagy and public health: a review of microbiological hazards. *Health*, *11*(10), 1272-1290.

Lee, H. L., Chandrawathani, P., Wong, W. Y., Tharam, S., & Lim, W. Y. (1995). A case of human enteric myiasis due to larvae of *Hermetia illucens* (Family: Stratiomyiadae): first report in Malaysia. *The Malaysian journal of pathology*, *17*(2), 109-111.

Leer, J., & Kjær, K. M. (2015). Strange Culinary Encounters: Stranger Fetishism in Jamie's Italian Escape and Gordon's Great Escape. *Food, Culture & Society*, *18*(2), 309-327.

Lewis, R. V. (1992). Spider silk: the unraveling of a mystery. *Accounts of chemical research*, *25*(9), 392-398.

Li, L., Xie, B., Dong, C., Wang, M., & Liu, H. (2016). Can closed artificial ecosystem have an impact on insect microbial community? A case study of yellow mealworm (*Tenebrio molitor* L.). *Ecological Engineering*, *86*, 183-189.

Ling, L. I., Yue, L. O. N. G., Xiaowen, D. I. N. G., & Xianzhi, H. U. A. N. G. (2017). The acute toxicity and genotoxicity of Cassava Silkworm Pupa. *Food and Machinery*, *33*(4), 49-53.

Liu, D. Y. (1991). The ancient Chinese edible insects and related issues. *Ancient and Modern Agriculture in China*, *9*, 1-8.

Looy, H., Dunkel, F. V., & Wood, J. R. (2014). How then shall we eat? Insect-eating attitudes and sustainable food ways. *Agriculture and human values*, *31*, 131-141.

Luo Ting Shun, L. T., Gao Meng Ting, G. M., Ma Fang Fang, M. F., Liu Guang Ming, L. G., & Zhang Cheng Gui, Z. C. (2012). Research advances in pharmacological action and clinical application of *Periplaneta americana*.

Mancini, S., Sogari, G., Menozzi, D., Nuvoloni, R., Torracca, B., Moruzzo, R., & Paci, G. (2019). Factors predicting the intention of eating an insect-based product. *Foods*, *8*(7), 270.

Mbabazi, M. (2011). *Dietary contribution of grasshoppers (Ruspolia nitidula) and white ants (Macrotermes bellicosus) and influence of processing methods on their nutrient composition* (Doctoral dissertation, Makerere University).

Messina, C. M., Gaglio, R., Morghese, M., Tolone, M., Arena, R., Moschetti, G., & Settanni, L. (2019). Microbiological profile and bioactive properties of insect powders used in food and feed formulations. *Foods*, *8*(9), 400.

Meyer-Rochow, V. B. (1979). *The diverse uses of insects in traditional societies*. Arbeitstelle für Ethnomedizin.

Meyer-Rochow, V. B. (2005). Traditional food insects and spiders in several ethnic groups of Northeast India, Papua New Guinea, Australia, and New Zealand. *Ecological implications of mini livestock-potential of insects, rodents, frog and snail. USA, Science Publishers Enfiel*, 389-414.

Meyer-Rochow, V. B., & Changkija, S. (1997). Uses of insects as human food in Papua New Guinea, Australia, and North-East India: Cross-cultural considerations and cautious conclusions. *Ecology of Food and Nutrition*, *36*(2-4), 159-185.

Meyer-Rochow, V. B., & Hakko, H. (2018). Can edible grasshoppers and silkworm pupae be tasted by humans when prevented to see and smell these insects? *Journal of Asia-Pacific Entomology*, *21*(2), 616-619.

Mozhui, L., Kakati, L. N., Kiewhuo, P., & Changkija, S. (2020). Traditional knowledge of the utilization of edible insects in Nagaland, North-East India. *Foods*, *9*(7), 852.

Mpuchane, S., Gashe, B. A., Allotey, J., Siame, B., Teferra, G., & Ditlhogo, M. (2000). Quality deterioration of phane, the edible caterpillar of an emperor moth *Imbrasia belina*. *Food Control*, *11*(6), 453-458.

Mpuchane, S., Taligoola, H. K., & Gashe, B. A. (1996). Fungi associated with *Imbrasia belina*, an edible caterpillar. *Botswana Notes and Records*, 193-197.

Murefu, T. R., Macheka, L., Musundire, R., & Manditsera, F. A. (2019). Safety of wild harvested and reared edible insects: A review. *Food Control*, *101*, 209-224.

Nations, U. (2012). The Millennium Development Goals Report 2012. *Millennium Development Goals Report.*

Nonaka, K. (2009). Feasting on insects. *Entomological research*, *39*(5), 304-312.

Nonaka, K., Sivilay, S., & Boulidam, S. (2008). The biodiversity of insects in Vientiane. *National Agriculture and Forestry Institute and Research Institute for Humanity and Nature. Nara, Japan.*

Oh, W. Y., Pyo, S., Lee, K. R., Lee, B. K., Shin, D. H., Cho, S. I., & Lee, S. M. (2003). Effect of *Holotrichia diomphalia* larvae on liver fibrosis and hepatotoxicity in rats. *Journal of Ethnopharmacology*, *87*(2-3), 175-180.

Osimani, A., Garofalo, C., Aquilanti, L., Milanović, V., Cardinali, F., Taccari, M., & Clementi, F. (2017). Transferable antibiotic resistances in marketed edible grasshoppers (*Locusta migratoria* migratorioides). *Journal of Food Science*, *82*(5), 1184-1192.

Parodi, A., Leip, A., De Boer, I. J. M., Slegers, P. M., Ziegler, F., Temme, E. H., & Van Zanten, H. H. E. (2018). The potential of future foods for sustainable and healthy diets. *Nature Sustainability*, *1*(12), 782-789.

Patel, S., Suleria, H. A. R., & Rauf, A. (2019). Edible insects as innovative foods: Nutritional and functional assessments. *Trends in Food Science & Technology*, *86*, 352-359.

Piha, S., Pohjanheimo, T., Lähteenmäki-Uutela, A., Křečková, Z., & Otterbring, T. (2018). The effects of consumer knowledge on the willingness to buy insect food: An exploratory cross-regional study in Northern and Central Europe. *Food quality and preference*, *70*, 1-10.

Pyo, S. J., Kang, D. G., Jung, C., & Sohn, H. Y. (2020). Anti-thrombotic, anti-oxidant and haemolysis activities of six edible insect species. *Foods*, *9*(4), 401.

Raheem, D., Raposo, A., Oluwole, O. B., Nieuwland, M., Saraiva, A., & Carrascosa, C. (2019). Entomophagy: Nutritional, ecological, safety and legislation aspects. *Food Research International*, *126*, 108672.

Ramos-Elorduy, J. (2005). Insects: a hopeful food source. *Ecological implications of mini livestock*, 263-291.

Ramos-Elorduy, J., Moreno, J. M. P., & Camacho, V. H. M. (2009). Edible aquatic Coleoptera of the world with an emphasis on Mexico. *Journal of Ethnobiology and Ethnomedicine*, *5*, 1-13.

Rozin, P., & Fallon, A. (1980). The psychological categorization of foods and non-foods: A preliminary taxonomy of food rejections. *Appetite*, *1*(3), 193-201.

Ruddle, K. (1973). The human use of insects: examples from the Yukpa. *Biotropica*, 94-101.

Sidali, K. L., Pizzo, S., Garrido-Pérez, E. I., & Schamel, G. (2019). Between food delicacies and food taboos: A structural equation model to assess Western students' acceptance of Amazonian insect food. *Food Research International*, *115*, 83-89.

Stork, N. E. (2018). How Many Species of Insects and Other Terrestrial Arthropods Are There on Earth? Annual Review of Entomology, 63(1), 31–45. https://doi.org/10.1146/annurev-ento-020117-043348.

Sungpuag, P. (1983). Nutritive value of unconventional protein source: insect [mole cricket, june beetle, buffalo dung beetle, short tail cricket, cricket, giant water bug, silkworm pupae, small locust, true water beetle and red ant are investigated for nutritive value].

Turner, J. S., & Soar, R. C. (2008). Beyond biomimicry: What termites can tell us about realizing the living building. In *First International Conference on Industrialized, Intelligent Construction at Loughborough University* (pp. 1-18).

Unicef. (2001). *Progress since the World Summit for Children: a statistical review*. Unicef.

Van Huis, A. (2016). Edible insects are the future? *Proceedings of the Nutrition Society*, *75*(3), 294-305.

Van Huis, A., & Oonincx, D. G. (2017). The environmental sustainability of insects as food and feed. A review. *Agronomy for Sustainable Development*, *37*, 1-14.

Van Huis, A., Van Itterbeeck, J., Klunder, H., Mertens, E., Halloran, A., Muir, G., & Vantomme, P. (2013). *Edible insects: future prospects for food and feed security* (No. 171). Food and Agriculture Organization of the United Nations.

Vandeweyer, D., Lievens, B., & Van Campenhout, L. (2020). Identification of bacterial endospores and targeted detection of foodborne viruses in industrially reared insects for food. *Nature Food*, *1*(8), 511-516.

Wang, C. H., 2011. Preliminary studies on the medicinal and dietary values of *Massicus raddei* Blessig. Dissertations Northeast Normal Univ.

Wang, Y. S., & Shelomi, M. (2017). Review of black soldier fly (*Hermetia illucens*) as animal feed and human food. *Foods*, *6*(10), 91.

WFP. (2015). Hunger Statistics. Hunger. http://www.wfp.org/hunger/stats.

Xiaoming, C., Ying, F., Hong, Z., & Zhiyong, C. (2010). Review of the nutritive value of edible insects. *Forest insects as food: humans bite back*, *85*.

Yen, A. L., Hanboonsong, Y. U. P. A., & Van Huis, A. (2013). *The role of edible insects in human recreation and tourism* (Vol. 11, pp. 169-185). Cambridge, UK: Cambridge University Press.

YF, H., XM, L., & WANG, Y. (2008). Research advance in medicinal value of *Periplaneta americana* [J]. *Medical Recapitulate*, *9*(14), 2822-2824.

Yhoung-Aree, J., & Viwatpanich, K. (2005). Edible insects in the Laos PDR, Myanmar, Thailand, and Vietnam. *Ecological implications of mini livestock: potential of insects, rodents, frogs and snails*, 415-440.

Chapter 17

The Use of Insects in the Cosmetic Industry

Kaushik Pramanik[1,*]
Thakur Mandar Vijay[2]
and Munj Sujal Suhas[2]

[1]Department of Agricultural Entomology, Bidhan Chandra Krishi Viswavidyalaya, Mohanpur, Nadia, West Bengal, India
[2]Department of Agricultural Entomology, University of Agricultural Sciences, Dharwad, Karnataka, India

Abstract

Most insect species have a tremendous impact on the global economy. The cost of crop injuries caused by different insect pests is one example of how insects can affect the financial system of a country. Though insects can have various effects on the economic sector, not all are pessimistic. Insects, the most diverse class of common creatures on earth, also provoke excessive concern about using them as food and feed resources. Apart from food, other crucial requirements for caring for human bodies are drugs and cosmetics. Insects may offer a sustainable way of generating special bio-built items that can be used for cosmetic purposes. Insects as a lipid source are a matter of incredible curiosity from economic and environmental perspectives. The cosmetics industry continually seeks innovative components to use in novel product formulations. The little-known pitch of 'economic entomology' delves into how insects or their by-products are mass-produced for social use. This chapter sheds light on several substances extracted from numerous insect species beneficial in cosmetics. We also try to show that insects have a bunch of bioactive ingredients that need to be harnessed for the good of humanity.

Keywords: bioactive substances, cosmetics, insect oils, eco-friendly, lipids

1. Introduction

Cosmetics are materials that pertain directly to the body of a human or inside their mouth to clean, protect, scent, enhance beauty, or improve odour (Anonymous, 2009). Cosmetics are a fascinating, unregulated group of over-the-counter commodities intended to enhance appearance and skin health. India ranks first in the world in the importation of raw materials for cosmetics. The era of modernisation is making people more health conscious. The fields of

* Corresponding Author's Email: kpramanik.kaushik@gmail.com.

In: Insects as Humans' Frenemies: From Friends to Foes
Editor: Istkhar Rao
ISBN: 979-8-89530-320-7

pharmaceuticals and cosmetics are increasing rapidly to make up for this demand. The cosmetics industry faces a major challenge in its quest for safe, nature-based raw materials. Natural-origin components that provide safety and excellent performance in their bioactives are already in high demand. Natural and responsibly sourced ingredients are highly needed to regulate the cosmetics market properly. Lipids and oils are extensively used in cosmetics, where they are a significant component of skin care products. Triglycerides are mollifying substances that smoothen the skin (Le, 1994). They indirectly moisturise the skin by decreasing transepidermal water loss (TEWL) (Stamatas et al., 2008). Lipids are also used in formulations to enhance viscosity or for their emulsifying qualities (Alander, 2007). Chitin and its compounds have antioxidant, antimicrobial, humectant, biodegradability, biocompatibility, cleaning, and protective functions that are applied in cosmeceutical and cosmetic materials for skin, nail, hair, dental, and lip care (Dutta et al., 2004).

Table 1. List of the commonly used insects and their derivatives/ parts in the cosmetic industry

Sl. No.	Name of the insects	Name of the insect's materials used	Usefulness
1.	Silkworm	Silkworm extract	Skin conditioner
		Silk	Hair/Skin conditioner
		Sericin	Hair/Skin conditioner, Smoothing.
2.	Spider	Web	Hair/Skin conditioner
		Peptides	Hair/Skin conditioner
3.	Beetle	Grubs	Skin conditioner
		Larvae of *Tenebrio molitor*	Skin conditioner, Humectant
4.	Bee	Venom	Skin conditioner, Astringent
		Wax	Emulsifying, Perfuming, Emollients
		Propolis	Skin conditioner
		Honey	Skin conditioner, Humectant
		Pollen	Skin conditioner, Emollients
		Royal jelly	Skin conditioner
		Nectar	Humectant, Skin conditioner
5.	Cochineal	Carmine dye	Pigmentation, Colouration
6.	Snail	Slime	Skin conditioner, Humectant
		Egg	Skin conditioner
		Allantoin	Skin conditioner, Soothing, Hair conditioning

Insects are the dominant organism among all living entities; over a million species have been identified to the present, with many more yet to be discovered. Insects have a high population with similarly high parameters of diversity (Chapman, 2009). The phrase "beneficial insect" was typically limited to the use of silkworms in the creation of silk, lac insects in the production of lac, and honeybees in the production of honey and wax. Apart from this, many other insects can be used as economic insects. Various insect species are considered food in many parts of the world. The beneficial role of insects requires more exploitation. Insects are used as a source of bioactive materials such as chitin, proteins, and fats. Especially lipid and chitin frame about 70–80 per cent of the dry weight in the insect. These various bioactive components may find a variety of applications in the cosmetics industry. Insects may become a viable option in the production of biopolymers by combining insect breeding with waste stream reduction (Diener, 2011). Insects are presently being extolled as a novel root for these valuable and helpful bioactive molecules, especially lipids and chitin, for a variety of uses

(Ganguly, 2018). Numerous organic residues can be utilised by larvae of various species, which bio-convert them into useful biomass with high protein and lipid content. Many synthetic chemicals are highly residual, which imposes tremendous stress on the environment, but bio-active materials such as lipids, fatty acids, and chitin don't have any ill effects on the ecosystem (Table 1). Thus, there is a wide range of opportunities for using insects as a source of raw materials in the cosmetics sector.

2. Use of Insects in Cosmetic Industries

2.1. Cochineal Scales (The Hemipterans)

Cochineal scales, *Dactylopius* spp., are cactus feeders, and they are now used in cosmetics, foods, medications, and textiles. One kilogram of cochineal dye is said to require 70,000 to 100,000 dead female cochineal beetles, which are used to make the red pigment (carmine). Due to their diet of red berries, the female of these beetles develops a red colour, which causes them to produce red carminic acid in response to danger. As it produces intense, vivid, and bright red colours, it is used to colour eye shadow, blush, and lipstick in the cosmetics industry (Figure 1). Today, cochineals are primarily harvested from prickly pear cacti on the plantations of the Canary Islands and Peru. Depending on the processing technique, the locals sun-dry the scales, quell them, and submerge them in an alcoholic liquid. Several studies have been ongoing on how to increase the yield of carmine from *Dactylopius coccus* (Rodríguez et al., 2005).

Figure 1. Red cochineal scales.

2.2. Lac Insects

Many products, containing sealants, shoe and floor polishes, printing inks, insulators, and varnish, contain lac as a key component. It is principally utilised as a nail polish remover in the cosmetics industry. The majority of lac is produced in India. The beginning of cultivation occurs when the farmer knots a stick with the trees that will be infested with hatching. Thousands of lacs gather on the host tree's branches and secrete the resin, which is then cut off and used to make stick lac (Figure 2). Following several washings and sieving operations to remove impurities of all types, the harvested stick lac is transformed into seed lac, a purer variety. However, shellac is produced by further removing impurities through heat treatment or solvent extraction from seed lac, which still contains 3–5% of impurities. Therefore, shellac is a refined oozing of the lacs (Figure 3).

Figure 2. Lac insects on host plant.

Figure 3. Shellac.

Shellac is used in the cosmetic industry for personal care items like mascara, eyeliners, and hair sprays. Das and Jacob (2011) noted that shellac is a common component of cosmetics like mascara, lipstick, nail polish, and hairspray. This is because shellac can keep the components of a condensed cake or tablet together. Additionally, it prevents an emulsion from separating into water-soluble and oil-soluble components, which aids in the maintenance of hairstyles by preventing moisture absorption by the hair (i.e., Shellac is non-hygroscopic). Seven cosmetic products with varying percentages of Shellac were patented between 1997 and 2002.

2.3. Uses of Insects' Oils

2.3.1. Hymenopteran Oils

Honey and beeswax have many cosmetic applications, including moisturising, tissue healing and softening (Ediriweera & Premarathna, 2012). Some of the uses cited are as follows:

- *As a facial scrub:* Mix 5g of almond seed powder with bee's honey, give it a little scrape, and then rinse it off (Needham, 2008).
- *As a face smoother:* A fantastic firming mask may be made with one tablespoon of honey, one egg white, one teaspoon of glycerin, and one cup of flour stirred together. Just apply it evenly, allow it to remain for fifteen minutes, and then rinse your face with warm water (Needham, 2008).
- *As facial softener:* Use a one-third cup of finely crushed oats and one or two tablespoons of honey as a facial softener. A dab of rose water is then combined after that. This is applied evenly across the face. Place them wherever, and give them a half-hour. After using a gentle cloth and lukewarm water to clean, rinse with cold water.
- *As a facial moisturising pack:* Mix two tablespoons of honey with two teaspoons of whole milk. This is applied to the face and let to sit there for fifteen minutes. It is then cleaned with both warm and cold water.
- *Pimples:* On pimples, honey is directly administered.
- *Cracked lips:* On chapped cheeks, honey is directly applied.
- *As Dry skin lotion:* To make a lotion for dry skin, mix 2.5 ml lemon juice with 5 ml olive oil and 5 ml bee's honey. It is removed from the skin after fifteen minutes.
- *Hair lustre:* Mix 5 ml of bee's honey with 4 glasses of warm water. It is applied as a hair treatment.
- *As Conditioner:* Before applying to the hair, mix 5 millilitres of bee's honey and 10 millilitres of olive oil. It's cleaned after fifteen minutes.

According to Burlando and Cornara (2013), honey has soothing, humectant, emollient, and hair-constraining effects in cosmetic preparations. It also maintains youthful skin, delays the appearance of wrinkles, balances pH, and guards against pathogenic infections.

2.3.2. Oil of Silkworms (Lepidoptera)

The caterpillars of the silkworms *Bombyx mori* participate in the creation of silk during the development of the cocoon. Sericin (20–30%) and fibroin (70–80%) are the two proteins that make up silk. The fibrous and structural component of silk is called fibroin, and the gooey covering that pastes the fibres and causes them to adhere to one another is called sericin. According to Voegeli et al. (1993), the annexation of sericin in creams causes the skin to become more hydrated and elastic, which has an anti-ageing and anti-wrinkle impact. Sericin in lotions prevents nails from becoming brittle and chapped. (Yamada et al., 2001). They discovered that sericin's action decreased impedance while increasing the concentration of hydroxyproline and hydrating epidermal cells. Sericin's occlusive effect, which stopped the transepidermal water loss that causes dry skin, was credited with the rise in hydration (Figures 4 & 5).

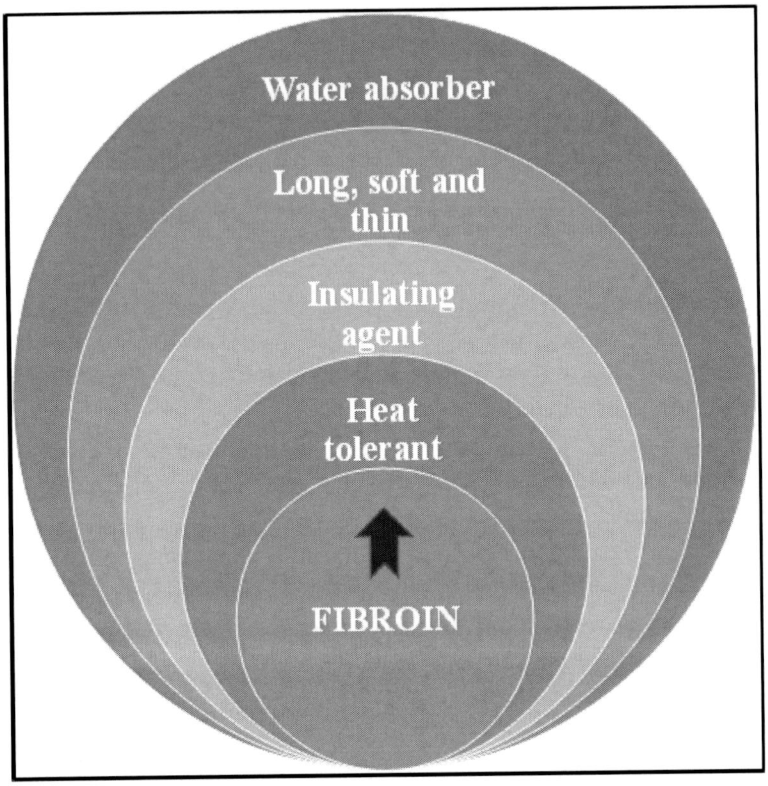

Figure 4. Properties of fibroin.

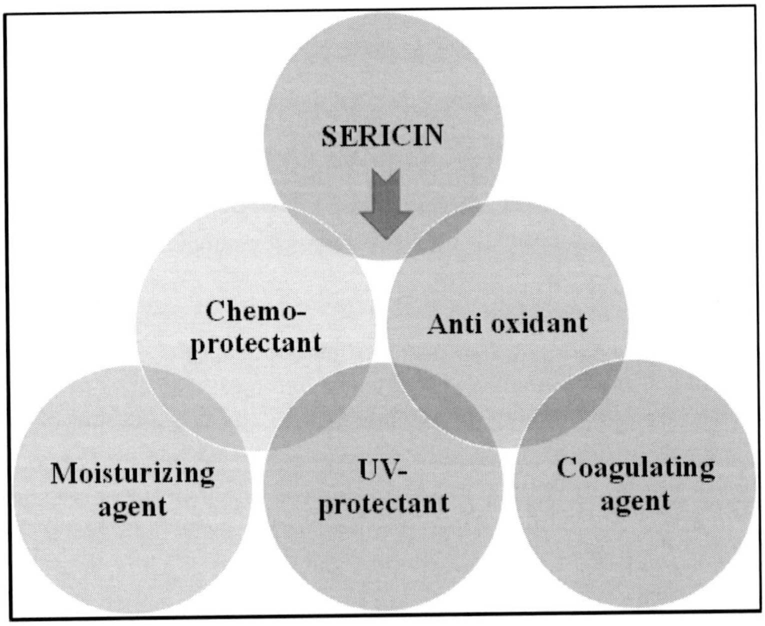

Figure 5. Properties of sericin.

3. Principal Bioactive Substances Used in the Cosmetic Industry

Lipids and chitin are common bioactive substances used in cosmetics. These materials can be procured from insect bodies in more or less quantity.

3.1. Lipids

Different lipids have varying roles in the composition of cosmetics. They are hydrating substances that prevent water loss via a variety of processes. The moisturising and soothing properties of lipids make the skin smooth and soft. Skin smoothness directly impacts skin radiance and brightness because it relies on the quantity of light imitated by the skin's surface, which enhances the skin's look (Draelos, 2018). Lipids can be used as emulsifiers and surfactants to keep water and oil in a product combined and to lower the surface tension between the product and the skin's surface. Due to the hydrophobic and hydrophilic moieties in their chemical structure, surfactants help liquids of various polarities develop emulsions and lower surface tension (Lourith & Kanlayavattanakul, 2009). The high commercial demand is typically met by plant-derived fats, especially those produced from oilseeds like palm, soy, rapeseed, and sunflower oils. Questions are being made about the rising lipid demand, including decreasing arable farmland and resource usage (De Luca et al., 2021). Insects have the potential to compensate for the gap created due to the reduced availability of plant-derived lipids. Insects use lipids as their main energy source during diapause, extended periods of flying, breeding, and other stressful situations (Duong & Nghia, 2014). Several variables affect the makeup of insect fatty acids and lipids, including species, life stage, nutrition, environmental factors, migratory movement, and sex. Female insects typically have higher levels of triglycerides than male insects.

3.2. Chitin

It has the same molecular makeup as cellulose, but instead of -OH groups at the C-2 positions, it has acetamide groups ($-NHCOCH_3$) (-OH). Chitin serves the same purpose in insects as collagen in mammals and cellulose in plants. It is an inelastic, white, and hard nitrogenous polysaccharide with minimal chemical acuteness (Casadidio et al., 2019; Dutta et al., 2004). Both cellulose and chitin have a crystalline structure that reduces solubility and is a significant restricting factor in their usage. Chitin has three variable crystalline configurations, i.e., α, β, and γ (Jang et al., 2014). Chitin is a biologically nontoxic substance; it is also biodegradable and biocompatible. Poor solubility is the main problem in applying chitin, so chemical groups are added to modify the structure, making it fit for use. They are magnificent moisturisers and rejuvenation agents that can shield the skin from possible hazards while enhancing vital skin processes like temperature control, defence, secretion, excretion, feeling, and absorption (Figure 6) (Sajna et al., 2015).

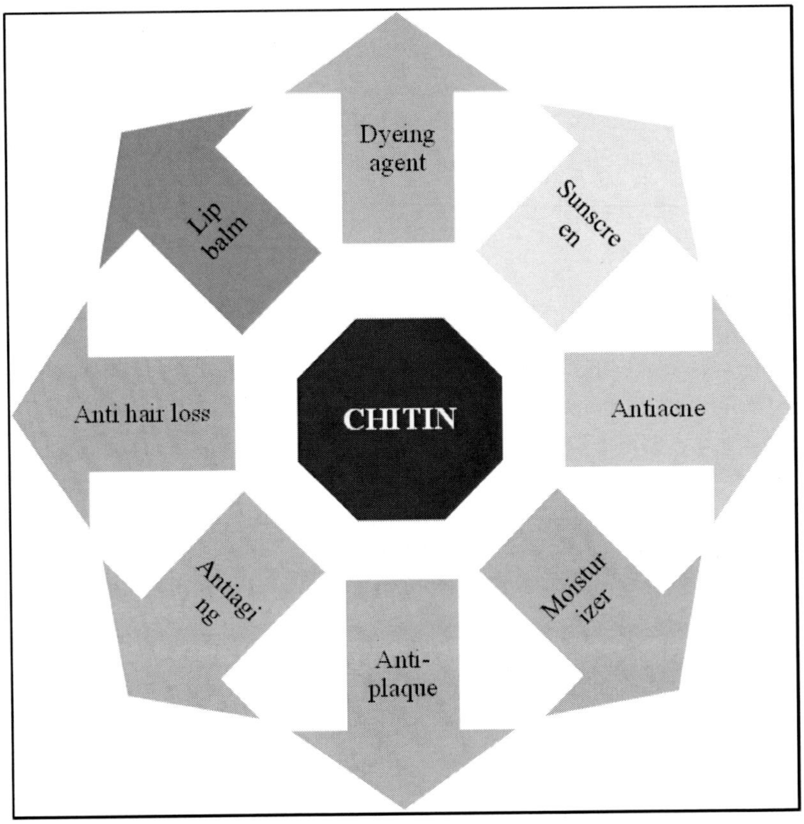

Figure 6. Uses of insect chitin in cosmetic industry.

4. Common Extraction Methods of Bioactive Materials Employed in the Cosmetic Industry

The extraction of bioactive materials in pure form is a key point. Different types of extraction methods are used depending on the kind of molecule to be extracted. Below are some common extraction methods:

4.1. Soxhlet Extraction

This is the classic lipid extraction technique. Petroleum ether, diethyl ether, hexane, ethyl acetate, and ethanol are some of the solvents that can be used in this extraction technique. So far, ethanol has provided the greatest extraction (Laroche et al., 2019). When the component to be taken out has regulated solubility in the particular solvent and the debasement is unsolvable, the Soxhlet extractor is employed for the extraction of liquids and solids. The extractor is used to transfer partially soluble components of a solid sample material to a liquid phase (solvent). This method makes use of non-polar liquids to produce neutral lipids.

4.2. Aqueous Extraction

The extraction technique is based on the aqueous technique outlined by Yi et al. (2014), with some changes, and is used for lipids. After blending frozen insects with demineralised water for one minute, the mixture is sonicated for 15 minutes. A stainless-steel filtration sieve with a 350-m pore size is used to sift the resulting insect suspension. The insect suspension is then centrifuged for 30 minutes at 4°C at 15,000 rpm. The supernatant is separated into three parts. The fatty component, supernatant, and particle are listed from top to bottom. The oil portion is then centrifuged for 30 minutes at 40°C using 15,000 rpm. A clear, dehydrated lipid extract is produced in the top layer of the centrifuged container following the second centrifugation. Two additional parts are obtained, i.e., a lower layer with water and water-soluble complexes and a thin sheet of cream. For further research and use, lipid isolates are kept in a nitrogen environment at -20ºC. This method gives the lowest yield of lipids among all methods.

4.3. Bligh and Dyer Method

The Bligh and Dyer technique combines lipid extraction, partitioning, and protein precipitation at the intersection of the two liquid stages. The solvent mixture content and ratios are distinct from those used in the Folch technique. The lipids in a cell homogenate are first removed with a 1:2 chloroform/methanol combination, and the chloroform phase (lipid-rich) is retrieved (Chandra et al., 2020).

4.4. TPP Extraction

This is the second-most efficient extraction method after the Soxhlet method. Three-phase partitioning can isolate up to 25 types of enzymes, proteins, and lipids. Ammonia sulphate and t-butanol are used for precipitation in this method. In this technique, purified water is combined with insect flour. With 0.1 M HCl, the pH of the fluid is brought down to 4.5. T-butanol and ammonium sulphate are then added, and the combination is agitated at 35°C for 1 hour. The combination is centrifuged at 2800 rpm for 5 minutes after being held at room temperature for 1 hour without movement. A rotating evaporator is used to evaporate the t-butanol while the supernatants are gathered in a round-bottom flask that has been pre-weighed. Oil samples were stored under a nitrogen environment at -20°C until use, and extractions were carried out in triplicate (Tzompa et al., 2014).

4.5. SC-C2 Method

The supercritical CO_2 extraction method is conducted at 35–40°C in the presence of CO_2 for lipid extraction. Oxidation and loss of heat-sensitive substances are prevented by this method. Organic solvents are not used in the complete process. Hence, the extractant obtained is pure, without any chemical residue. High efficacy and low energy requirements make it a cheaper method of extraction. Insect meal and glass pebbles are added in equal amounts to a reactor that has been operating for 75 minutes at 325 pressure and 55°C at a flow rate of 10 g/ml. Once the lipid has been extracted, the defatted waste is gathered and measured. The oil portion is stored at 20°C in a nitrogen atmosphere until it is used (Tzompa et al., 2014).

4.6. Chitin Extraction Method

Numerous chemical methods are used for the extraction of chitin from insect bodies. Demineralisation and deproteinisation are common steps followed in chitin extraction for crustacean shells, but delipidation is also used in the case of insects. The quantity of lipids in the insect body may have an impact on the output during extraction and the amount of chitin and chitosan present. Delipidation/defatting is done before the deproteinisation stage because 80 per cent of fats are in the triacylglycerol state. During the delipidation procedure, a $CHCl_3$:CH_3OH (7:3) combination is used as a reagent at 20°C for 4 hours (Mohan et al., 2020). The second crucial stage is deproteinisation because chitin and proteins are linked by bonds. In general, a variety of compounds have been used to deproteinise industrial insect chitin. Additionally, NaOH is the preferred chemical base and is used in a range of values from 0.125 to 5.0 M (Kaya et al., 2014). Alternative techniques for protein extraction requiring the use of enzymes such as alkaline proteases are developing and constitute a newfound approach (Duong & Nghia, 2014).

Demineralisation is the term for the removal of minerals, primarily calcium carbonate. Sulfuric, oxalic, nitric, hydrochloric, acetic, and formic acids are frequently used in acid treatment to accomplish this procedure. HCl is better than all of these other acids in the separation of chitin from insects. In the demineralisation process, calcium carbonate is broken down into calcium chloride, and carbon dioxide is released (Mohan et al., 2020).

Conclusion

Insects are a better option for providing raw materials in the cosmetics industry. As insects are the most dominant animals on earth, availability is very easy, with almost negligible cost. Most of the extraction methods are cheaper as compared to synthetic raw materials. Rearing of insects is easy and has a low cost as it can be reared on waste materials. The fecundity of insects is very high, and fast breeding is also possible. All these factors indicate an unlimited supply chain at a very low cost. The use of plants or any other related material is not at all eco-friendly or economically superior. Insects are an unlimited source of raw materials. Deriving more economic benefits from insect pests to compensate for the losses up to a certain limit caused by them is possible.

References

Alander, J. (2007). Cosmetics in Vegetable Oils and Fats (Aarhuskarlshamn, Karlshamn and Sweden). Pp: 214-234.

Anonymous (2009). Regulation (CE) No. 1223 of the European Parliament and of the Council of 30 November on cosmetic products. *Official Journal of the European Union, 342*, 59-209.

Burlando, B., & Cornara, L. (2013). Honey in dermatology and skin care: a review. *Journal of Cosmetic Dermatology, 12(4),* 306-313.

Casadidio, C., Peregrina, D. V., Gigliobianco, M. R., Deng, S., Censi, R., & Di Martino, P. (2019). Chitin and chitosans: Characteristics, eco-friendly processes, and applications in cosmetic science. *Marine Drugs, 17(6),* 369.

Chandra, R., Vishal, G., Sánchez, C. E. G., & Uribe, J. A. G. (2020). Bioreactor for algae cultivation and biodiesel production. In *Bioreactors* (pp. 289-307). Elsevier.

Chapman, A. D. (2009). Numbers of Living Species in Australia and the World, 1st ed.; Australian Biological Resources Study: Canberra, Australia. Pp: 01-78.

Das, S., & Jacob, S. E. (2011). Shellac. *DERM, 22(4),* 220-222.

De Luca, M., Pappalardo, I., Limongi, A. R., Viviano, E., Radice, R. P., Todisco, S., ... & Vassallo, A. (2021). Lipids from microalgae for cosmetic applications. *Cosmetics, 8(2),* 52.

Diener, S., Zurbrügg, C., Gutiérrez, F. R., Nguyen, D. H., Morel, A., Koottatep, T., & Tockner, K. (2011). Black soldier fly larvae for organic waste treatment-prospects and constraints. *Proceedings of the WasteSafe, 2,* 13-15.

Draelos, Z. D. (2018). The science behind skin care: moisturizers. *Journal of Cosmetic Dermatology, 17(2),* 138-144.

Duong, N. T. H., & Nghia, N. D. (2014). Kinetics and optimization of the deproteinization by pepsin in chitin extraction from white shrimp shell. *Journal of Chitin and Chitosan Science, 2(1),* 21-28.

Duong, N. T. H., & Nghia, N. D. (2014). Kinetics and optimization of the deproteinization by pepsin in chitin extraction from white shrimp shell. *Journal of Chitin and Chitosan Science, 2(1),* 21-28.

Dutta, P. K., Dutta, J. & Tripathi, V. S. (2004). Chitin and chitosan: Chemistry, properties and applications. *Journal of Scientific and Industrial Research, 63,* 20-31.

Ediriweera, E. R. H. S. S., & Premarathna, N. Y. S. (2012). Medicinal and cosmetic uses of Bee's Honey–A review. *Ayu, 33(2),* 178.

Ganguly, S., Neog, P., Gogoi, M., Bordoloi, P. L. & Para, P. A. (2018). Edible Insects as Sources of Novel Bioactive Compounds. In Recent Research Trends in Veterinary Sciences and Animal Husbandry; AkiNik Publications, Delhi, India. Pp: 55-69.

Jang, M. K., Kong, B. G., Jeong, Y. I., Lee, C. H., & Nah, J. W. (2004). Physicochemical characterization of α-chitin, β-chitin, and γ-chitin separated from natural resources. *Journal of Polymer Science Part A: Polymer Chemistry, 42(14),* 3423-3432.

Kaya, M., Baran, T., Mentes, A., Asaroglu, M., Sezen, G., & Tozak, K. O. (2014). Extraction and characterization of α-chitin and chitosan from six different aquatic invertebrates. *Food Biophysics, 9,* 145-157.

Laroche, M., Perreault, V., Marciniak, A., Gravel, A., Chamberland, J., & Doyen, A. (2019). Comparison of conventional and sustainable lipid extraction methods for the production of oil and protein isolate from edible insect meal. *Foods, 8(11),* 572.

Le Po ole, H. A. C. (1994). Natural Oils and Fats Multifunctional Ingredients for Skin Care (Dekker, Wormerveer). *The Netherlands,* 47-56.

Lourith, N., & Kanlayavattanakul, M. (2009). Natural surfactants used in cosmetics: glycolipids. *International Journal of Cosmetic Science, 31(4),* 255-261.

Mohan, K., Ganesan, A. R., Muralisankar, T., Jayakumar, R., Sathishkumar, P., Uthayakumar, V., ... & Revathi, N. (2020). Recent insights into the extraction, characterization, and bioactivities of chitin and chitosan from insects. *Trends in Food Science & Technology, 105,* 17-42.

Needham, A. W. (2009). Health benefits of honey, 2008.

Rodríguez, L. C., Faúndez, E., Seymour, J., Escobar, C. A., Espinoza, L., Petroutsa, M., ... & Niemeyer, H. M. (2005). Factores bióticos y concentración de ácido carmínico en la cochinilla (*Dactylopius coccus* Costa) (Homoptera: Dactylopiidae). *Agricultura Técnica, 65(3),* 323-329.

Sajna, K. V., Gottumukkala, L. D., Sukumaran, R. K. & Pandey, A. (2015). White Biotechnology in Cosmetics. In Industrial Biorefineries and White Biotechnology; Elsevier: Amsterdam. The Netherlands. Pp: 607-652.

Stamatas, G. N., de Sterke, J., Hauser, M., von Stetten, O., & van der Pol, A. (2008). Lipid uptake and skin occlusion following topical application of oils on adult and infant skin. *Journal of Dermatological Science, 50(2)*, 135-142.

Tzompa-Sosa, D. A., Yi, L., van Valenberg, H. J., van Boekel, M. A., & Lakemond, C. M. (2014). Insect lipid profile: aqueous versus organic solvent-based extraction methods. *Food Research International, 62,* 1087-1094.

Voegeli, R., Meier, J., Blust, R., & Hofsteter, R. (1993). Sericin silk protein: unique structure and properties. *Cosmetics and Toiletries, 108(12),* 101-108.

Yamada H., Yamasaki K. & Zozald, I. C. (2001). Nail Cosmetics containing sericin. Patent EP 1632214 A1.

Yi, H. Y., Chowdhury, M., Huang, Y. D., & Yu, X. Q. (2014). Insect antimicrobial peptides and their applications. *Applied Microbiology and Biotechnology, 98,* 5807-5822.

Chapter 18

The Role of Insects in the Biodegradation of Plastics

Sakshi Singh
Ayushi Chaudhary
Parul Chauhan
and Istkhar Rao*

Department of Bioscience and Biotechnology, Banasthali Vidyapith, Banasthali, Rajasthan, India

Abstract

Plastic is an artificial polymer obtained from conservative petrified oil that is resistant to biodegradation. Of the total plastic, about 92% is PE (polyethylene) and PP (polypropylene), both of which are widely used in wrapping and covering. Lately, the manufacturing of plastic has increased rampantly; hence, innovative and ecological solutions for plastic degradation are crucially needed. At all times, a surge of soil pollution, water pollution, and the shutting of landfill plots have generated distress about plastics. With the uncontrolled application of plastics and rising insistence on methods convenient for plastic wither disposal, the call for the biodegradation of plastic and biodegradable plastic wastes has assumed growing significance lately. Recognition of the waste issues and their effect on the surroundings has sparked the latest curiosity in the field of atrophy polymers. The heed in ecological matters is rising, as there is growing pressure for an evolutionary matter that does not load nature. The importance of biodegradation is for water-dissoluble or water-unmixable polymers, since these, in time, launch into watercourses that can neither be reprocessed nor burnt up. While sucking, insects confront a large variety of hydrocarbon polymers in their meals, and many insects accommodate microbial symbionts in their alimentary canal that assist in the mortification of these polymers. Hence, this gave a sense of insects as synthetic polymer bio-degraders. In a twinkling, numerous insects, like mealworms, rice meal worms, super worms, etc., especially those belonging to the orders Coleoptera and Lepidoptera, were recognised to possess the exceptional capacity to ingest and mortify a broad domain of artificial polymers (polypropylene, polyethylene, polyethylene terephthalate, polyoxymethylene, and polyvinyl chloride) in lower molar masses, uncomplicated, as well as harmless molecules that are finally egested as facula. The wax worms digest complex polyethylene plastic films into simple, degradable compounds known as ethylene glycol. This exceptional capacity to digest matter, traditionally viewed as indigestible, may have begun with the wax worm's ability to digest beeswax. The current chapter enlightens the plastic degradation capabilities of some of the important insects and their future uses in plastic management.

---

* Corresponding Author's Email: istkharrao@gmail.com; istkhar@banasthali.in.

In: Insects as Humans' Frenemies: From Friends to Foes
Editor: Istkhar Rao
ISBN: 979-8-89530-320-7

Keywords: plastic, artificial polymers, biodegradation, insects, microbes, degradable compounds

1. Introduction

Plastic is an artificial material constructed from numerous sources, like natural polymers such as polyethylene, PVC, polystyrene, and nylon, or from fossil fuels, which include coal, natural gas, and petroleum, and from renewable sources like corn starch and sugarcane. It is an extremely multifaceted material that is globally employed in countless distinct factories because of its longevity, featherweight, and cheapness (Ahmed et al., 2018). Plastic can be configured into distinct shapes and structures, and it can be built into an array of products, such as bottles, bags, containers, toys, and electronics. It is an adaptable substance that can be cast into various forms and structures and has advantages in covering manufacturing, shipping, automotive, medical management, and wiring systems. However, plastic has caused great ecological trouble in the present day, as it is indestructible and can take hundreds of years to degrade. Plastic squandering has caused pollution of oceans, rivers, and landfills, creating a threat to fauna and mortal fitness (Krueger et al., 2015). To lower the environmental influence of plastic, there are numerous attempts to decrease plastic application and encourage reprocessing, such as disallowing single-use plastics, executing a plastic sag levy, boosting the use of reusable vessels, and implementing plastic recycling programs. Additionally, there has been a push towards developing new, sustainable materials to replace plastic in certain applications. Varieties of plastics are being used globally (Table 1); for example, polyethylene (PE) is the most commonly used plastic all over the world. Based on the density of plastic and degree of branching, it is divided into various types, such as UHMWPE (ultra-high molecular weight polyethylene), HDPE (high-density polyethylene), XLPE (cross-linked polyethylene), MDPE (medium density polyethylene), and LDPE (low-density polyethylene) (Muhamad et al., 2015; Yoshida et al., 2016).

Table 1. Utility, types and application of polymers

S. No.	Polymer types	Utility %	Formula	Application
1	Polyethylene	29.7	$(CnH_2n)n$	Covering film, waste and keister bags, electric wires and cables insulation, squeeze bottles, toys, and housewares.
2	Polypropylene	19.3	$(C_3H_6)n$	Used in automobiles, packaging, construction products, consumer goods, and more.
3	Polyvinyl chloride	10	(C_2H_3Cl)	Pipes and fittings. Power, data and telecoms wiring and cables. Cable and services ducting.
4	Polyethylene terephthalate	1.1	$(C_{10}H_8O_4)n$	Manufacturing of shopping bags, water bottles, videotapes. · For manufacturing, containers and bags
5	Acrylic	<1%	CH_2=CHCOOH	used for lighting, electronics screens, automotive components and outdoor glazing
6	Polystyrene	<1%	$(C_8H_8)n$	disposable plastic vessels and dinnerware, CD "jewel" cases, smoke detector housings, license plate frames, plastic model
7	Acrylonitrile-butadiene-styrene	<1%	$(C_8H_8)x\cdot(C_4H_6)y\cdot(C_3H_3N)z)$	used to make light, rigid, moulded products such as pipe, automotive body parts, wheel covers, enclosures, and protective headgear.

The second most manufactured commodity plastic worldwide is polypropylene, followed by polyvinyl chloride (third most utilised), polyethylene terephthalate (fourth most created synthetically worldwide), and several others.

2. Plastic Waste Management

In the initial years, around 1950, the world production of plastic was only two million metric tonnes per year. But after that, annual production kept on increasing and hence reached four hundred sixty (460) million metric tonnes now (https://ourworldindata.org/plastic-pollution). Despite the constant acceleration in production, there was a slight decrease in yearly production around 2009 and 2010, which proved to be primarily due to the 2008 financial crisis that happened globally, during which a crater in resource production and consumption, including energy, was seen. Before 1980, there was no clue about the reprocessing and burning of plastic, so 100 per cent of it was thrown away. Later, in 1980, the incineration method was used, and in 1990, recycling came into use (Lebreton et al., 2019). In 2015, on average, fifty-five per cent of world plastic waste was thrown away, twenty-five per cent was burnt, and twenty per cent was recycled. The cumulative production of synthetic fibres, polymers, and additives was 8300 million tons. Since 1950, ~5800 M tonnes of primary plastic have been discarded, which is no longer in use, and only nine per cent have been recycled (Glaser et al., 2019). By 2015, a total of 8,300 metric tonnes of primary plastic were produced globally. Out of that, 500 metric tonnes (~6%) were recycled, 700 metric tonnes (~9%) were burnt, 2,500 metric tonnes (30%) were still in use, and 4600 metric tonnes (~55%) were estimated to have been discarded, ultimately accumulating in landfills and natural environments (Geyer et al., 2017). Research suggests that plastic waste was landfilled and only 18% recycled. Plastic production reached 368 million metric tonnes around the world in 2019 (Ekanayaka et al., 2022). In a business-as-usual scenario where the production, waste generation, and waste management of plastics keep their current trends, the mass of mismanaged plastic waste that is left in uncontrolled landfills and is more likely to leach into the environment than being properly processed is estimated to double by 2050, reaching 12,000 metric tonnes.

However, projections of various scenarios designed to reduce plastic pollution around 2016 and 2040 estimate that substantial amounts of plastic waste will remain mismanaged worldwide, even under a "system change" scenario where all the currently available interventions are immediately implemented. Estimates show that in 2017, 32% of the 25 metric tonnes of plastic waste produced in Europe was collected for recycling (Chamas et al., 2020). A significant amount of the collected waste was then exported to other non-EU countries, where different environmental standards for plastic waste management are applied. For the past quarter of the century, it was found that globally, China was the largest importer of plastic waste, handling approximately half of the globally produced plastic waste meant for recycling. However, starting in 2018, the Chinese government enacted a "National Sword" policy, banning the ship from carrying most of this plastic waste inside the country. The ban has posed an important challenge to global plastic waste management, disrupting the global trade network (Haider et al., 2019). None of the other large economies involved in the network can take on the short-term role of China, increasing the risk of illegal dumping and incineration of plastic waste and redirecting the flow of waste towards other countries, which are usually developing countries not well prepared to manage their plastic waste stream (Emadian et al., 2017). The challenge posed by the Chinese ban can also prove to be an opportunity, particularly given the low prices of plastic waste, pushing investment and innovation towards a more robust and complete recycling and recovery system that is regionally developed.

3. Biodegradation

The decomposition of biological substances (Figure 1) by microbes such as bacteria, fungi, and other pathogens is known as biodegradation. These microbes convert complex compounds into simpler ones like CO_2, H_2O, and other biomolecules (Ali et al., 2021). For smooth ecological interaction, biodegradation plays an important role in the continuous availability of nutrients by naturally cycling them and is also accountable for the degradation of many organic materials, including dead plant and animal matter, sewage, and waste products. It potentially conserves the intrinsic value of the materials used and incorporates them into biological cycles. Microorganisms require specific conditions to work, such as some requiring an oxygen-rich environment, i.e., aerobic, and others requiring an oxygen-poor environment, i.e., anaerobic. Depending on the microbes taking part and the nature of the material to be degraded, biodegradation can occur under both conditions (Wierckx et al., 2018). For example, in landfills, biodegradation by microbes is done under anaerobic conditions, like the process of biogas formation. Biodegradation can work as a panacea for today's big problem of plastic mismanagement and decrease white pollution, as after degradation, it integrates materials, conserves their intrinsic value, and finally includes them in organic cycles (Lucas et al., 2008). Plastic biodegradation can be described as the degradation of plastic films by microbes present in the guts of some insect species belonging to the Coleopteran and Lepidopteran orders. It is fundamentally separated into steps such as fragmentation and deterioration; depolymerisation; assimilation; and mineralisation. This has lately caught the attention of synthetic plastic degradation experts (Kundungalet et al., 2019; Lin et al., 2022).

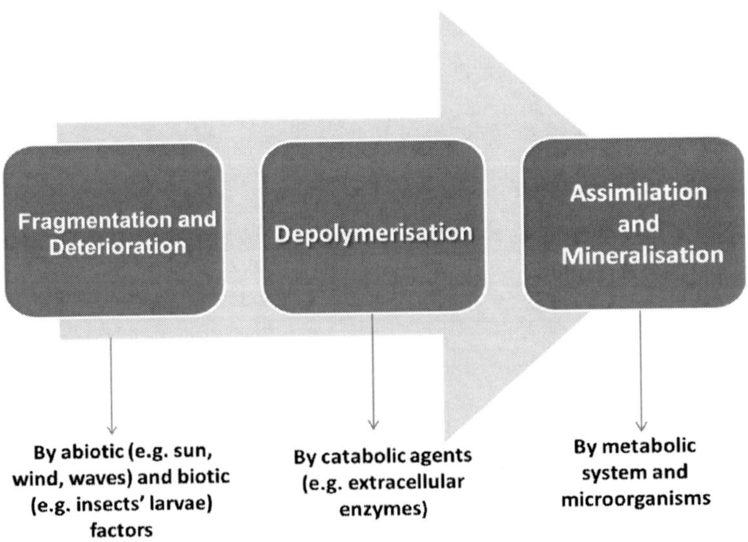

Figure 1. Basic steps in biodegradation process.

4. Insects

Insects occupy positions in the animal kingdom under invertebrates, which are subclassified as phylum Arthropoda and class Insecta. Arthropoda consists of 70% of the world's total species and is known as the largest phylum (Gullan et al., 2014). There are over 1 million known species of insects, with a total of over 30 million species estimated by scientists, and many have vanished from the earth before they could be identified. Some of the most common groups of insects include ants, bees, butterflies, beetles, flies, grasshoppers, and mosquitoes (Rumpold et al., 2017). They have distinct characteristics that make them adaptable to almost all kinds of climatic conditions and help them to inhabit various locations and survive. Their body is divided into 3 different regions (the abdomen, thorax, and head), 3 pairs of jointed legs, typically 1-2 pairs of wings, antennae overhead, compound eyes, and a chitinous exoskeleton. They also have wings, although not all insects can fly. These features make insects available in virtually every habitat on Earth, from land to water to air, and they participate in ecosystems as pollinators, decomposers, and prey for other animals (Jiang et al., 2019). Insects feed on varieties of meals, as they have different types of mouthparts like sucking, syphoning, sponging, etc., which help species feed on nectar, others on plant matter, and still others on insects or animals (Sheikh et al., 2017). They use a variety of methods to locate food, including visual cues (dancing style), scents known as pheromones, and vibrations. Insects reproduce through both sexual and asexual means, depending on the species. Many insects undergo metamorphosis, with distinct larval, pupal, and adult stages. While many insects are beneficial to humans and the environment, others can be harmful, either through direct damage to crops or property or by transmitting diseases. Some common examples of harmful insects include mosquitoes, ticks, and termites. Insects are also important pests, causing damage to crops and spreading disease.

4.1. Insects in Plastic Biodegradation

Many substances, such as petroleum products, plastics, and pesticides, are not easily broken down by natural processes, and their accumulation can have harmful effects on the environment. Biodegradation technologies can help mitigate these effects by breaking down these substances into harmless compounds. The rate and efficiency depend on several factors of biodegradation, which include the type as well as the amount of organic matter, the presence of microorganisms capable of degrading the substance, and environmental factors such as temperature, moisture, and pH (Lau et al., 2020; Wang et al., 2020). The degradation of plastic itself is a multistep process (Figure 2). Understanding these steps and conditions is important for the development of effective biodegradation strategies. Insects play important roles in many ecosystems, including decomposing organic matter, pollinating plants, and serving as a food source for different animals. However, some insects can also be pests, causing damage to crops or transmitting diseases to humans and other animals. In addition to their ecological and economic significance, insects have also played a cultural role throughout human history. They have been depicted in art, literature, and mythology and have been used for food, medicine, and even as a source of dye for textiles.

Insects have been demonstrated to recreate a function in the degradation of plastics, although their specific impact and effectiveness can vary depending on the kind of plastic and the species of insect involved (Berasategui et al., 2016). Coleopteran and lepidopteran are two important groups of insects having insects of plastic degradation capabilities (Table 2). One well-known example of an insect that can degrade plastic is the wax moth, Galleria mellonella. Wax moths are known to consume polyethylene, which is a common type of plastic used in products such as shopping bags and packaging materials (Cassone et al., 2020). The discovery of wax worms eating plastic was by chance and unexpected. In Spain, a scientist, Federica Bertocchini, who belonged to the Institute of Biomedicine and Biotechnology of Cantabria and was also an amateur beekeeper, was disappointed when she observed one of her hives of bees invaded by Galleria. Then she cleaned the hive of bees and kept the resin worms in an ordinary shopping bag that was made from plastic. Then the lady enclosed that bag with the insects and kept it somewhere in her house. After some time, when she came back to the place, she noticed the worms were creeping all over the place and was shocked to see that. There were holes in the plastic bag, and with this, she concluded that the insects had eaten that plastic and come out of that bag. Their objective was to find out whether these wax worms could be used to tackle today's biggest issue, the international crisis caused by plastic, as publicised in the Journal of Current Biology on April 24, 2017 (Bombelli et al., 2017). It was in Japan that, for the first time, researchers identified plastic-degrading bacteria. They encountered these plastic-degrading bacteria in a residue model accepted from a plastic recycling establishment (Yoshida et al., 2016). Studies have shown that the wax moth larvae can break down polyethylene into smaller pieces and that they can consume up to 92 mg of polyethylene per day.

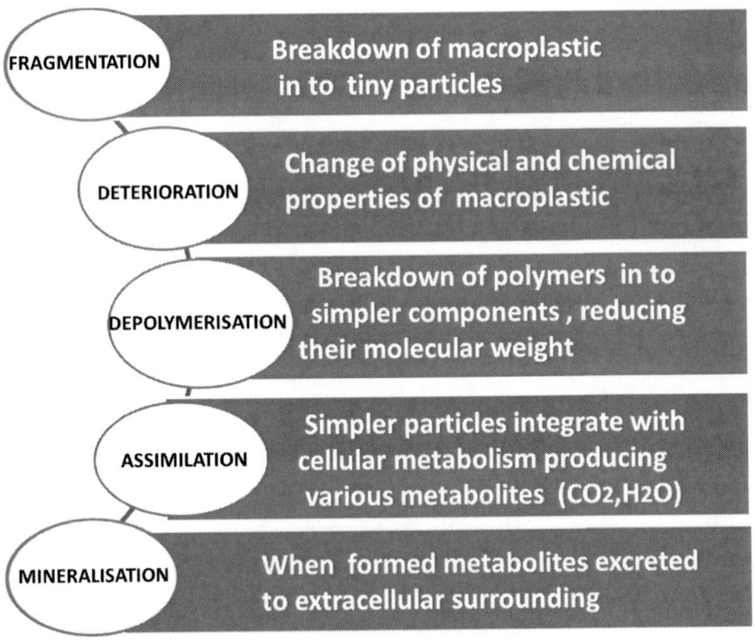

Figure 2. Steps involved in the plastic biodegradation process.

Another insect that has been shown to have a potential role in plastic degradation is the mealworm, Tenebrio molitor (Brandon et al., 2018). Mealworms can consume polystyrene, which is another common type of plastic used in products such as foam insulation and disposable cups. Research has shown that mealworms can break down polystyrene into smaller fragments and that they can survive and grow on a diet of polystyrene alone. While insects have shown some potential for plastic degradation, it is important to note that they are not a silver bullet solution for the plastic pollution problem. Insects can only consume a limited amount of plastic, and their effectiveness can be influenced by abiotic factors such as temperature, humidity, and the type of plastic involved (Mallar et al., 2020). Additionally, it is still unclear what the long-term environmental impacts of insects consuming plastic might be. Therefore, efforts to reduce plastic waste and increase recycling and biodegradability of plastics should continue to be a priority.

Table 2. The life cycles of plastic-eating insects

S No.	Insect	Order	Life cycle			
			Egg	Larvae	Pupa	Adult
1	*Tenebrio molitor*	Coleoptera	4-19 days	2-6 months	7-21 days	A few months to a year
2	*Zophobus atratus*	Coleoptera	2-4 weeks	3-6 months	2-3 weeks	4-8 months
3	*Tribolium confusum*	Coleoptera	5-12 days	30 days	1-3 weeks	Weeks to several months
4	*Tribolium casteneum*	Coleoptera	2-5 days	2-6 weeks	5-18 days	Weeks to several months
5	*Plodia interpunctella*	Lepidoptera		2-14 days	4-10 days	1-2 months
6	*Galleria mellonella*	Lepidoptera	7-22 days	1-6 months	1-9 weeks	Male-21 days Female-12 days
7	*Achroia grisella*	Lepidoptera	5-8 days	6-7 weeks	1-2 months	Male-7 days
8	*Corcyra cephalonica*	Lepidoptera	4-19 days		3-30 days	
9	*Spodoptera frugiperda*	Lepidoptera		2-4 weeks	7-37 days	1-3 weeks

5. Selected Plastic Eating Insects

Several insects belonging to the orders Coleoptera and Lepidoptera are reported to have plastic-eating capabilities and are listed in Table 2. Not only the insects but in many cases, their bacterial symbiont also play their role in plastic degradation and are listed in Table 3. The descriptions of some of these selected insects are given below:

5.1. Coleopterans

Coleopterans, also known as beetles, are a diverse group of insects that make up the order Coleoptera. They are one of the largest orders of insects, with over 0.4 million known species, and are found in almost every habitat on Earth (Rees et al., 2018). Beetles have two pairs of wings, with the front pair (elytra) being hard and protective, covering the hind wings and the body. This characteristic distinguishes them from other insects, like butterflies or flies, which have only one pair of wings. Beetles can vary in size, colour, and shape. Some of the most common types of beetles include ladybirds, fireflies, and scarab beetles (Crowson et al., 2013).

They play important roles in many ecosystems, including pollination, decomposition, and food webs. Some beetles are also important pests, causing damage to crops, stored grains, and wood. Beetles have a complex life cycle, starting as an egg and then moving through several larval stages, pupation, and finally, adulthood. They have a wide variety of feeding habits, with some species being herbivores, others being predators, and still others being scavengers. Their mouthparts are adapted for chewing, and they have a complete metamorphosis, going through egg, larval, pupal, and adult stages. Habitats of beetles occupy a wide range, from terrestrial to freshwater and even marine environments (Crowson et al., 2013). They play important roles in ecosystems, serving as pollinators, decomposers, predators, and prey. Beetles have been studied extensively by entomologists due to their ecological importance and their economic impact as agricultural pests and vectors of disease. The diversity of Coleopterans is staggering, with beetles ranging in size from less than 1 mm to over 17 cm in length. Some beetles are important agricultural pests, while others play important roles in pollination and decomposition (Rees et al., 2018). Many species of beetles have important cultural and economic significance, such as the European honeybee, which is a critical pollinator of crops, and the mountain pine beetle, which is a destructive forest pest. The coleopteran insects that can break down artificial polymers are discussed and detailed below:

5.1.1. Mealworm

Tenebrio molitor (mealworm) is a species of beetle that is commonly used as a food source for birds, reptiles, and other animals. The larvae of mealworms are known to consume a variety of organic materials, including grains. Some studies have suggested that the larvae of mealworms can consume and break down certain types of plastic, such as polystyrene foam, into smaller fragments by chewing, which are then further broken down in their gut by the microbiome (Billenet et al., 2020). The mealworm larvae are capable of chewing and eating polystyrene (Styrofoam) plastic when given as their only diet, and in 14–15 hours, they were able to degrade approximately half of the ingested plastic in their gut. The bacterial strain (YT2) present in the gut of mealworms was observed to alter the surface topography of PS materials. Polystyrene samples were injected with the Exiguobacterium sp. bacterial strain (YT2) (Urbanek et al., 2020). After two months of incubation, it was found that polystyrene samples lost seven per cent of their weight. As a result, it was discovered that yellow mealworms fed separately on PS and PE plastic caused mass loss of both polymers. The yellow mealworms decompose both polyethylene and polystyrene, whereas, in solely plastic diets, the breakdown of PE was seen to be significantly higher (48%) than that of PS (32%) (Genta et al., 2006). It is important to keep in mind that not only can they eat various kinds of plastics, but they can also ingest potentially harmful plastic additives without suffering negative consequences.

5.1.1.1. Life Cycle Stages

As shown in Figure 3, the female darkling beetle lays tiny white or yellowish eggs in soil or substrate, and each egg is about 1mm in length. In 4 to 19 days, eggs hatch and convert into mealworms, influenced by temperature and humidity. The newly hatched mealworm is a small, white, and slender worm-like creature that is about 1-2 mm long. During this stage, the mealworm grows in size by periodically shedding its skin, known as moulting. It can take anywhere from 2 to 6 months for the mealworm to reach its full size, which can vary from 2.5 cm to 3.5 cm long (Wu et al., 2019). When the mealworm has reached its full size, it stops eating and moulting and enters the pupa stage. The pupa is a non-feeding, non-moving stage where the mealworm transforms into an adult darkling beetle. The pupa stage lasts from 7 to 21 days. After the pupa stage, the darkling beetle emerges as an adult. The adult beetle is typically black or brown and has two pairs of wings, although it is unable to fly. It is capable of reproducing and laying eggs, and its lifespan can vary from a few months to a year. Overall, the lifecycle of a mealworm can take from 2 to 6 months, depending on environmental conditions (Yang et al., 2020).

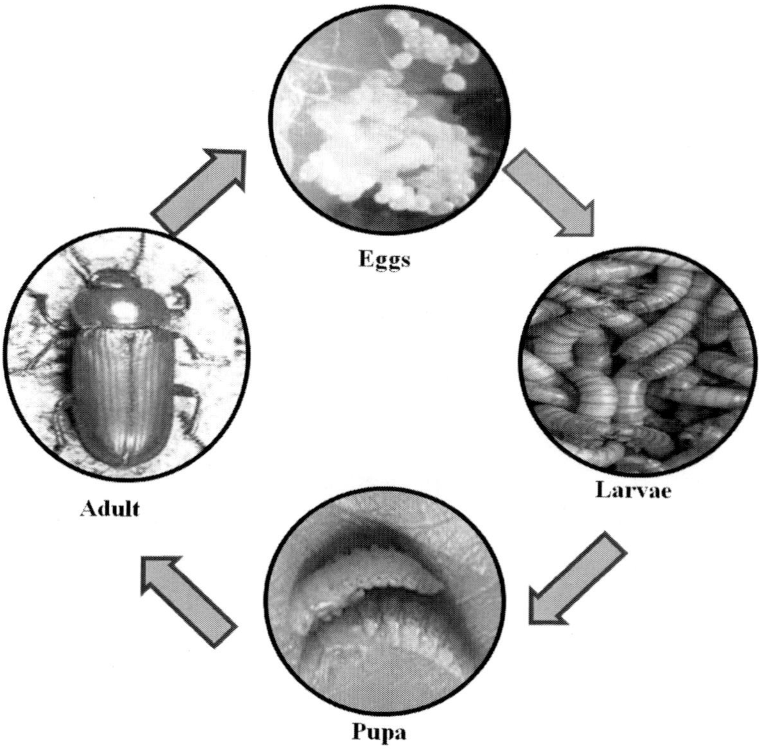

Figure 3. Life cycle of *Tenebrio molitor.*

5.1.2. Superworm

Superworms are the larval stage of the beetle *Zophobas atratus* (Family: Tenebrionidae). They are also known by the names of kingworms or blind click-beetles. Their front wing, known as the elytra present on their back, is much darker, and once they reach maturity, they become darker and are then called "black beetles." These are capable of chewing and eating plastic because they have mandibular mouthparts. Super worms are found to feed on Styrofoam as their sole diet. It was observed that when Styrofoam samples were inoculated with super worms, they instantly started to ingest and penetrate through the blocks, making hollows in the blocks within an hour (Luo et al., 2021). Further investigation led to the discovery of three bacterial strains, which were *Aeromonas* sp. and *Klebsiella pneumonia* from Enterobacteriaceae and *Pseudomonas* sp. from Pseudomonadaceae, respectively, from the guts of super worms (Kim et al., 2020). The presence and isolation of these bacterial strains confirmed their potency to degrade polystyrene.

5.1.2.1. Life Cycle Stage

The female *Zophobas* beetle lays tiny white or yellowish eggs in soil or substrate, and each egg is about 1.5mm in length. In a period of two to four weeks, eggs convert into adults, influenced by the surrounding temperature and humidity (Kim et al., 2015). The newly hatched super worm is a small, white, and slender worm-like creature that is about 1-2 cm long. During this stage, the super worm grows in size by periodically shedding its skin. It can take anywhere from 3 to 6 months for the super worm to reach its full size, which can vary from 5 to 6 cm long. When the superworm has reached its full size, it stops eating and moulting and enters the pupa stage. The pupa is a non-feeding, non-moving stage where the superworm transforms into an adult *Zophobas* beetle. The pupa stage lasts from 2 to 3 weeks. After the pupa stage, the *Zophobas* beetle emerges as an adult. The adult beetle is typically dark brown or black and has two pairs of wings, although it is unable to fly. It is capable of reproducing and laying eggs, and its lifespan can vary from a few months to a year (Quennedey et al., 1995). Overall, the lifecycle of a superworm can take from 4 to 8 months, depending on environmental conditions (Figure 4).

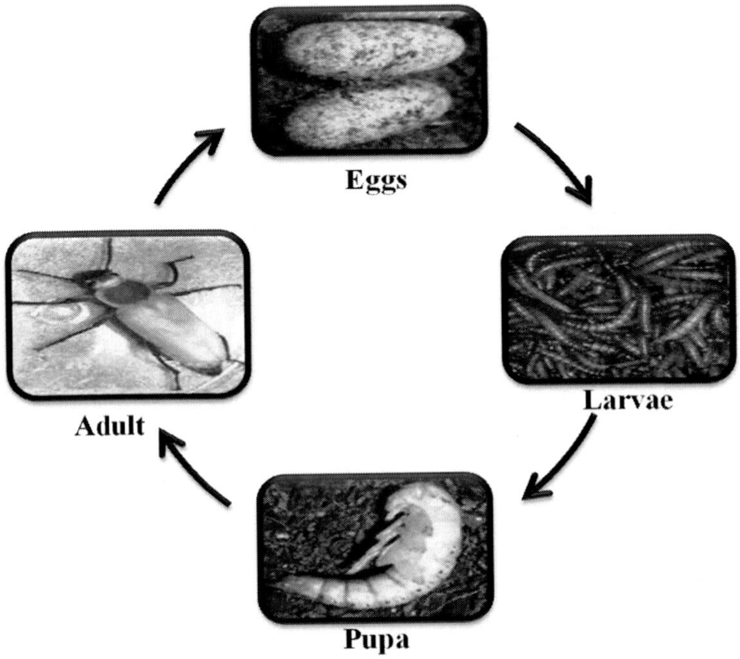

Figure 4. Life cycle of *Zophobas atratus.*

5.1.3. Confused Flour Beetle

Tribolium confusum (confused flour beetle) is a kind of darkling beetle which is, known as a flour beetle, that is a common insect pest renowned for infesting and attacking stored grain and flour. They are tiny, measuring only 3-6 mm (1/8–1/4 inch) long, and have a reddish-brown hue. Due to its similarity to the red flour beetle, the word "confused" appears in the insect's name. During the study, it was observed that its larvae degraded PE film, PS, and ethylene vinyl acetate. After degradation, when the larvae's weight was observed, there was a loss in their weight (Abdulhay, 2020), indicating that PE, PS, and EVA are not effective sources of energy for the beetles. Further oxidation of the ingested Styrofoam was demonstrated using Fourier-transform infrared spectroscopy (FTIR). This resulted in the discovery of *Serratia* species on the PS film isolated from the gut (Woo et al., 2020). This species is responsible for the degradation of PS, PS, and EVA.

5.1.3.1. Life Cycle

A female may lay up to thousands of eggs in a lifetime, and within five to twelve days, the eggs hatch. The grub is worm-like, with a whitish cream colour and slender, cylindrical, faint stripes at the end segment. Two spine-like appendages are present; during development, they undergo five to twelve instar stages and reach about 0.18 inches in length within thirty days. Pupa stays generally lying in the grain and is uncovered. Adults are active and move unevenly. They are oval, flat, and brown and ingest meal (Baldwin et al., 2003). There is a significant difference between the two species of *Tribolium*: in *T. confusum*, the compound eyes are completely curved or grooved, and the antennae are not gently stiffened, whereas in *T. castaneum*, the notch is incomplete and the antennae have a clear 3-segmented club (Baldwin et al., 2003). Figure 5 shows the life cycle stages of *T. confusum*.

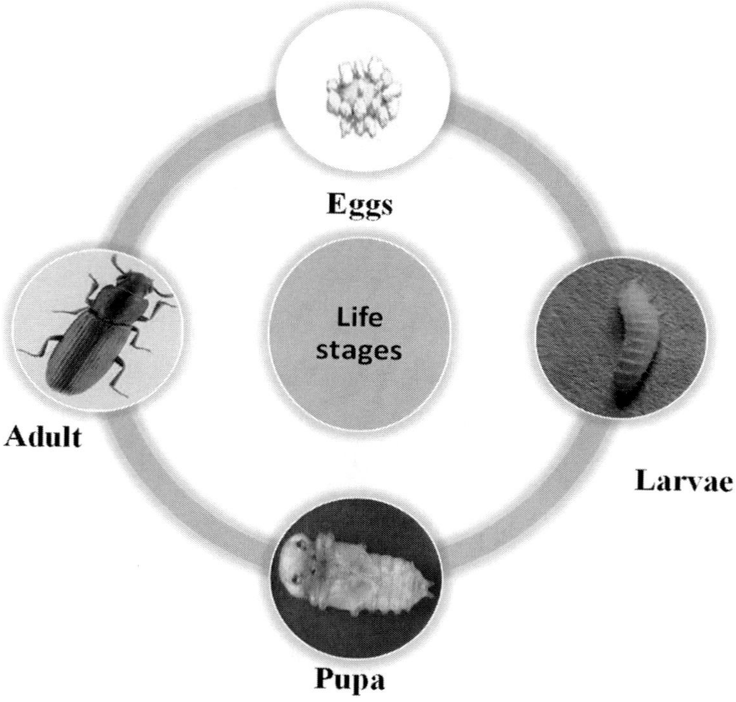

Figure 5. Life cycle of *Tribolium confusum*.

5.1.4. Red Flour Beetle

Tribolium castaneum (red flour beetle) is commonly found in stored grains and cereals, such as flour, rice, and oats. The adult reaches approximately three to four millimetres in length and has a reddish-brown hue. The body is flat and curl-sided; the wing cases (elytra) are completely clumped, present on the head and initial region of the thorax. The head also possesses enlarged antennae with reddish-black eyes. During a study, the Red Flour Beetle samples fed on polystyrene films showed holes and pits on them, whereas the control did not show any changes

(Wang et al., 2018). For further investigation, the gut tissue of beetles fed on PS was isolated for metagenomic analysis, and its result was compared with the gut tissue of beetles that were fed a normal diet. As per the result, the number of microbial genes responsible for complex compound metabolism showed that the number of microbial genes related to the metabolism of complex compounds was remarkably higher in the larvae who fed on PS compared to the larvae who did not (the control group), showing that bacteria responsible for polystyrene degradation increased with the advance of PS feeding. Finally, a gut tissue bacterial strain was identified named *Acinetobacter* sp. (strain AnTc-1, which has a high potency for degrading PS). On incubation of PS powder with AnTc-1 for 60 days, weight loss of up to 12 per cent of the starting weight was registered. With this conclusion, it was concluded that red flour beetles can biodegrade plastics.

5.1.4.1. Life Cycle

The life cycle of a red-fleshed beetle consists of 4 stages: egg, larva, pupa, and adult (Figure 6). The female red flour beetle lays tiny white or cream-coloured eggs in flour or other grain products, and each egg is about 0.5 mm in length. Within two to five days, eggs are formulated into larvae, influenced by the surrounding temperature and humidity (Brown et al., 2009). The newly hatched larva is a small, white, and worm-like creature that is about 2 mm in length. During this stage, the larva feeds on the flour or grain product and grows in size by moulting. It can take anywhere from 2 to 6 weeks for the larva to reach its full size, which can vary from 4 to 5 mm in length. When the larva has reached its full size, it stops eating and moulting and enters the pupa stage. The pupa is a non-feeding, non-moving stage and lasts from 5 to 18 days. After the pupa stage, the red-fleshed beetle emerges as an adult. The adult beetle is typically reddish-brown and has two pairs of wings, although it is unable to fly (Devi et al., 2015). It is capable of reproducing and laying eggs, and its lifespan can vary from a few weeks to several months. Overall, the lifecycle of a red flour beetle can take from 4 to 12 weeks and is affected by surrounding abiotic factors like humidity and temperature (Brown et al., 2009). It is important to properly store and handle flour and grain products to prevent infestations of red flour beetles.

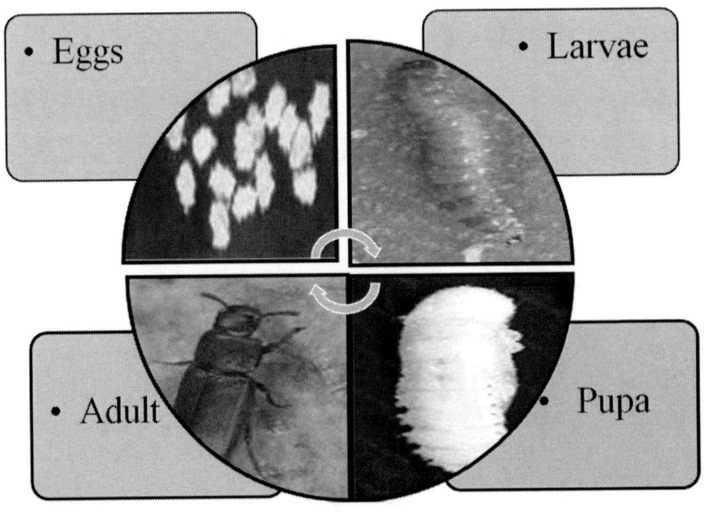

Figure 6. Life cycle of *Tribolium castaneum*.

5.2. Lepidoptera

Lepidoptera is a group of insects commonly known as butterflies and moths. They are characterised by their large, scale-covered wings, which are often brightly coloured and patterned. Lepidoptera is one of the most varied groups of insects, with over one lakh eighty thousand described species found throughout the world (Powell et al., 2009). The four unique life stages of Lepidopteran insects include egg, larva (caterpillar), pupa (chrysalis), and adult. The caterpillar stage is characterised by rapid growth and feeding, while the adult stage is focused on mating and reproduction (Smith et al., 1965). Many species of butterflies and moths are important pollinators, while others are agricultural pests. Some common examples of lepidopterans include monarch butterflies, silk moths, and sphinx moths. The diverse patterns and colours of their wings have made lepidopterons popular art subjects, and they have been used as symbols of beauty and transformation in many cultures throughout history. The life cycle of a lepidopteron starts with an egg, which hatches into a caterpillar. The caterpillar goes through several stages of growth and eventually forms a pupa, inside which it undergoes metamorphosis to become an adult butterfly or moth (Powell et al., 2009). Lepidopterans have a wide range of feeding habits. Some species are herbivores, feeding on plants, while others are predators, feeding on other insects. Some species of butterflies and moths are important pollinators, while others are considered pests, damaging crops and stored food products (Bernays et al., 1997). Below is a discussion and description of the Lepidopteran insects capable of decomposing man-made polymers.

5.2.1. The Indian Meal-Moth

Plodia interpunctella (the Indian meal moth), commonly known as the pantry moth, is a pest of the common household that infests stored food items such as grains, cereal, nuts, dried fruits, and pet food. The caterpillar has a small head that is black to brown in colour and feeds on cereals, fruits, and grains. The adult moths are small, typically measuring about 8-10 mm in length, with a wingspan of around 16–20 mm, and have black-tipped feet (Silhacek et al., 1972). The adult moths lay eggs on food products, and the larvae that hatch from these eggs are responsible for most of the damage to stored foods. The larvae are creamy-white with brownish heads, and they spin silken threads as they move through the food, leaving behind webbing and frass (insect excrement) that can contaminate the food and make it unappetising (Subramanyam, B. 2011). The caterpillar resides in bee colonies as a parasite and feeds on beeswax and other food gathered by bees, like flower pollen, cocoons, and grain mixes. During the study, meal moths were fed with PE films, and it was observed that there was some destruction in the physical properties of PE, like a change in surface topography, as many small pits and cavities were found on the surface of biofilms through scanning electron microscopy (SEM) and atomic force microscopy (AFM) (Graham et al., 1997). The gut tissue of moths fed on PE was investigated, and two intestinal bacteria were isolated and recognised as gram-positive strain YP1, *Bacillus* sp., and gram-negative strain YT1, *Enterobacter asburiae* (Graham, 1997; Yang et al., 2015). Separately, these bacterial strains were grown on PE sheets, and it was found that they attached to PE films instantly, and within three to four hours, both the bacterial strains YP1 and YT1 formed biofilms, indicating that they are set for biodegradation. Wax worm gut bacterial strains YT1 and YP1 are effective at decomposing PE, as seen by the physical and chemical alterations of inoculated PE samples. Thus, Indian meal moth larvae that consume plastic may offer a promising source of insects that degrade plastic (Yang et al., 2023).

5.2.1.1. Life Cycle

The eggs have a greyish-white hue and are 0.3–0.5 mm (0.01–0.023 inches) in length. Eggs lay singly or in groups of between twelve and thirty on the food source, and at a time, 100 to 300 eggs are laid by adult females. Within 2–14 days, eggs hatch into larvae, which undergo 5-7 instars (Silhacek et al., 1972). Newly hatched larvae and mature larvae feed on grain and grain germs, respectively. The larvae usually have an off-white colour but can be of other colours like pink, brown, or green. They reach up to twelve mm in length and possess pro-legs for motion. Adult larvae can weave webs and leave silk strands behind as they move through the environment. After 4–10 days, the pupa turns into an adult. Adults mate, and the life cycles are reprocessed. This species' complete life cycle lasts between 30 and 300 days (Subramanyam, B. 2011). A life cycle typically lasts 50 days. A life cycle can be as brief as 28 days under ideal circumstances, but the colder winter months prevent this. Moths can have seven to nine generations every year (Figure 7).

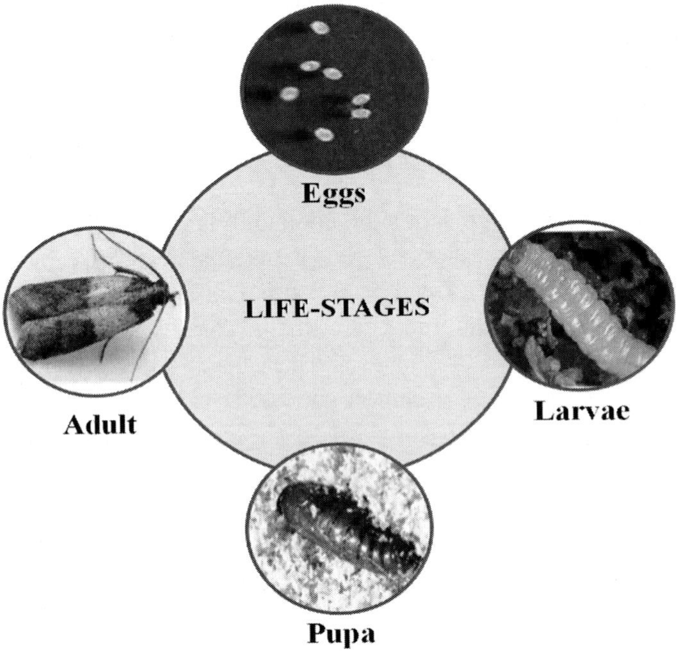

Figure 7. Life cycle of *Plodia interpunctella*.

5.2.2. The Greater Wax Moth

Galleria mellonella (the greater wax moth) is a species of moth that is known for its ability to consume beeswax and honeycomb, which are natural substances produced by honeybees. Its larval stages feed on bees and beehives and cause high damage because of its feeding habit. Recently, some studies suggest that the larvae of *G. mellonella* can also break down certain types of plastic, particularly polyethylene (Lou et al., 2020). During a study, when microbes were isolated from *G. mellonella* caterpillar guts and grown in a liquid C-free medium with PE for sixty weeks, it was observed that these gut microbes played an important role in the biodegradation of polyethylene. The discovered microbes, *Acinetobacter* species ACT126, as well as *Enterobacter* sp. and the fungus *Aspergillus flavus*, were assisting in the process of PE degradation (Khyade et al., 2018). Further studies show that they not only degrade PE but also polystyrene. As larvae were found to ingest PS microbes helping in PS degradation. The gram-positive lactic acid-producing bacteria, *Enterococcus* sp., the facultative anaerobic gram-positive bacteria, *Bacillus cereus*, and the gram-negative rod-shaped bacteria, *Serratia marcescens* were separated from the intestines of larvae and suspected of contributing to the degradation of PS (Khyade et al., 2018).

5.2.2.1. Life Cycle

The greater wax moth undergoes four stages in its life: egg, larvae, pupae, and adult (Figure 8). The egg shape is plain and globular, reaching 0.4 to 0.5 mm in size (Chandel et al., 2003). It is found in a diverse range of colours, from pink to cream to white. Females lay eggs in groups in small fissures and clefts of the beehive, and they hatch between seven and twenty-two days. Larva varies in length from three to thirty mm and is found in white and dull grey. Their meals include honey, pollen, the cast-off skin of honeybee larvae, and the midrib of the waxcomb; deficiency of food also shows cannibalism. They eat vigorously in the initial instar stages. The period of larval stages may range from twenty-eight days to six months, during which they undergo 8 to 10 stages of moulting (Kwadha et al., 2017). They spun silk during all instar stages, but at the final stage of instar, a silk cocoon is spun by larvae for itself and gets into the pupa stage. The pupa stage is inactive and immovable; it does not eat and stays inside the cocoon for one to nine weeks until it comes out as an adult. They vary in size, reaching from 14-16 mm. Pupae at first are a light brown colour, which slowly becomes dark brown just before they are prepared to come out. Adult males and females are grey and brown. They vary in size from 8-10 mm, with a wingspan of about 30-40 mm (Chandel et al., 2003). Compared to males, females are heavier and bigger, with beak-like heads. The front wing is plain at the external periphery, and the labial palp is extended forward. The semilunar notch is an identifiable feature of males. The lifespans of males and females are twenty-one and twelve days, respectively.

5.2.3. The Lesser Wax Worm

Achroia grisella (the lesser wax moth) is a species of moth that is known for its ability to consume beeswax, which is a natural substance produced by honeybees. While the larvae of lesser wax moths can break down beeswax, studies have shown that while the larvae of lesser wax moths can consume and break down certain types of plastic, such as polyethylene and polypropylene, their ability to degrade these materials is limited (Cepeda-Aponte et al., 2002). Naturally, *A. grisella* larvae consume beeswax, which shows similarity with polyethylene in structural formula. When *A. grisella* caterpillars were inoculated with PE films, then these PE films that were inoculated showed a high change in polyethylene weight and lessened the remaining polyethylene, recommending that most larvae ingest PE either by breaking down or absorbing the PE. The biodegradation process that auspiciously occurs in *A. grisella* larvae was approved by FTIR and NMR analyses (Chalup et al., 2018). The analysis report shows new carbonyl and alcoholic groups with a high amount of unsaturated hydrocarbons, which confirms the biodegradation process of PE in the lesser wax worm.

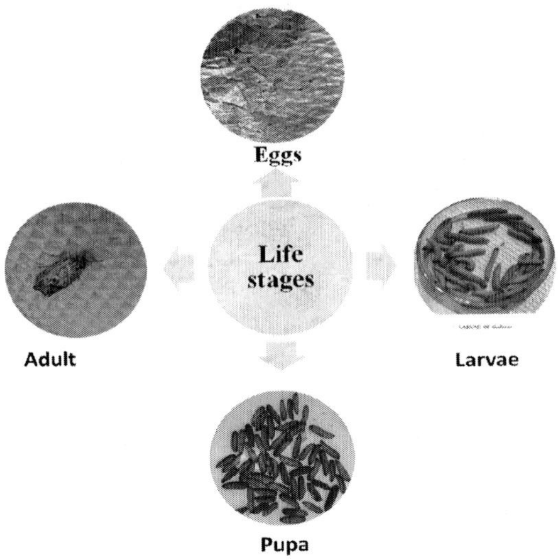

Figure 8. Life cycle of *Galleria mellonella.*

5.2.3.1. Life Cycle

Having four life stages of holometabolous insects (Figure 9), their eggs are similar to those of greater wax moths. They are globular in shape and creamy white. In between 5-8 days, eggs hatch, but this can be influenced by warmer temperatures as it decreases the hatching time. In about 6-7 weeks, larvae fully develop, but it sometimes takes more time, up to 5 months (Chandel et al., 2003). They grow up to 20 mm in length and possess slim white bodies with a brown head. It's the only feeding stage, as after this, the pupae grow up to 11 mm in size and are a yellow-tan colour. The silk cocoon is white but generally enveloped in grass. It takes about 37 days to emerge as an adult, but pupation can last up to 60 days (Chandel et al. 2003). Adults are found in different colours, such as silver, grey, or beige, with a yellow head. They are slim and grow up to half an inch in size with a wingspan of half an inch. Compared to females, males are smaller. The lifespan of an adult is 7 days, including female oviposition and mating (Mahgoub et al., 2020).

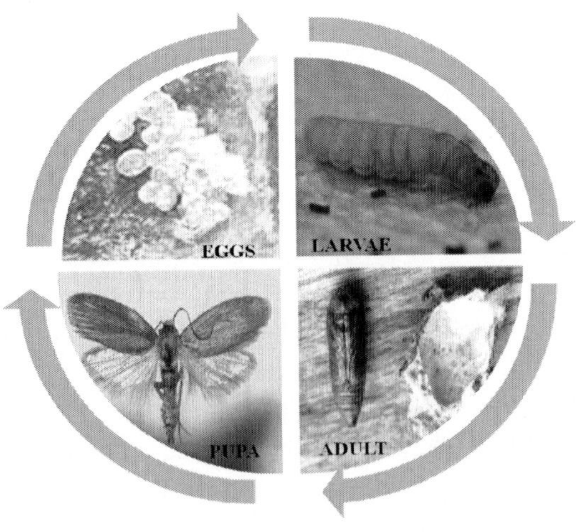

Figure 9. Life cycle of *Achroia grisella.*

5.2.4. Rice Mealworm

The rice mealworm, scientifically called *Corcycra cephalonica*, is a species of darkling beetle that is commonly used as a food source for reptiles, birds, and other animals. While the larvae of rice mealworms are known to consume a variety of organic materials, including grains, it is important to note that they are not efficient plastic degraders. An experiment (Kesti et al., 2019) was done to find the role of the rice moth larvae gut microbiome in PE degradation by comparing a group of untreated rice moths with a group of gut-microbiome suppressed rice moths pre-treated with antibiotics. Regardless of the pretreatment, both groups of rice moths were able to chew and consume similar quantities of PE after 20 days (20% and 25% of the initial substrate for the control and antibiotic groups, respectively), which led the researchers to conclude that the *C. cephalonica* gut microbiome may not play a major role in PE consumption (Kesti et al., 2019) but some other microbes and enzymes may help in degrading PE.

5.2.4.1. Life Cycle

There are four stages in the life cycle: eggs, larvae, pupae, and adults (Figure 10). The adult female lays about five hundred eggs. The eggs hatch between four and nineteen days into the larval stages, and they feed on crops and dead insects. Moulting occurs, and the eggs undergo nine to twenty instars. After the last moult, they pupate (Kesti et al., 2018). The pupa initially is white but gradually, with time, turns brown. After three to thirty days, influenced by the surrounding temperature, it comes out as an adult beetle.

5.2.5. Fall Armyworm

The fall armyworm, *Spodoptera frugiperda*, is a species of moth that is known for its destructive impact on crops, particularly maize, rice, and sorghum. While the fall armyworm is not known for its ability to degrade plastic, it is considered a major pest in many parts of the world and has a significant impact on food security (Sabra et al., 2022). During a study, *S. frugiperda* larvae were fed on PVC, and the intestinal tissues of these treated larvae were taken for genomic analysis. The PVC-degrading *Klebsiella* strain EMBL-1, which depends on PVC for its growth, was isolated from the gut. Later, multigenomic analysis integrating genome-, transcriptome-, protein-, and metabolite-level insights into the strain confirmed the discovery of some functional genes, enzymes, and metabolic pathways closely related to PVC degradation (Zhang et al., 2022). Ingesting PVC film notably altered the larval intestinal microbiota through selective enhancement of *Enterococcus*, *Ochrobactrum*, *Falsochrobactrum*, *Microbacterium*, *Sphingobacterium*, and *Klebsiella*. *Klebsiella* sp. EMBL-1 was confirmed as the first *Klebsiella* bacteria known to actively degrade and utilise PVC by various classic physicochemical and morphological analyses. Overall, both *S. frugiperda* and strain EMBL-1 were first found to survive effectively on PVC film by using the polymer as the sole energy source (Zhang et al., 2022).

5.2.5.1. Life Cycle

The fall armyworm's life span varies with the seasons; they complete their life cycle (Figure 10) in one month during summer and take two months in spring and autumn. The longest time taken during winter is eighty to ninety days. The armyworm's egg is arch-shaped and is approx. 0.4 mm (0.015 in) in length and 0.3 mm (3⁄256 in) in height (Ashok et al., 2020). Around 1,500 eggs are laid by females in their lifetime. Influenced by warm weather, the eggs will hatch into larvae in a short time. The larvae pass through six distinct instars, each of which differs a bit in morphology. Based on temperatures, it may go from fourteen days to one month (Sabra et al., 2022). Mature larvae measure about 30-51 mm in length. The larvae have a distinctive inverted "Y" suture on the forehead. The larvae stay inside the cocoon for about seven to thirty-seven days and form silk. Based on the surrounding temperature, the pupation period may vary. Adults live for 1-3 weeks and are seen at night when the climate is humid and warm (Ashok et al., 2020).

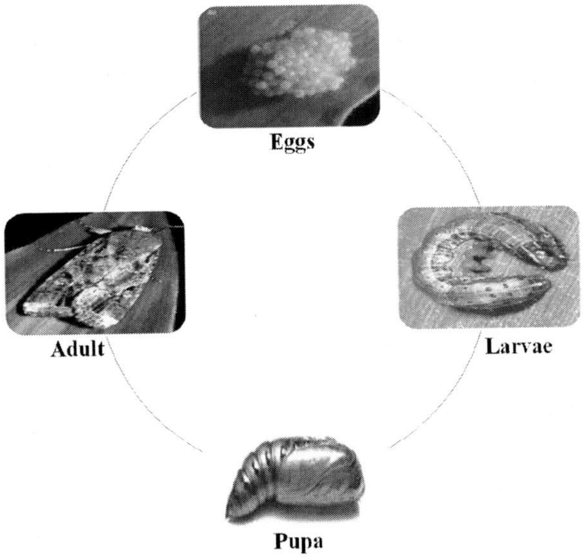

Figure 10. Life cycle of *Spodoptera frugiperda.*

Table 3. Some insects with their common name, the ability to eat different forms of plastic and their bacterial symbionts

S. No.	Common name	Scientific name	Order	Plastic they eat	Bacterial symbiont
1	Confused flour beetle	*Tribolium confusum*	Coleoptera	Ethylene-vinyl acetate	*Acinetobacter sp.*
2	Fall armyworm	*Spodoptera frugiperda*	Lepidoptera	Polyvinyl chloride	*Klebsiella strain*
3	Greater wax moth	*Galleria mellonella*	Lepidoptera	Polyethylene	*Enterobacter asburiae*
4	Indian meal moth	*Plodia interpunctella*	Lepidoptera	Polyethylene	*Enterobacter tabaci Bacillus subtilis*
5	Lesser wax moth	*Achroia grisella*	Lepidoptera	Polyethylene	
6	Mealworm	*Tenebrio molitor*	Coleoptera	Polyvinyl chloride Polystyrene Polyethylene	*Exiguobacterium sp. Pseudomonas*
7	Red flour beetle	*Tribolium castaneum*	Coleoptera	Polystyrene	*Acinetobacter sp. AnTc-1*
8	Rice mealworm	*Corcyra cephalonica*	Lepidoptera	Polyethylene	*Staphylococcus saprophyticus*
9	Super worm	*Zophobas atratus*	Coleoptera	Polyphenylene sulfide Polyethylene	*Pseudomonas aeruginosa*

Conclusion

The collection of plastic waste in the surroundings is creating an ever-increasing environmental ultimatum. Biodegradable plastics are viewed as eco-friendly; they have a wide range of potential to be used in various fields, like packaging, for a long time. Polyethylene (PE) has been viewed as non-biodegradable. Plastic biodegradation is regarded as a particularly promising waste management strategy, especially for single-use and short-in-use phase products used for agricultural purposes and packaging, which are widely recognised as the main drivers of global plastic waste mismanagement and environmental pollution. The biodegradation of various synthetic polymers and plastic products has been studied under different environmental conditions. Numerous microbial strains and enzymes isolated from diverse environments (e.g., soil, compost, landfills, seawater) were reported to show the degrading capabilities of synthetic polymers that commonly make up plastic packaging products. The degradation processes and pathways of these plastic products have been thoroughly reviewed, pointing out the methodological challenges and knowledge gaps that are still to be resolved to attain conclusive evidence of complete plastic biodegradation. Recent experiments have revealed that some insects and their microbial symbionts possess plastic biodegradation capabilities, raising interest in their untapped potential as a bioresource for new plastic waste management alternatives. Interactions between insects and plastic, particularly plastic packaging, have been described as early as the 1950s. The large majority of reports aimed at testing and improving the resistance of plastic packaging to the attack of insect "pests," ensuring the integrity of the packaged product (i.e., food). This reflects on one of the major purposes and benefits of using and developing plastic packaging technologies: reducing food waste. However, it was also noted that historical trends show rising rates of food waste as well as waste from plastic packaging in affluent nations like Europe, supporting the claim that plastic packaging falls short of addressing significant underlying causes of food waste. Moreover, many environmental impact analyses and life cycle assessments that advocate for the use of plastic packaging may heavily rely on assumptions that ignore or oversimplify the complex reality of current plastic packaging waste management.

Acknowledgments

The authors are thankful to the Vice Chancellor, Banasthali Vidyapith and the Head of the Department of Bioscience and Biotechnology for providing the necessary facilities to facilitate the work. Facilities provided by DST-FIST to the University are highly acknowledged.

References

Ahmed, T., Shahid, M., Azeem, F., Rasul, I., Shah, A. A., Noman, M., & Muhammad, S. (2018). Biodegradation of plastics: current scenario and future prospects for environmental safety. *Environmental Science and Pollution Research*, *25*, 7287-7298.

Ali, S. S., Elsamahy, T., Al-Tohamy, R., Zhu, D., Mahmoud, Y. A. G., Koutra, E., & Sun, J. (2021). Plastic wastes biodegradation: Mechanisms, challenges and future prospects. *Science of The Total Environment*, *780*, 146590.

Ashok, K., Kennedy, J. S., Geethalakshmi, V., Jeyakumar, P., Sathiah, N., & Balasubramani, V. (2020). Lifetable study of fall army worm *Spodoptera frugiperda* (JE Smith) on maize. *Indian Journal of Entomology*, 82(3), 574-579.

Baldwin, R., & Fasulo, T. R. (20202003). Confused Flour Beetle, Tribolium confusum Jacquelin du Val and Red Flour Beetle, *Tribolium castaneum* (Herbst) (Insecta: Coleoptera: Tenebrionidae). UF/IFAS Extension.

Berasategui, A., Shukla, S., Salem, H., & Kaltenpoth, M. (2016). Potential applications of insect symbionts in biotechnology. *Applied Microbiology and Biotechnology*, *100*, 1567-1577.

Bernays, E. A. (1997). Feeding by lepidopteran larvae is dangerous. *Ecological Entomology*, 22(1), 121-123.

Bombelli, P., Howe, C. J., & Bertocchini, F. (2017). Polyethylene bio-degradation by caterpillars of the wax moth *Galleria mellonella. Current bBiology*, *27*(8), R292-R293.

Brandon, A. M., Gao, S. H., Tian, R., Ning, D., Yang, S. S., Zhou, J., & Criddle, C. S. (2018). Biodegradation of polyethylene and plastic mixtures in mealworms (larvae of *Tenebrio molitor*) and effects on the gut microbiome. *Environmental Science & Technology*, *52*(11), 6526-6533.

Brown, S. J., Shippy, T. D., Miller, S., Bolognesi, R., Beeman, R. W., Lorenzen, M. D., ... & Klingler, M. (2009). The red flour beetle, *Tribolium castaneum* (Coleoptera): a model for studies of development and pest biology. Cold Spring Harbor Protocols, 2009(8), pdb-emo126.

Cassone, B. J., Grove, H. C., Elebute, O., Villanueva, S. M., & LeMoine, C. M. (2020). Role of the intestinal microbiome in low-density polyethylene degradation by caterpillar larvae of the greater wax moth, *Galleria mellonella. Proceedings of the Royal Society B*, *287*(1922), 20200112.

Cepeda-Aponte, O. I., Imperatriz-Fonseca, V. L., & Velthuis, H. H. (2002). Lesser wax moth Achroia grisella: first report for stingless bees and new capture method. *Journal of aApicultural Rresearch*, *41*(3-4), 107-108.

Chalup, A., Ayup, M. M., Monmany Garzia, A. C., Malizia, A., Martin, E., De Cristóbal, R., & Galindo-Cardona, A. (2018). First report of the lesser wax moth *Achroia grisella* F. (Lepidoptera: Pyralidae) consuming polyethylene (silo-bag) in northwestern Argentina. *Journal of Apicultural Research*, *57*(4), 569-571.

Chamas, A., Moon, H., Zheng, J., Qiu, Y., Tabassum, T., Jang, J. H., & Suh, S. (2020). Degradation rates of plastics in the environment. *ACS Sustainable Chemistry & Engineering*, *8*(9), 3494-3511.

Chandel, Y. S., Sanjeev Sharma, S. S., & Verma, K. S. (2003). Comparative biology of the greater wax moth, *Galleria mellonella* L., and lesser wax moth, *Achroia grisella* F. *Pest Management and Economic Zoology*, *11(1)*, 69-74.

Crowson, R. A. (2013). The biology of the Coleoptera. Academic press.

Devi, M. B., & Devi, N. V. (2015). Biology of rust-red flour beetle, *Tribolium castaneum* (Herbst) (Coleoptera: Tenebrionidae). In Biological Forum (Vol. 7, No. 1, pp. 12-15). Satya Prakashan.

Emadian, S. M., Onay, T. T., & Demirel, B. (2017). Biodegradation of bioplastics in natural environments. *Waste Management*, *59*, 526-536.

Genta, F. A., Dillon, R. J., Terra, W. R., & Ferreira, C. (2006). Potential role for gut microbiota in cell wall digestion and glucoside detoxification in *Tenebrio molitor* larvae. *Journal of Insect Physiology*, *52*(6), 593-601.

Geyer, R., Jambeck, J. R., & Law, K. L. (2017). Production, use, and fate of all plastics ever made. *Science Aadvances*, *3*(7), e1700782.

Glaser, J. A. (2019). Biological degradation of polymers in the environment. *Plastics in the Environment*, *1*, 13.

Graham, B. T. (1997). Penetration of Polyvinyl Chloride and Polypropylene Packaging Films by Ephestia cautella (Lepidoptera: Pyralidae) and Plodia interpunctella (Lepidoptera: Pyralidae) Larvae, and Tribolium confusum (Coleoptera: Tenebrionidae) Adults. *Journal of Economic Entomology*, *90*, 1028–1031.

Gullan, P. J., & Cranston, P. S. (2014). *The insects: an outline of entomology*. John Wiley & Sons.

Haider, T. P., Völker, C., Kramm, J., Landfester, K., & Wurm, F. R. (2019). Plastics of the future? The impact of biodegradable polymers on the environment and on society. *Angewandte Chemie International Edition*, *58*(1), 50-62.

Jiang, L., Hua, Y., Hu, G. L., & Hua, B. Z. (2019). Habitat divergence shapes the morphological diversity of larval insects: insights from scorpionflies. *Scientific Reports*, *9*(1), 12708.

Kesti, S. S. K., & Thimmappa, S. C. T. (2019). First report on biodegradation of low density polyethylene by rice moth larvae, *Corcyra cephalonica* (Stainton). *The Holistic Approach to Environment*, *9*(4), 79-83.

Kesti, S. S., & Shivasharana, C. (2018). The role of insects and microorganisms in plastic biodegradation: A comprehensive review. *International Journal of Scientific Research in Biological Sciences*, *5*, 6.

Khyade, V. B. (2018). Review on biodegradation of plastic through waxworm (Order: Lepidoptera; Family: Pyralidae). *International Academic Journal of Economics*, *5*(4), 31-38.

Kim, H. R., Lee, H. M., Yu, H. C., Jeon, E., Lee, S., Li, J., & Kim, D. H. (2020). Biodegradation of polystyrene by Pseudomonas sp. isolated from the gut of superworms (larvae of *Zophobas atratus*). *Environmental Science & Technology*, *54*(11), 6987-6996.

Kim, S. Y., Kim, H. G., Song, S. H., & Kim, N. J. (2015). Developmental characteristics of of *Zophobas atratus* Fab. (Coleoptera: Tenebrionidae) under crowded or isolated conditions and effects of juvenile hormone analogue applications. *Journal of Insect Physiology*, 41(2), 143-152.

Krueger, M. C., Harms, H., & Schlosser, D. (2015). Prospects for microbiological solutions to environmental pollution with plastics. *Applied Microbiology and Biotechnology*, *99*, 8857-8874.

Kwadha, C. A., Ong'amo, G. O., Ndegwa, P. N., Raina, S. K., & Fombong, A. T. (2017). The biology and control of the greater wax moth, *Galleria mellonella*. *Insects*, 8(2), 61.

Lau, W. W., Shiran, Y., Bailey, R. M., Cook, E., Stuchtey, M. R., Koskella, J., & Palardy, J. E. (2020). Evaluating scenarios toward zero plastic pollution. *Science*, *369*(6510), 1455-1461.

Lebreton, L., & Andrady, A. (2019). Future scenarios of global plastic waste generation and disposal. *Palgrave Communications*, *5*(1), 1-11.

Lou, Y., Ekaterina, P., Yang, S. S., Lu, B., Liu, B., Ren, N., & Xing, D. (2020). Biodegradation of polyethylene and polystyrene by greater wax moth larvae (*Galleria mellonella* L.) and the effect of co-diet supplementation on the core gut microbiome. *Environmental Science & Technology*, *54*(5), 2821-2831.

Lucas, N., Bienaime, C., Belloy, C., Queneudec, M., Silvestre, F., & Nava-Saucedo, J. E. (2008). Polymer biodegradation: Mechanisms and estimation techniques–A review. *Chemosphere*, *73*(4), 429-442.

Luo, L., Wang, Y., Guo, H., Yang, Y., Qi, N., Zhao, X., & Zhou, A. (2021). Biodegradation of foam plastics by *Zophobas atratus* larvae (Coleoptera: Tenebrionidae) associated with changes of gut digestive enzymes activities and microbiome. *Chemosphere*, *282*, 131006.

Mahgoub, M. O., Wei, H. L., Omar, D. B., & Naim, A. (2020). Life table and demographic parameters of the lesser wax moth, *Achroia grisella*, reared on natural honey bee wax. *WJAR*, *8*(1), 12-15.

Mallar, C. N. R. (2020). A review on the role of insects and their gut microbiota in plastic biodegradation.

Muhamad, W. N. A. W., Othman, R., Shaharuddin, R. I., & Irani, M. S. (2015). Microorganism as plastic biodegradation agent towards sustainable environment. *Advances in Environmental BiologyAdv Environ Biol*, *9*(13), 8-13.

Powell, J. A. (2009). Lepidoptera: moths, butterflies. In *Encyclopedia of iInsects* (pp. 559-587). Academic Press.

Quennedey, A., Aribi, N., Everaerts, C., & Delbecque, J. P. (1995). Postembryonic development Postembryonic Development of Zophobas atratus Fab. (Coleoptera: Tenebrionidae) under Crowded or Isolated Conditions and Effects of Juvenile Hormone Analogue Applications. *Journal of Insect Physiology*, 41, 143-152. Beetle,

Rees, D. P. (2018). Coleoptera. In *Integrated management of insects in stored products* (pp. 1-39). CRC Press.

Ren, L., Men, L., Zhang, Z., Guan, F., Tian, J., Wang, B., & Zhang, W. (2019). Biodegradation of polyethylene by *Enterobacter* sp. D1 from the guts of wax moth *Galleria mellonella*. *International Journal of eEnvironmental Research and Public Health*, *16*(11), 1941.

Rumpold, B. A., Klocke, M., & Schlüter, O. (2017). Insect biodiversity: underutilized bioresource for sustainable applications in life sciences. *Regional Environmental Change*, *17*(5), 1445-1454.

Sabra, I. M., Kandil, M. A., El-Shennawy, R. M., & Ahmed, A. F. (2022). Effect of Temperature on Development and Life Table Parameters of Fall Armyworm *Spodoptera frugiperda* (Lepidoptera: Noctuidae). *Journal of Plant Protection and Pathology*, 13(11), 267-271.

Sheikh, A. A., Rehman, N. Z., & Kumar, R. (2017). Diverse adaptations in insects: A review. *Journal of Entomology and Zoology Studies*, *5*(2), 343-350.

Silhacek, D. L., & Miller, G. L. (1972). Growth and development of the Indian meal moth, *Plodia interpunctella* (Lepidoptera: Phycitidae), under laboratory mass-rearing conditions. *Annals of the Entomological Society of America*, 65(5), 1084-1087.

Smith, T. L. (1965). External morphology of the larva, pupa, and adult of the wax moth, *Galleria mellonella* L. *Journal of the Kansas Entomological Society*, 287-310.

Subramanyam, B. (2011). Characteristics of Indian meal moth. *Milling Journal*, 19(1), 38-41.

Urbanek, A. K., Rybak, J., Wróbel, M., Leluk, K., & Mirończuk, A. M. (2020). A comprehensive assessment of microbiome diversity in *Tenebrio molitor* fed with polystyrene waste. *Environmental Pollution*, *262*, 114281.

Wang, S., Wang, L., Fan, X., Yu, C., Feng, L., & Yi, L. (2020). An insight into diversity and functionalities of gut microbiota in insects. *Current Microbiology*, *77*, 1976-1986.

Wierckx, N., Narancic, T., Eberlein, C., Wei, R., Drzyzga, O., Magnin, A., & Blank, L. M. (2018). Plastic biodegradation: Challenges and opportunities. *Consequences of microbial interactions with hydrocarbons, oils, and lipids: biodegradation and bioremediation*, 1-29.

Wu, Q., Tao, H., & Wong, M. H. (2019). Feeding and metabolism effects of three common microplastics on *Tenebrio molitor* L. *Environmental Geochemistry and Health*, *41*, 17-26.

Yang, S. S., & Wu, W. M. (2020). Biodegradation of plastics in *Tenebrio* genus (mealworms). *Microplastics in Terrestrial Environments: Emerging Contaminants and Major Challenges*, 385-422.

Yang, X. G., Wen, P. P., Yang, Y. F., Jia, P. P., Li, W. G., & Pei, D. S. (2023). Plastic biodegradation by in vitro environmental microorganisms and in vivo gut microorganisms of insects. *Frontiers in Microbiology*, *13*. https://doi.org/10.3389/fmicb.2022.1001750.

Yang, Y., Chen, J., Wu, W. M., Zhao, J., & Yang, J. (2015). Complete genome sequence of Bacillus sp. YP1, a polyethylene-degrading bacterium from waxworm's gut. *Journal of biotechnologyBiotechnology*, *200*, 77-78.

Yoshida, S., Hiraga, K., Takehana, T., Taniguchi, I., Yamaji, H., Maeda, Y., ... & Oda, K. (2016). A bacterium that degrades and assimilates poly (ethylene terephthalate). *Science*, *351*(6278), 1196-1199.

Zhang, Y., Pu, S., Lv, X., Gao, Y., & Ge, L. (2020). Global trends and prospects in microplastics research: A bibliometric analysis. *Journal of Hazardous Materials*, *400*, 123110.

Zhang, Z., Peng, H., Yang, D., Zhang, G., Zhang, J., & Ju, F. (2022). Polyvinyl chloride degradation by a bacterium isolated from the gut of insect larvae. *Nature Communications*, 13(1). https://doi.org/10.1038/s41467-022-32903-y.

Chapter 19

Forensic Entomology: Insects as Forensic Helpers

Aby Joseph*
and Nrashant Singh
Department of Forensic Science, Amity University, Dubai, United Arab Emirates

Abstract

Forensic applications of entomology in criminal and civil cases investigate insects and arthropods, estimating postmortem intervals in criminal investigations, and managing insect pests in urban areas in civil cases investigations. Forensic entomology studies have been used for various applications in identifying food contamination, analysing toxic substances in insects, investigating traffic accidents and neglect cases, and aiding in blood spatter pattern analysis and wildlife crime investigations. It also serves as a valuable investigation tool for identifying offenders and providing crucial evidence in certain civil cases. In criminal investigations, forensic entomologists utilise knowledge of various insect lifecycles and behaviours to estimate the time since death, determine the stage of decomposition, and provide valuable information to investigators in criminal investigations, analysing and presenting insect evidence to contribute to successful convictions in the court of law. They also play a vital role in cases of animal abuse or wildlife crimes by examining insect evidence to determine the cause and time of death. Forensic entomology investigations require specific equipment, specific sampling tools, and packaging materials to collect and preserve insect evidence from crime scenes while maintaining proper documentation and following the chain of custody protocol. Forensic entomology laboratory procedures involve estimating postmortem intervals using insect temperature development models of life cycles and accumulated degree hours (ADH) methods, while entomotoxicology focuses on extracting and analysing toxins from insects using different chromatography and spectroscopy techniques, which aid in identifying the cause of death. In urban and food entomology, species identification techniques include visual screening, microscopic characteristic examination, and DNA barcoding techniques. In conclusion, forensic entomology has an important role in solving complex cases and ensuring justice is served.

Keywords: forensic science, entomology, criminal investigation, evidence, entomotoxicology

* Corresponding Author's Email: abyjsph@gmail.com.

In: Insects as Humans' Frenemies: From Friends to Foes
Editor: Istkhar Rao
ISBN: 979-8-89530-320-7

1. Introduction

Forensic entomology primarily deals with insects and other arthropods and interprets their evidence in criminal investigations (Byrd & Castner, 2001). Despite their use in criminal cases, entomological techniques can be used in civil or medical cases as well. Entomological investigations are generally divided into three fields: forensic entomology, urban entomology, and stored food and non-food product pests. The most prevalent application of forensic entomology is medico-legal entomology. Medico-legal entomology is one of the ways to determine the postmortem interval of the deceased. A forensic entomologist can estimate the post-mortem interval based on insect evidence and life cycle analyses. During the process of colonising human and animal remains, insects and arthropods can provide important information about a person's death as well as other forensic data. An early insect to colonise a corpse is the blowfly, which is useful to forensic entomologists in estimating how long has passed since the corpse was buried. Furthermore, beetles, moths, and mites provide valuable information for forensic investigations (Anderson, 2001; Dadour et al., 2001).

A large majority of the world's population lives in urban areas, which pose many challenges, including overcrowding, substandard housing, pollution, and vector-borne diseases. Urban areas can benefit from insect management and control programmes that are effective, economical, and environmentally friendly through entomological studies. Those who study urban entomology concentrate on insects that cause significant economic damage to buildings, and forensic entomologists often take part in civil litigation related to these pests. There are criminal and civil aspects to this issue, as urban pests feed on the living as well as the dead. There are some circumstances where pests, such as bed bugs or cockroaches, can be used to identify a suspect or determine the location of a crime. The marks and wounds that have been caused by the insects' mandibles on the skin of the living or deceased may also be used as indicators of the insect's activity (Hagstrum et al., 2013). The presence of these pests can cause significant damage to stored products and also provide valuable information for forensic investigations. In food safety and product liability cases, the presence of certain pests in stored products, such as the Indian meal moth, can be used to determine the product's origin and age. A related field of forensic investigation is food entomology, which involves the identification of insects and arthropods found in grains, seeds, and food products. A forensic entomologist can act as an expert witness in civil and criminal proceedings concerning food contamination (Hagstrum & Phillips, 2017).

A branch of forensic entomology, forensic entomotoxicology studies how insects and other arthropods produce toxic and poisonous substances as well as their effects on both humans and animals. An important function of forensic entomotoxicologists is to investigate poisonous insect bites and stings. If an injury or death is caused by a toxin or venom, they have to determine the source and nature of the poison. A forensic investigation using entomotoxicology can provide valuable information about the identity, timing, and circumstances of an insect or arthropod that was involved in the incident. Numerous research experiments have shown that drugs and toxic substances can be detected, identified, and quantified in insects. A combination of molecular biology, chemical analysis, and advanced instrumental techniques have been used to identify the specific insects involved in the incident as well as the toxins responsible for the allergic reactions (Chopi et al., 2019). There are many practical applications of insect toxicology, including the development of new insecticides, the assessment of environmental toxins, and the study of insect resistance to pesticides. It also sheds light on how toxic substances interact with humans. Observing the effects of toxins on insects can assist researchers in determining whether their exposure poses a health risk to humans and other animals.

The forensic application of entomology extends beyond the above applications; it also helps in the investigations of traffic and transportation accidents, child abuse, and elderly neglect. The presence or absence of bee stings may be the cause of several single-occupant car accidents that lack a clear cause. According to some accident studies, insects are the cause of motor vehicle accidents (Hamilton, 2008). Besides causing motor vehicle crashes, insects can also cause aircraft crashes by obstructing fuel lines and interfering with essential instruments. The remains of insects lodged on the car's radiator, front fascia, and windshield must also be examined by forensic entomologists. Identifying the location and travel path of an automobile by analysing these remains is especially valuable when pinpointing its location. An infestation of insects on someone living, usually a child or elderly person, is called myiasis, and this type of neglect can be detected by a forensic entomologist (Amendt et al., 2011).

There is also the possibility that insects can influence blood spatter pattern analysis in a crime scene investigation. Fly and flea blood tracks might mislead a crime scene technician if blood was found in unexpected places, like ceilings and cracks or in pools of blood or spattered blood. Similarly, flies can feed on blood and pass partially digested blood through their faeces, known as flyspecks." Bloodstain analysis may also be muddled by fly regurgitation and possibly blood droplets dropped on distant surfaces. Fleas feed on living organisms, leaving behind undigested blood on household surfaces. It would be confusing for crime scene technicians if faecal drops already present in an indoor crime scene detected human blood. To identify and properly document cockroaches, flies, and flea natural artefacts, their presence, feeding, and defecation must be recognised. Insects that feed on living, decaying, or dried plant materials are identified by forensic entomologists. The importance of this is especially relevant when dealing with vegetative material, such as cannabis imported from abroad (Rivers & Dahlem, 2022). Wildlife crime investigations can be aided by forensic entomology by analysing insect evidence found at crime scenes. The developmental stage of the insects found on the carcass can be used as evidence in a wildlife crime investigation. In this way, the entomologist can estimate the time since death and determine the timeline for the wildlife crime investigation. In addition, toxins present in the insects, or the carcass can be used to prosecute the perpetrators (Magni et al., 2016).

To summarise, forensic entomology has wide applications to provide valuable information for investigating criminal and civil cases because it can aid in identifying offenders and prosecuting them for various crimes. Forensic entomology has been used to solve many criminal cases, providing insight into the timing, cause of death, and movements of victims, witnesses, and suspects.

1.1. Forensic Entomology: Case Studies

The infamous "John George Haigh Acid Bath Murders" in England in the 1940s serve as a classic example of forensic investigations. To destroy evidence, suspect John George Haigh dissolved the bodies of his victims in sulphuric acid. Keith Simpson, an entomologist with a speciality in forensics, discovered insect pupae in the sludge, indicating that the victims had decomposed before being submerged in acid. It was this evidence that helped establish Haigh's guilt and enabled him to be convicted (Vermeij et al., 2015). Entomological evidence was used to estimate the post-mortem interval (PMI) in five violent deaths in Southern Brazil. Various species of blowflies were collected and identified, and a PMI was calculated using the development rates of these species and the ambient temperature in which bodies and insects were found. The insights provided by the collected larvae helped redirect the investigation, identifying potential perpetrators and shedding light on the nature of the crime. It is evident from these findings that entomological evidence can be applied to investigations of violent deaths in a very promising way (Thyssen et al., 2018).

Forensic entomology is also useful in missing person investigations, where determining the time of death is critical. Insect evidence can provide insights into the postmortem interval when a person has been missing for extended periods by providing insight into the possible time of death or disappearance. In the case of Natalee Holloway, forensic entomology was used to estimate her death date. Holloway was last seen near the location where blowfly larvae were found, which aided investigators in establishing a timeline for her disappearance (Douglas et al., 2013). Italian investigators have been unable to locate the two missing children seventeen months after the father of those two children was charged with homicide, kidnapping, and concealing two bodies. Three months after the father's conviction, firefighters found the two victims in an abandoned place, and their dental records confirmed their identities. Dehydration rates of beetles were used as a gauge by forensic entomology to determine their approximate PMI. This entomological study ultimately cleared the father of the homicide charge (Introna et al., 2011).

An array of case studies demonstrates how forensic entomology can be applied to criminal investigations. Insect evidence can provide insight into a variety of crimes, from estimating the time of death to assisting in missing person investigations. Forensic entomologists can assist investigators with understanding events, establishing timelines, and collecting critical evidence for resolving criminal cases.

1.2. Forensically Important Insects

Many insects are attracted to decomposing bodies in a predictable pattern, generally including flies (Diptera), beetles (Coleoptera), butterflies and moths (Lepidoptera), wasps, bees, and ants (Hymenoptera) (Jens et al., 2011).

Blowflies (Diptera) are mainly used to determine the early stages of time since death through their life cycle. Blowflies (Family: Clliphoridae) and fleshflies (Family: Sarcophagidae) feed on flesh and are the first in line to arrive after death within minutes or the first couple of hours. Their eggs hatch into larvae, which later develop into three instars before pupating and emerging as adults. Each blowfly larva develops through every instar at different rates at which it hatches. A forensic entomologist will be able to determine the time since a person died by analysing the development stages of blowfly larvae types found on the copse (Greenberg, 1991).

Beetles (Coleoptera) are the next group of insects that have a significant role in forensic entomology. In contrast to blowflies, beetles are attracted to decomposing bodies in the second phase of the decomposition process. They can also provide valuable information about how long has passed since death. When these beetles are found on a decaying body, it indicates that the larvae of the blowfly have finished developing and left the body. The carrion beetles (Coleoptera: Silphidae) are attracted when soft tissue is still present on the decaying body. The presence of these beetles indicates that the body hasn't been decomposing for more than a week. During the final stages of decomposition, when only bone and hair remain on the body, another kind of beetle named the burying beetle *Nicrophorus* sp. (Coleoptera: Silphidae) is attracted to it. If they are present, they can indicate that the body has been decomposing for several weeks or more. Blowfly larvae, along with other insects, are the main source of food for these beetles. A body that is decomposed with beetles can also suggest that it has been decomposing for more than a few months (Gennard, 2012).

Butterflies and moths (Lepidoptera) can also serve as forensic entomological evidence. Insects that are present in the body's final stages of decomposition could indicate an extended duration of decomposition. Moths or butterflies may also indicate the geographical location where the body was found, as some species are restricted to specific regions. Ants, wasps, and other insect species of the hymenopteran families often inhabit decomposing bodies. In addition to butterflies and moths, Lepidoptera plays a crucial role in forensic entomology. Generally, they are not insects that colonise dead bodies, but they can provide valuable information during the later stages of decomposition. The location of the body may be able to be determined by some lepidopteran species that are native to specific geographical regions. A long period of decomposition might be indicated if these species are present (Benecke, 2001). In a forensic entomological research study, the carcasses of piglets exposed to insects lost 90% of their body weight in six days, while carcasses not exposed to insects lost 30%. The life cycle of each category of insect is used to estimate the length of time since death in the stages of decomposition (Dawson et al., 2020).

1.3. Stages of Decomposition and Insect Activity

There are the following stages of insect decomposition:

1. *Fresh:* Fresh decomposition occurs within 24 to 48 hours after death when blowflies feed and lay eggs on the body.
2. *Bloat:* This is the stage when gas accumulates within the body following death, causing it to swell and discolour. The first insects to colonise the body during this time were blowflies (Diptera: Calliphoridae) and fleshflies (Diptera: Sarcophagidae). Microbes in the human body produce bloated gases when they produce them rapidly. This causes the feeding larvae to have a local increase in body temperature. Normally, soil invertebrates are repelled from soil-based bodies due to fluid leakage and ammonia formation.
3. *Active Decay:* As tissues break down and fluids are released during active decay, decomposition occurs approximately five to eleven days after death. During this stage, a wide range of insects are active, such as beetles (Coleoptera), ants (Hymenoptera: Formicidae), as well as blowflies and fleshflies. As the larvae feed, decay openings form in their bodies, allowing gases to escape, resulting in deflation of the body. In this stage, immature and adult beetles are also feeding on the tissues of the body and preying on the fly larvae, which are leaving the body to pupate.

4. *Post-decomposition:* Decomposition progresses as soft tissues are lost, leaving bones and hair behind. This stage is mainly dominated by beetles and other insects, including the burying beetle (Coleoptera: Silphidae) and dermestid beetle (Coleoptera: Dermestidae). The majority of soft tissues have been consumed by insects in this stage, leaving mummified skin, cartilage, and bones.
5. *Skeletal:* Approximately 25 days to months after death, the dry stage of decomposition occurs, characterised by the absence of soft tissues. This stage of development is characterised by moths (Lepidoptera) and beetles (Coleoptera) that feed mainly on dry tissue. The bones and hair remain as the only remains of the body, without any or a few carrion insects.
6. The stages of decomposition of a body as well as the insect activity associated with them must be understood by forensic entomologists in order to estimate the post-mortem interval and to provide valuable information to investigators (Goff, 1993).

1.4. The Role of Forensic Entomologists in Criminal Investigations

To determine when a victim died, where the murder occurred, and other key information about the crime scene, forensic entomologists use their understanding of insect lifecycles and behaviour to assist criminal investigators. It is one of the primary tasks of the police to obtain and analyse insect evidence at a crime scene. Based on the species and developmental stages of insects present on or near the body, forensic entomologists can determine the stage of decomposition and estimate the time since death when collecting insects. Ensure that the evidence is accurate and up-to-date. It is the responsibility of the forensic entomologist to ensure that the evidence has been collected, killed, and preserved according to the approved standard operating procedures. Following a systematic written and photographic record of the insect evidence collected at the crime scene, the entomologist will perform a detailed analysis, resulting in a report and court appearance. A forensic entomologist may also investigate cases of animal abuse or wildlife poaching as part of the work they perform. Insect evidence can be used to determine what caused an animal's death and when it died. In general, forensic entomologists play an important role in criminal investigation by offering valuable information about insects to investigators. Forensic entomologists contribute to successful convictions in criminal cases by analysing and presenting insect evidence (Smith, 1986).

2. Forensic Entomology: Sampling Materials and Methods

It is preferable to wear protective clothing when collecting entomological evidence from a crime scene, especially from the corpse, to avoid contamination. It is strongly recommended that you wear overalls, gloves, and shoes or boots. To collect insects and additional information, the following equipment is recommended at the scene:

- Tools kit
- Protocol sheets for recording the types of specimens collected

- Graphite pencil or pen with a water- and alcohol-resistant tip (avoid standard ink, which goes bad in wet conditions or if splashed with ethanol)
- Identifying labels
- Collection forceps (with different spring tensions for adults and less fragile immatures)
- Collecting buckets and spoons for maggots
- Egg collection brush (after moistening)
- Storage containers for living and dead insects of various sizes
- Trowels and shovels are used to find larvae and puparia buried in soil and leaf litter
- A strong plastic or paper bag with two compartments is ideal for soil samples and leaf litter
- Egg storage and handling of tissue paper
- Temperature measuring apparatus for measuring body temperatures, ambient temperatures, and larval mass temperatures
- An ethanol solution (70-95%) is suitable for storing dead specimens
- Video/camera for documenting samples
- ABFO Scales
- Ice packs and reusable cooler bags
- Insect reference samples
- Insect net for handheld use
- A data logger that measures the temperature of the environment at scenes

The following materials and methods are used in forensic entomology investigations:

- several insects can be collected from the body, the soil, or the surroundings.
- Forceps, tweezers, and aspirators will be used to collect the insects without damaging them.
- Specimens that are already dead, whether immature or adult, should be stored in 70–95% ethanol. Formalin and formaldehyde should not be used.
- Collect eggs on moistened tissue paper and place them in vials. It is important to prevent maggots from escaping from the vials while air is being allowed inside. As soon as possible, hand over the samples to an expert for rearing. To preserve and kill all specimens, place them in 70–95% ethanol.
- In most cases, larvae should be kept in vials at controlled temperatures and humidity; cool temperatures (2–6°C) are preferred. Line the containers with tissue paper to absorb any fluids produced by the maggots. Live specimens should be transferred immediately to an expert in entomology for rearing. Take the remaining specimens and immerse them in extremely hot water (>80°C) for 30 seconds, but do not immerse them in boiling water. Following the draining of samples, ethanol should be used as a rinse before being stored in a vial with 70–95% ethanol. Once a procedure is completed, it must be documented as soon as possible. Larvae can also be killed in a deep freezer (preferably at 20°C). Putrefaction will cause the larvae to darken and shrink within a few days, so do not put them directly into ethanol. There can be some morphological characteristics obscured when there is shrinkage, making it difficult to estimate real length as a result.

- It is important to transfer pupae within 24 hours for rearing; otherwise, they should be stored in vials under controlled conditions. To allow air to enter the vials, small holes should be punched in the lids. Live adults should be placed in a vial and frozen (ideally at 20°C) for one hour; dead specimens should be stored in 70–95% ethanol. Pins can be used for adult specimens on a form board. When an adult emerges from a puparium, it should be killed after its wings have hardened and its colour has developed (around 6 hours after emergence).
- Specimens of insects, such as pupal skins, puparia, and beetle faeces, should be stored in vials or in 70%–95% ethanol when completely dehydrated.

2.1. Chain of Custody

- It is important to identify the instructing authority, the principal contact, as well as the type of approach and the time of the approach.
- It is important to document when and where samples are collected.
- Every sample must have a unique code (number and/or name); this code should be placed on all samples so they can be retrieved in the future.
- A protocol sheet should be used to label each vial and note the sampling location (e.g., head, leg, carpet).
- To ensure a chain of custody, separate seals should be used for each sample.

2.2. Application

2.2.1. Forensic Entomology: Postmortem Interval Estimation

To estimate an insect's age, developmental landmarks such as length, weight, and life cycle stage are used to correlate insect growth and temperature. Instead of measuring weight and length, the life cycle stage is preferred due to confounding factors such as diet, competition, and specimen preservation techniques (Goff, 1993). A large part of age determination is based on the relationship between temperature and different developmental stages, such as the early stages of development, larval stages, pupae, and hatchlings. The temperature-dependent durations of the stages of the forensic investigation can be obtained from reference data, which are compared with different modelling techniques based on the thermal history of the specimens collected. Postmortem intervals (PMI) can be calculated using insects using the Accumulated Degree Hours (ADH) method. A linear model describes the relationship between temperature and insect development in the ADH method (Amendt et al., 2007). Based on temperature data obtained from the location where the body was found, accumulated degree hours (ADH) are calculated. Every hour that an insect develops, ADH is calculated by adding the degree hours (DH) above a certain temperature threshold (usually 10°C). It is possible to estimate the time of colonisation using the ADH value as a measure of the thermal history of the developing insects. ADH indicates the difference between the mean temperature and the threshold temperature during an hour. Insect development rates are compared to a database containing developmental data for insects found on the body to show how temperature affects insect development. Calculate how long the insect will take to reach the developed stage based on the temperature data for the insect with the help of this database. There are limitations associated with ADH methods for estimating PMI and the potential influence of factors such as humidity, diet, and environmental contamination. Therefore, when using this method in conjunction with other approaches, it is important to consult an experienced forensic entomologist.

2.2.2. Forensic Entomotoxicology

In entomotoxicology, insects' toxins are identified and analysed using a variety of methods, including toxins that must be extracted from the insects. Insect toxins are extracted from crushed insects using solvents or chemicals. Once the toxins have been extracted, chromatography can be used to separate them from other compounds. Toxins are separated based on their physical and chemical properties after passing the extract through a stationary phase, such as column or paper chromatography. To identify the toxic substances after they have been extracted, mass spectrometry or nuclear magnetic resonance spectroscopy is used. It is possible to determine the molecular structure and concentration of a toxin in an insect by using these techniques. In the final step, entomologists interpret and formulate the results to solve the case. Toxin detection in insects can provide valuable information about the cause, timing, and location of death. The court of law can use it to identify suspects and build cases against them in insect poisoning and abuse cases (Introna et al., 2001).

2.2.3. Urban and Food Entomology: Species Identification

Urban entomologists study the various insects found in human dwellings, including homes, buildings, and public spaces. Specific methods are employed by urban entomologists to identify insects. A visual inspection of the collected samples involves examining the insect and identifying it according to its physical characteristics, including its body shape, colour, and size. Cockroaches, ants, and bedbugs can be identified with the help of magnifiers and stereoscopes. Microscopic examination is the practice of increased magnification and examination for the detailed physical characteristics of insects. A microscopic examination helps to detect insects that are too small to see with the naked eye, particularly mites or lice. Microscopy can also be used to detect insect eggs and other signs of infestation. DNA barcoding can be used to identify insects that are difficult to differentiate based on their physical characteristics in a visual examination. In urban environments, these methods can be combined to accurately identify insects at different stages of their life cycle, such as larvae and pupae. It is crucial to identify pests accurately to control and manage them effectively.

Conclusion

In summary, entomology provides a multifaceted forensic discipline that plays an important role in a variety of forensic investigative procedures, making it an essential discipline within the legal context. The main purpose of forensic entomology is to determine the time since death by analysing insects' life cycles found on or near deceased individuals, establishing a comprehensive timeline of events, determining the location of the crime, and assisting law enforcement agencies in identifying suspects based on their meticulous analysis. Further, certain insects, including blowflies and beetles, can be used to identify drug-related activities in urban environments, making it easier to detect and investigate drug-related crimes.

Forensic entomology also plays an important role in revealing pesticide and toxic substance exposure in both humans and animals. Forensic entomologists can provide crucial insights into potential poisoning or environmental contamination cases by examining insects that have come into contact with the suspected contamination in the environment. The study of the contaminant's information not only assists in determining the cause of death but also in investigating suspected foul play or hazardous environmental conditions. In addition, forensic entomology provides substantial evidence about toxic exposure in toxicology investigations. Insects that feed on contaminated plants and animals can accumulate toxins in their systems, providing valuable information about the levels and duration of exposure to toxins. It is especially valuable when toxic exposure is suspected, as it assists in establishing the cause of death and aids in investigating potential criminal activity or environmental hazards.

The field of forensic entomology has an important impact on food safety investigations for legal health and safety compliance. Forensic entomologists can identify infestations and determine the source and extent of contamination by meticulously analysing insects of various species. This information plays a critical role in identifying and mitigating potential health risks, thus safeguarding public health and preventing future outbreaks.

In conclusion, the entomological applications are not limited to the estimation of postmortem intervals but also in urban investigations related to drug detection, food safety enquiries, and toxicology assessments, which illustrates its importance within the field of forensic science. The variety of applications provides valuable evidence and enhances the efficiency and effectiveness of investigations in the legal context. In the coming decades, forensic entomology will play an increasingly important role in solving complex cases and ensuring justice is served.

References

Amendt, J., Campobasso, C. P., Gaudry, E., Reiter, C., LeBlanc, H. N., & J. R. Hall, M. (2007). Best practice in forensic entomology—standards and guidelines. *International Journal of Legal Medicine*, 121, 90-104.

Amendt, J., Richards, C. S., Campobasso, C. P., Zehner, R., & Hall, M. J. (2011). Forensic entomology: applications and limitations. *Forensic science, medicine, and pathology*, *7*, 379-392.

Anderson, G. (2001). Forensic Entomology in British Colombia: A brief history. *Society of British Columbia* 98, pp 127-135.

Benecke, M. (2001). A brief history of forensic entomology. *Forensic Science International*, *120*(1-2), 2-14.

Byrd, J. H., and Castner, J. L. eds. (2001). *Forensic Entomology. The Utility of Arthropods in Legal Investigations*. Boca Raton, Florida: CRC Press.

Chophi, R., Sharma, S., Sharma, S., & Singh, R. (2019). Forensic Entomotoxicology: current concepts, trends and challenges. *Journal of Forensic and Legal Medicine*, *67*, 28-36.

Dadour, I. R., et al. (2001). Forensic Entomology: application, education and research in Western Australia, *Forensic Science International,* 120, 48-52.

Dawson, B. M., Barton, P. S., & Wallman, J. F. (2020). Contrasting insect activity and decomposition of pigs and humans in an Australian environment: a preliminary study. *Forensic Science International*, *316*, 110515.

Douglas, J. E., Burgess, A. W., Burgess, A. G., & Ressler, R. K. (2013). *Crime classification manual: A standard system for investigating and classifying violent crime*. John Wiley & Sons.

Gennard, D. (2012). *Forensic entomology: An introduction*. John Wiley & Sons.

Goff, M. L. (1993). Estimation of Postmortem Interval Using Arthropod Development and Successional Patterns. *Forensic Science Review, 5*, 81-94.

Greenberg, B. (1991). Flies as forensic indicators. *Journal of Medical Entomology*, 28(5), 565-577.

Hagstrum, D. W., & Phillips, T. W. (2017). Evolution of stored-product entomology: protecting the world food supply. *Annual Review of Entomology*, *62*, 379-397.

Hagstrum, D. W., Flinn, P. W., & Reed, C. (2013). *Stored-product insects and other pests. In Forensic Entomology: The Utility of Arthropods in Legal Investigations* (2nd ed., pp. 299-318). John Wiley & Sons, Lt.

Hamilton, S. L. (2008). *Forensic Entomology: Bugs & Bodies: Bugs & Bodies*. ABDO.

Introna, F., Campobasso, C. P., & Goff, M. L. (2001). Entomotoxicology. *Forensic Science International*, *120*(1-2), 42-47.

Introna, F., De Donno, A., Santoro, V., Corrado, S., Romano, V., Porcelli, F., & Campobasso, C. P. (2011). The bodies of two missing children in an enclosed underground environment. *Forensic Science International*, *207*(1-3), e40-e47.

Jens, Amendt., Cs, Richards., Carlo, Pietro, Campobasso., Richard, Zehner., Martin, J., R., Hall. (2011). Forensic entomology: applications and limitations. *Forensic Science Medicine and Pathology*, 7(4):379-392.

Magni, P. A., Dadour, I. R., & Nansen, C. (2016). Forensic entomology in wildlife crimeinvestigations. *Forensic Science, Medicine, and Pathology*, 12(4), 458-474.

Rivers, D. B., & Dahlem, G. A. (2022). *The science of forensic entomology*. John Wiley & Sons.

Smith, G. V. (1986). A Manual of Forensic Entomology. British Museum (Natural History) and Cornell University Press, Oxford, London.

Thyssen, P. J., Aquino, M. F., Purgato, N. C., Martins, E., Costa, A. A., Lima, C. G., & Dias, C. R. (2018). Implications of entomological evidence during the investigation of five cases of violent death in Southern Brazil. *Journal of Forensic Science Research*, *2*, 001-008.

Vermeij, E., Zoon, P., van Wijk, M., & Gerretsen, R. (2015). Microscopic residues of bone from dissolving human remains in acids. *Journal of Forensic Sciences*, 60(3), 770-776.

Chapter 20

A Threat to the Insect World: The Changing Climate

Pawan Kumar[1]
and Suman Tanwar[2,*]
[1]Dharm Samaj PG College, Aligarh, Uttar Pradesh, India
[2]Government College for Girls, Sector-14, Gurugram, Haryana, India

Abstract

Insects are a diverse group of invertebrates that play an essential role in the ecosystems of our planet, significantly influencing both human life and the broader animal kingdom. Their impact on the environment is extensive, contributing to processes such as pollination, nutrient cycling, and serving as food sources for a wide range of species. The distribution and characteristics of these important organisms are largely shaped by climatic conditions, which serve as a fundamental determinant in the configuration of both natural and anthropogenic ecosystems. Climate factors greatly affect various components of the environment, including hydrological cycles, cryospheric conditions, as well as terrestrial and aquatic ecosystems. This encompasses critical areas such as agriculture and forestry, which are directly impacted by the ecosystem services provided by insects. Even minor shifts in climate can produce rapid changes in the distribution and population dynamics of insect species due to several inherent traits. Insects possess relatively short life cycles, high mobility, extensive reproductive capacities, and sensitivity to temperature variations, all of which enable them to respond quickly to environmental changes.

This chapter delves into the intriguing world of insects, exploring the profound effects of climate change on their biodiversity, population levels, and geographical dispersal. It examines how rising global temperatures, increased carbon dioxide concentrations, and altered precipitation patterns impact critical ecological parameters such as insect behaviour, reproductive rates, and survival strategies. By understanding these relationships, we can better appreciate the vital role insects play in ecosystem health and stability in the face of a changing climate.

Keywords: insects, climate, biodiversity, abundance, temperature, carbon dioxide

---

* Corresponding Author's Email: tanwarsuman111@gmail.com.

In: Insects as Humans' Frenemies: From Friends to Foes
Editor: Istkhar Rao
ISBN: 979-8-89530-320-7

1. Introduction

Climate change encompasses changes in environmental factors such as temperature, precipitation, and humidity over a long period. Because of elevated temperatures, increased carbon dioxide, and other hazardous gases, universal food production and irregular rainfall have been severely affected. The universal temperature has been regularly increasing since 1900, with an elevation of around 1°C since then (Shrestha, 2019). The climate plays a significant role in deciding the distributions and different characteristics of natural and manmade ecosystems, including water measures, hydrology, cryology, aquatic and terrestrial ecosystems, and agriculture and forestry (Rosenzweig et al., 2007).

Alterations in climate could cause alterations in geographical distribution patterns, enhanced overwintering, modifications in population growth rates, extensions in the number of generations, a rise in the development season, variations in crop-pest concurrency of phenology, alterations in interspecific interactions, and a rise in the risk of annexation by migrant pests (Porter et al., 1991; Parmesan, 2007; Memmott et al., 2007). Climate and weather can have a strong influence on the distribution and development of insects (Ladányi & Horváth, 2010). A prolonged or drastic drought might make trees more vulnerable to insects (Hanson & Weltzin, 2000). Over decades, climate change has had effects on the schedule and rigidity of insect outbreaks that could change species distributions.

Earlier studies have shown a spatial modelling technique to understand the correlation between continued variation in climate and the geographic distribution of the species. With the use of simulations, a series of maps were generated that show anticipated shifts in zones where the studied species might amplify its range if patterned climatic surroundings at decadal and annual intervals were prolonged (Coops et al., 2005).

2. Effects of Increasing Temperature

Temperature can influence insect species in many ways. As insects are ectotherms, their life cycles, growth, and development are primarily influenced by temperature. Therefore, an alteration in temperature is noticeable enough to modify insect dynamics. Species that cannot acclimatise and flourish in rising temperatures have a tough time regulating their populations, while other species may flourish and increase their numbers faster. Temperature has an impact on mobility, metamorphosis, metabolism, and host availability, which regulate the probability of differences in insect pest population and energy (Shrestha, 2019). All these effects have an eventual influence on food security and agriculture, which further creates new challenges for pest handling. Long-term observation of insect behaviour and population size, especially in identifiably sensitive areas, could be an indicator of biological response to climate alteration (Shrestha, 2019).

Elevated temperatures can have a significant effect on insect survival, population size, development, and geographic range. Temperature can have different effects on insect species, depending on the development strategy (Bale et al., 2002). Insects, which take many longer periods to complete one life cycle, will prefer moderate temperature changes. In temperate areas, the majority of insects have their growth time in the warmer months, due to which species whose niche space is regulated by environmental variations will react more predictably to changing climates, while those whose niche is affected by other abiotic or biotic factors will be less certain (Bale et al., 2002). The general prognosis is that if global temperature rises, species will transfer their geographical ranges to the poles or higher altitudes and expand their population size (Bale et al., 2002; Samways et al., 2005). Gender ratios of some pest species, like thrips, may be affected by temperature (Lewis et al., 1997) and possibly influence their reproduction rates. Warmer winter temperatures cause lower mortality of insects in winter, which may be crucial to rising insect populations (Harrington et al., 2001).

Insects that spend significant parts of their life cycles in the soil could be more slowly affected by temperature fluctuations than those on land, as soil acts as an insulating medium that will regulate temperature fluctuations more than the air (Bale et al., 2002). The species diversity of insects per area tends to decline with higher latitude and elevation (Williams, 1996), indicating that increasing temperatures may result in more insect species assaulting more hosts in moderate climates (Bale et al., 2002).

3. Impacts of Elevated Levels of Carbon Dioxide

Carbon dioxide is a well-known gas that is often associated with climate change. Elevated levels of carbon dioxide have always been the subject of research on climate change. CO_2 plays an important role in rising insect populations by interfering with the plant's nutritional configuration. Plants consist mainly of carbon, and increasing carbon dioxide levels make them grow faster as they can assimilate carbon more quickly. A rise in CO_2 is often understood as an elevation in photosynthetic rate, and it is accepted that plants acquire good drought tolerance with rising carbon dioxide concentrations, which further enhances the accessibility of hosts for insects even in adverse environments. As a result, it produces conditions where pest occurrences are plentiful. More importantly, the prime cause of the elevated level of destruction at a higher CO_2 level is the reduced nutrient amount in plants. Plants subjected to higher CO_2 suggested a reduced amount of nitrogen, causing deficient nutrient levels in host plants. An increase in CO_2 in the atmosphere normally elevates the carbon-to-nitrogen ratio caused by the aggregation of non-structural carbohydrates in plant tissues. As a result, the nutritional standard for protein-deficient insects decreases, diminishing the nitrogen amount by 15–25% in the tissues (Coviella & Trumble, 1999).

The predicted rise in atmospheric carbon dioxide and universal mean temperature may affect insect-plant interactions. Plant characteristics prime for insect herbivores, like nitrogen content, could be directly influenced by increased carbon dioxide and temperature, while insect herbivores are probably directly influenced only by temperature. The populations of insects did not vary significantly with elevated carbon dioxide but tended to rise moderately; mean weights were reduced at elevated temperatures. Plant biomass and height were not significantly changed by the carbon dioxide, but growth rates before affliction were increased by raised carbon dioxide. These observations show that the combined effects of both increased temperature and carbon dioxide may intensify pest destruction in plants that respond poorly to elevated atmospheric carbon dioxide (Flynn et al., 2005). Moreover, variations in climate elevate the possibility of insect transport from one area to another (Whinam et al., 2005).

4. Effects of Change in Precipitation Patterns

Alterations in frequency, intensity, and level of precipitation are also measures of climate change. As seen in most circumstances, the frequency of rainfall and the intensity of rainfall have an inverse relationship, which has induced the prevalence of floods and droughts. Soil-dwelling species are directly influenced by the intersecting precipitation. The intensity of precipitation, influencing the diapause, floods because of heavy rainfall, and extended stagnation of water intimidate the existence of insects. Furthermore, the eggs and larvae of insects may be wiped out by floods and heavy rainfall. Under heavy rains, small pests like whiteflies, aphids, mites, jassids, etc. may decline (Pathak et al., 2012).

5. Impact of Changing Climate

Climate affects insects in different ways; a few of them are mentioned in Figure 1. Some of the important impacts of climate change and their brief descriptions are listed below.

5.1. Changes in Geographical Dispersal, Diversity, and Abundance of Insects

Insects encompass the largest group of animals and play a significant role at different ecosystem levels (Kremen et al., 1993; Kannan et al., 2009). Insects are fine measures of climate position. Insect homogeneity in a habitat stipulates the health position of an ecosystem, as they are fine measures of environmental variation and thereby play a dominant role in food chains (Gregory et al., 2009). The variations in climate may have an impact on the correlative abundance of different species of insects, and those species that are incapable of adjusting to the alterations could be lost with time (Thomas et al., 2014).

Climate change can result in reduced activity of decomposers, and predators, and enhanced herbivory, which may have detrimental effects on agriculture. Climate change is susceptible to habitat shattering, which may be responsible for the formation of smaller subunits of species. Such shattering modifies the microenvironment at the shattered edges, and edge effects comprise alterations in microclimate, temperature, humidity, and light, and each of these can create a strong effect on the vitality and configuration of the species in the fragment (Bebber et al., 2013).

For species survival in modifying climates, they must either adjust *in situ* to a new environment or change their dispersal according to favourable surroundings. Insects with large population sizes and short generation times can react swiftly to climate alteration due to the quick response of their phenology, habitat use, fecundity, survival, and selection. These alterations to insect life history may, in succession, generate fast changes in their abundance and dispersal. Plenty of insect pests will be increased by elevated temperatures, and changing temperatures have an effect on about every single species of insect (Bale, 2002). In-season effects of warming comprise the capacity for elevated levels of feeding and growth, including the probability of further generations in a specific year, which has detrimental effects on crop growth and pest management (Cannon, 1998). Elevated universal temperatures will also affect the phenology of insects, like the untimely appearance of insect pests in their agricultural habitats and the emergence time of a variety of insect pests (Roderick and Watt, 1992).

Not all insect pests can sustain themselves and adjust to the changing climate. Climate alteration can confront accessibility for food, space for oviposition, mobility, sex expression, etc. Moreover, climate change implies an alteration in vegetation abundance in the environment, which may cause a reduction in the host population, forcing insect pests to suffer from shortages of hosts for food and oviposition, etc. (Shrestha, 2019).

5.2. Alterations in Insect Phenology

The study of how an insect's life cycle, especially timings, adapts to variations in the weather and environment is known as insect phenology. Alterations in insect phenology are the most noticeable measure of climate change. It is easier to observe climate change by studying the alteration in the phenology of insects. With elevated temperatures, it is anticipated that insects will cross their larval stages rapidly and become adults prematurely. Anticipated responses in insects could include an advance in the timing of larval and adult appearance and a rise in the extent of flight time (Menéndez et al., 2007).

5.3. Alteration in Diapause Duration

Insects have developed different behavioural strategies, like diapause (a physiological situation in which unfavourable environmental circumstances cause an animal's growth to be slowed), to live in adverse environments. Diapause is influenced by many factors, including temperature, photoperiod, humidity, etc. Diapause plays an important role in maintaining the population of insect pests and the seasonal control of insect life cycles, due to which the insects have a better advantage in sustaining themselves in harsh environmental conditions. Aestivation and hibernation are two types of diapauses that allow insects to sustain themselves at higher and lower temperatures, respectively (Shrestha, 2019).

5.4. Geographical Range Extension

The dispersal of insects relies on the interconnection of insects with climatic factors. An altering climate also implies the expansion of climate zones favourable for particular species. It is speculated that a 1°C increase in temperature would cause a speed of 200 km northwards (in the northern hemisphere) or 40 m skywards (in altitude). Formerly unfitting regions may support the survival of newer pests. Minimal temperature instead of maximum temperature plays a significant role in deciding the universal dispersal of insect species; hence, any elevation in temperature will create a higher efficiency to survive at higher altitudes, eventually transferring pest strength from south to north. As many insect pests' abundance is directly determined by the temperature instead of vegetation, rising temperatures can cause an elevation in pest abundance. With an increase in temperature, insect pests are anticipated to increase their geographic span from the tropics and subtropics to temperate zones at higher altitudes, together with transfers in farming regions of their host plants. Most polar zones are restricted from insect outbursts due to low temperatures and regular frosts. In the future, estimated climate warming (Carroll et al., 2004) and enhanced occurrence of dry periods (Logan et al., 2000) are anticipated to result in more regular insect outbursts in temperate zones as well.

5.5. Impact on Pollinators

Pollinators can be defined as valuable insects that, during their life-supporting actions, deploy the pollen from another to the stigma of identical or non-identical flowers. Honeybees, butterflies, bumblebees, wasps, flies, etc. are examples of familiar pollinators. About 73% of the world's farming is pollinated by bees, 19% by flies, 6.5% by bats, 5% by beetles, 5% by wasps, 4% by butterflies and moths, and 4% by birds (Abrol, 2009). The quality and quantity of pollination have numerous connections for species diversity, food security, ecosystem balance, and durability to climate alteration (FAO, 2008).

5.6. Enhanced Overwintering Survival

Being cold-blooded, insects have a finite capacity for homeostasis with exterior temperature variations. Hence, insects have evolved different procedures, like physiological adjustments, e.g., diapause, to favour life in thermally traumatic environments and behavioural escape through migration (Bale & Hayward, 2010). Looking at the past 100-year climate frame of India, warming was more pronounced during the winter, where a significant increase was observed as compared to the maximum temperature (IMD, 2010). Consequently, insects undergoing hibernation are likely to encounter the most remarkable changes in their thermal environment (Bale and Hayward, 2010). Stimulated metabolic rates at inflated temperatures reduce the period of insect diapause because of the rapid exhaustion of stored nutrient resources (Hahn & Denlinger, 2007).

5.7. Increases in the Number of Generations

It has been speculated that with a 2°C temperature rise, insects might encounter 1–5 auxiliary life cycles per season (Yamamura & Kiritani, 1998). Warming might reduce the experience of harsh cold episodes, which could in turn extend the overwintering region for insect pests (Patterson et al., 1999). Thus, a universal rise in temperature within a specific favourable span may facilitate the proportion of reproduction, development, and survival in tropical and subtropical insects. Eventually, insects will be proficient at finishing a greater number of generations per year (Yamamura & Kiritani, 1998; Petzoldt & Seaman, 2010).

5.8. Introduction of Invasive Alien Species

Climate alteration can also encourage the appearance and formation of alien species. The risk of the inception of invasive alien species (species that have major detrimental effects on their new environment when they are unintentionally or purposefully introduced into a natural area where they are not typically present) is enhanced by universal climate change. Nonetheless, the reasons for biological invasions are numerous and heterogeneous; modifications in abiotic and/or biotic factors of the environment (climate alteration, biological regulation) are accepted as the main operators of species invasion (Dukes & Mooney, 1999; IPCC, 2007). Modernisation and development of world agricultural business integrated with fast transport and transmission methods nowadays have considerably and effectively enhanced the possibility of exotic introductions.

5.9. Enhanced Prevalence of Insect-Transmitted Plant Diseases

Climate change may result in an increased occurrence of insect-vectored plant diseases through range extension and the fast proliferation of insect vectors (Petzoldt & Seaman, 2010). Elevated temperatures, especially in the prior season, have been observed to enhance the occurrence of viral diseases in potatoes due to the premature colonisation of virus-bearing aphids, the prime vectors for potato viruses in Northern Europe (Robert et al., 2000).

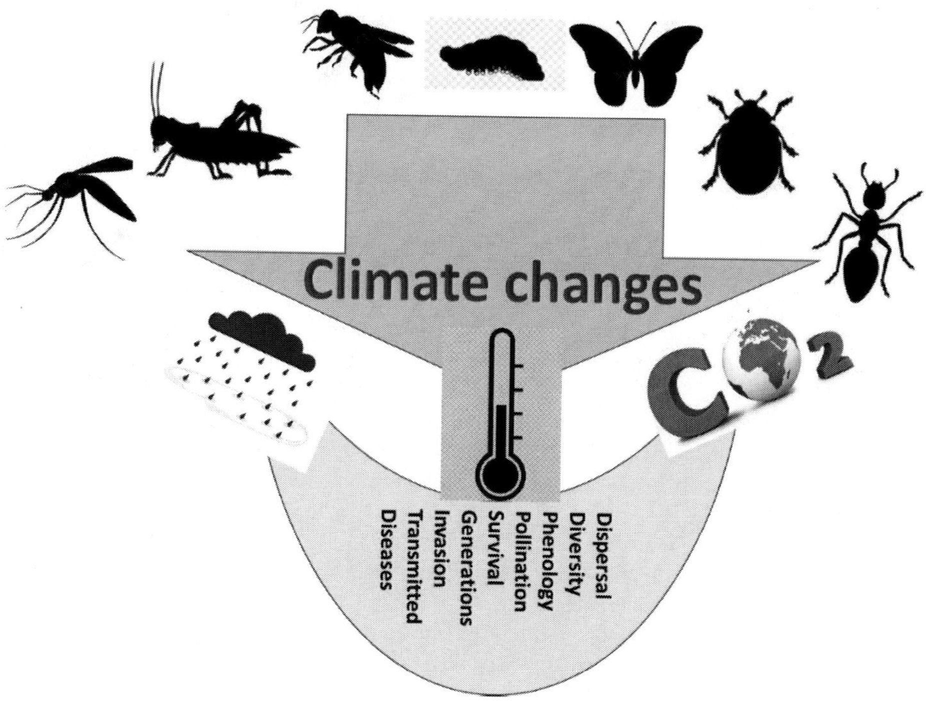

Figure 1. Effects of climate changes on insects.

Conclusion

Insects respond to changes in climate in a variety of ways, exhibiting alterations in their life cycles, movement patterns, and interactions within their communities. These responses can lead to shifts in the timing of important biological events, such as reproduction and migration, as well as changes in how different species coexist and compete for resources. To combat the challenges posed by climate change, it is crucial to enhance the functional diversity within agroecosystems. By doing so, we can increase the resilience of these systems, allowing them to better withstand disturbances and minimize the risks associated with insect pest damage.

References

Abrol, DP. (2009) "Pollination – Basic Concepts." In *D. P. Abrol, Pollination Biology.*

Bale, J. S., & Hayward, S. A. L. (2010). Insect overwintering in a changing climate. *Journal of Experimental Biology*, *213*(6), 980-994.

Bale, J. S., Masters, G. J., Hodkinson, I. D., Awmack, C., Bezemer, T. M., Brown, V. K., ... & Whittaker, J. B. (2002). Herbivory in global climate change research: direct effects of rising temperature on insect herbivores. *Global Change Biology*, *8*(1), 1-16.

Bebber, D. P., Ramotowski, M. A., & Gurr, S. J. (2013). Crop pests and pathogens move polewards in a warming world. *Nature Climate Change*, *3*(11), 985-988.

Cannon, R. J. (1998). The implications of predicted climate change for insect pests in the UK, with emphasis on non-indigenous species. *Global Change Biology*, *4*(7), 785-796.

Cleland, E. E., Chuine, I., Menzel, A., Mooney, H. A., & Schwartz, M. D. (2007). Shifting plant phenology in response to global change. *Trends in Ecology & Evolution*, *22*(7), 357-365.

Coops, N. C., Waring, R. H., & Law, B. E. (2005). Assessing the past and future distribution and productivity of ponderosa pine in the Pacific Northwest using a process model, 3-PG. *Ecological Modelling*, *183*(1), 107-124.

Coviella, C. E., & Trumble, J. T. (1999). Effects of elevated atmospheric carbon dioxide on insect-plant interactions. *Conservation Biology*, *13*(4), 700-712.

Dukes, J. S., & Mooney, H. A. (1999). Does global change increase the success of biological invaders? *Trends in Ecology & Evolution*, *14*(4), 135-139.

FAO. (2008). "Rapid Assessment of Pollinators Status."

Flynn, D. F., Sudderth, E. A., & Bazzaz, F. A. (2006). Effects of aphid herbivory on biomass and leaf-level physiology of Solanum dulcamara under elevated temperature and CO2. *Environmental and Experimental Botany*, *56*(1), 10-18.

Gaston, K. J. (1996). Biodiversity-latitudinal gradients. *Progress in Physical Geography*, *20*(4), 466-476.

Gregory, P. J., Johnson, S. N., Newton, A. C., & Ingram, J. S. (2009). Integrating pests and pathogens into the climate change/food security debate. *Journal of experimental botany*, *60*(10), 2827-2838.

Hahn, D. A., & Denlinger, D. L. (2007). Meeting the energetic demands of insect diapause: nutrient storage and utilization. *Journal of insect physiology*, *53*(8), 760-773.

Hanson, P. J., & Weltzin, J. F. (2000). Drought disturbance from climate change: response of United States forests. *Science of the Total Environment,* 262(3), 205-220.

Harrington, R., Fleming, R. A., & Woiwod, I. P. (2001). Climate change impacts on insect management and conservation in temperate regions: can they be predicted? *Agricultural and Forest Entomology*, *3*(4), 233-240.

IMD. (2010). Annual Climate Summary India *Meteorological Department, Pune. Government of India, Ministry of Earth Sciences, pp.* 27.

IPCC In: (Eds.: Parry, M. L., Canziani, O. F., Palutikof, J. P., van der Linden, P. J., Hanson, C. E.) (2007). Climate Change- Impacts, Adaptation and Vulnerability. *Cambridge University Press, Cambridge, UK,* pp. 976.

Kannan, R., & James, D. A. (2009). Effects of climate change on global biodiversity: a review of key literature. *Tropical Ecology*, *50*(1), 31.

Kremen, C., Colwell, R. K., Erwin, T. L., Murphy, D. D., Noss, R. A., & Sanjayan, M. A. (1993). Terrestrial arthropod assemblages: their use in conservation planning. *Conservation biology*, 796-808.

Ladányi, M., & Horváth, L. (2010). A review of the potential climate change impact on insect populations-general and agricultural aspects. *Applied Ecology and Environmental Research*, *8*(2), 143-152.

Lewis, W. J., Van Lenteren, J. C., Phatak, S. C., & Tumlinson Iii, J. H. (1997). A total system approach to sustainable pest management. *Proceedings of the National Academy of Sciences*, *94*(23), 12243-12248.

Logan, J. A., Régnière, J., & Powell, J. A. (2003). Assessing the impacts of global warming on forest pest dynamics. *Frontiers in Ecology and the Environment*, *1*(3), 130-137.

Menéndez, R. (2007). How are insects responding to global warming? *Tijdschrift voor Entomologie*, *150*(2), 355.

Parmesan, C. (2007). Influences of species, latitudes and methodologies on estimates of phenological response to global warming. *Global Change Biology*, *13*(9), 1860-1872.

Patterson, D. T., Westbrook, J. K., Joyce, R. J. V., Lingren, P. D., & Rogasik, J. (1999). Weeds, insects, and diseases. *Climatic Change*, *43*(4), 711-727.

Petzoldt, C., & Seaman, A. (2010). Climate Change Effects on Insects and Pathogens. New York State Agricultural Extension Station, Geneva, NY 14456.

Porter, J. H., Parry, M. L., & Carter, T. R. (1991). The potential effects of climatic change on agricultural insect pests. *Agricultural and Forest Meteorology*, *57*(1-3), 221-240.

Robert, Y., Woodford, J. T., & Ducray-Bourdin, D. G. (2000). Some epidemiological approaches to the control of aphid-borne virus diseases in seed potato crops in northern Europe. *Virus Research*, *71*(1-2), 33-47.

Rosenzweig, C., Major, D. C., Demong, K., Stanton, C., Horton, R., & Stults, M. (2007). Managing climate change risks in New York City's water system: assessment and adaptation planning. *Mitigation and Adaptation Strategies for Global Change*, *12*, 1391-1409.

Samways, M. J. (2005). *Insect diversity conservation*. Cambridge University Press.

Sehgal, V. K., Singh, M. R., Chaudhary, A., Jain, N., & Pathak, H. (2013). Vulnerability of agriculture to climate change: District level assessment in the Indo-Gangetic Plains.

Shrestha, S. (2019). Effects of climate change in agricultural insect pest. *Acta Scientific Agriculture*, *3*(12), 74-80.

Taylor, S. W., & Carroll, A. L. (2003, October). Disturbance, forest age, and mountain pine beetle outbreak dynamics in BC: A historical perspective. In *Mountain pine beetle symposium: Challenges and solutions* (Vol. 3031). Natural Resources Canada, Canadian Forest Service, Pacific Forestry Centre Victoria.

Thomas C., et al. (2014). "Extinction risk from climate change Extinction risk from climate change." *Nature*.

War, A. R., Taggar, G. K., War, M. Y., & Hussain, B. (2016). Impact of climate change on insect pests, plant chemical ecology, tritrophic interactions and food production. *International Journal of Clinical and Biological Sciences*, *1*(02), 16-29.

Whinam, J., Chilcott, N., & Bergstrom, D. M. (2005). Subantarctic hitchhikers: expeditioners as vectors for the introduction of alien organisms. *Biological Conservation*, *121*(2), 207-219.

Yamamura, K., & Kiritani, K. (1998). A simple method to estimate the potential increase in the number of generations under global warming in temperate zones. *Applied Entomology and Zoology*, *33*(2), 289-298.

Chapter 21

Insects as Experimental Models

Nishi Keshari[1,*]
Sethu Lakshmi S.[2]
and Alagesan Keerthana[3]
[1]Department of Entomology, Dr. Rajendra Prasad Central Agricultural University, Pusa, Samastipur, Bihar, India
[2]Division of Nematology, Indian Agricultural Research Institute, Pusa Campus, New Delhi, India
[3]Department of Agricultural Entomology Tamil Nadu Agricultural University, Coimbatore, India

Abstract

Complex biodiversity and different factors can negatively or positively influence our thriving world. For a better understanding of these matters, biological, medical, and environmental researches are conducted using model organisms as per experimental requirements. Model organisms are species that have been extensively investigated by scientists, are well-known, and have had their biology thoroughly analysed, even to the extent to which scientists are aware of the sequential growth of each cell over time, as in the case of *Caenorhabditis elegans*. Model organisms are frequently used in research because they are small, affordable, easily bred in the lab, have well-understood genetics, and may have particular uses for humans. Each of these elements is significant because it facilitates study that would not typically be possible at the human level. These species may all multiply quickly due to their quick generation rates. Also, many animals have applications that are particular to them and help with biological studies. Among insects, the fruit fly (*Drosophila melanogaster*), known for over 100 years, has been an efficient aid in genetic, behavioural, and molecular biology investigations. Other insects, like *Galleria mellonella*, *Apis mellifera*, *Bombyx mori*, *Periplaneta americana*, and *Locusta migratoria*, are often used in different experimental studies as model organisms. A growing number of insect species have been proposed as model animals in life sciences studies in recent years due to their global distribution, environmental significance, and similarities in their structural elements, energy metabolism, and signalling pathways in mammals. Insects are novel model species for studying human disorders and determining the toxicity of chemicals. Furthermore, in the early stages of drug development and physiological processes, it is effective. In recent studies, beetles have aroused scientists' interest because of their important function in ecosystems that are both natural and dominated by humans, with a high survival capacity and a 300 million-year-old history.

---

* Corresponding Author's Email: nishinema@gmail.com

In: Insects as Humans' Frenemies: From Friends to Foes
Editor: Istkhar Rao
ISBN: 979-8-89530-320-7

Keywords: insects, model organisms, *Galleria mellonella*, *Scathophaga stercoraria*, *Tribolium castaneum*, *Bombyx mori*

1. Introduction

The study of model and experimental organisms provides relevant information about heredity, development, physiology, and the underlying cellular and molecular processes. Despite the diversity of life, these organisms provide information not only to their related but to some extent the knowledge about unrelated life forms on earth and can help in retrieving significant analytical results in experimental research as forms of universal applicability. With the growing technology and evolution of many species, biological science has evolved to include the investigation of more complex systems with the help of living organisms except for the human species, which have been widely studied, are simple to uphold and reproduce in a laboratory setup, and provide unique clinical advantages, which are known as model organisms (Hunter, 2008). The use of these organisms in research has resulted in important discoveries, which are simple yet effective methods of gene control that were initially established in *Escherichia coli*, and work in mice has contributed significantly to our knowledge of cancer and metabolic illnesses. The essential mechanisms of heredity were uncovered in *Drosophila* sp. (Levy & Currie, 2015). They provide useful insight into the biological world, but they are chosen rather for their simplicity and convenience in research studies. Model organisms serve as hypothesis generators because the information they provide leads to theories that can be evaluated in other species. Species that are simple to work with within the lab and out in the field; the study of these species should be integrated with other sub-disciplines so that more substantial insight can be gathered, and the study of these species should be relevant to addressing issues facing the world. These experimental organisms are frequently employed to investigate potential reasons for remedies to diseases in both humans and animals, as they contain a plethora of biological information. Model organisms are, therefore, desirable to study and can serve as an example for challenging direct study or complex understanding in higher animals.

When studying chromatin, meiosis, and other eukaryotic-specific components of the central dogma, for instance, more complex model systems like yeast must be used because the simplest systems, in some cases, are so simple that they lack basic features of interest. For instance, eukaryotes employ a wider spectrum of regulatory systems than do bacteria and bacteriophages. These systems were referred to as model organisms, which meant "an organism that is inherently convenient to study a particular area of biology" (Russell et al., 2017).

For a better and more comprehensive understanding of certain scientific issues, animal models are frequently utilised in the life sciences and other disciplines. Higher creatures, such as mammals, are frequently used in the research. However, widespread usage of animals could result in problems with animal rights, emotions, and many other bioethical difficulties, in addition to biosafety concerns (Levy, 2012). As a result, it becomes a problem to find a suitable model organism to substitute for or limit the use of mammals in ongoing life science research. Research showed that about 80% of pathogen infection trials using mammals could be substituted with insects; several specialists have endorsed and advocated utilising lower organisms instead of higher ones (Renwick & Kavanagh, 2007; Meng, 2017). Taking the above-mentioned areas in life science into account, research insects have served as useful "model systems" in biology for almost a century. Just a couple of illustrations: fruit flies are used in genomic and biological research; insects and their larvae are used in hormonal investigations; houseflies, blowflies, and locusts are used in neuroscience; silk moths are used in pheromone-related research; and neuroethology studies are done with the help of crickets and honeybees.

Nonetheless, fresh model organisms continue to emerge. Due to their widespread distribution and importance in the ecosystem, their ability to extend the study to vertebrates, and their comparatively inexpensive cost of rearing in recent biological sciences studies, an increasing number of insect species have been proposed as model organisms. *Tribolium castaneum*, coming under the largest insect order of beetles, was the first species to have its entire genome sequenced. Beetles are also showing signs of becoming more and more popular as model organisms in numerous studies. Several species representing different coleopteran families have been employed in addition to the red flour beetle, including the burying beetle, the Colorado potato beetle, the seven-spot ladybird, the mealworm, and many more. They are being incorporated more frequently into two important study areas: biomedical and environmental investigations. Using beetles as models, biomedical research primarily focuses on revealing the mechanics of fundamental life activities, including eating, neurotransmission, and biological defence system activity, as well as the various mechanisms of numerous disorders such as neurodegenerative, cardiovascular, biochemical, and immunological barriers (Adamski, 2019). Signalling pathways, energy metabolism, and structural factors of insects are analogous to those of mammals. Insects are interesting target species for analysing human ailments and estimating pharmaceutical toxicity. In the initial stages of medication development and pharmacological process research, insects are also inexpensive, simple to rear, and time-efficient. *Drosophila melanogaster* has been proven to be a useful research tool in the fields of genetics, behaviour, and molecular biology for more than a century. In addition to well-known insect species, researchers have recently looked for novel insect species that can serve as models for research in the physiological and ecological sciences (Adamski, 2019).

2. The Typical Characteristics of Model Organisms

- Model organisms were specifically chosen as research materials because they were thought to be simple and reasonably priced to collect, transport, maintain, and modify experimentally (Ankeny, 2011).
- More specifically, model organisms generally have high fecundity, small anatomical and genomic sizes, quick reproduction durations, brief life cycles, and high mutation rates or responsiveness to basic genetic alteration approaches. These characteristics, as genetic tools, are directly correlated with these traits.
- They were created utilising intricate standardisation procedures that enable the creation of a standardised strain, which then serves as the basis for more research.
- They serve as the foundation for clarifying processes supposed to be shared by many different species, and they are always taken to represent a larger group of organisms outside of themselves.

3. Insects as Model Organisms

There has always been interest in using insects as experimental models. From the fundamental to the applied disciplines of biology, they have served as experimental models (Table 1). Here is a list of several significant insects, along with a brief description, that are used as model animals.

3.1. *Drosophila Melanogaster* (Order: Diptera; Family: Drosophilidae)

Drosophila melanogaster, commonly known as the fruit fly, has been frequently used for more than a hundred years in both conventional and contemporary genetics. The life cycle of a fruit fly is brief. At 25°C, a single fertile mating pair can give birth to hundreds of progenies in around 10 days. Compared to the typically used rodent models, this is far faster (Pandy & Nichols, 2011). They are utilised in an extensive spectrum of biological systems, including genetics and heredity, learning, behaviour, and ageing.

3.2. *Manduca Sexta* (Linnaeus, 1763), (Order: Lepidoptera; Family: Sphingidae)

The Carolina sphinx moth (*Manduca sexta*) is an economically important pest of the tobacco plant; its larval stage is known alternatively as the tobacco hornworm or goliath worm. The facultative specialist goliath worm can consume various types of plant leaves and is specifically adapted to metabolise the neurotoxic nicotine in tobacco leaves and use it as a host defence against predators (Kumar et al., 2013). *M. sexta* has attracted the attention of scientists for a variety of reasons. Its large size, ease of rearing, short life cycle (30 to 50 days), and accessible nervous system have made it a model organism in neurobiology. Additionally, tobacco hornworms are susceptible to attacks by parasitoid wasps, specifically *Cotesia congregata* Say, 1836 (Diptera: Braconidae), which uses a symbiotic polydnavirus (PDV) to suppress the metamorphosis and immune system of *M. sexta* larvae (Beckage et al., 1994; Lavine & Beckage, 1996). *M. sexta* has been employed as an effective biochemical model to investigate biochemical signalling processes in insect cellular immunity and examine several facets of innate immunity (Jiang et al., 2010; Cao et al., 2015).

3.3. *Galleria Mellonella* (Order: Lepidoptera; Family: Pyralidae)

Galleria mellonella is an economically important pest of beehives and is distributed over the world where there is an economic production of honey by feeding on the midrib of the honeycomb and small quantities of honey and propolis (Warren & Huddleston, 1962). The greater wax moth has drawn interest from the scientific community for various reasons. The greater wax moth was employed as an experimental model organism for evaluating the virulence of microbes and understanding their ecotoxicology. It is a model for research on insect immunology, including studies on pathogen detection, haemocyte activation, and wound healing (Diehl-Jones et al., 1996; Dunphy & Downer, 1994; Theopold et al., 2004). *G.*

mellonella larvae, as well as the larvae of *Plodia interpunctella* (the lesser wax moth), have been discovered to be able to digest polyethylene (Khyade, 2018). This is most likely because of symbiotic bacteria in the insects' midguts (Bombelli et al., 2017), which leads us to a promising field of science, i.e., the biodegradation of plastic, making it one of the most exciting models for scientific research.

3.4. *Tribolium Castaneum* (Order: Coleoptera; Family: Tenebrionidae)

The red flour beetle, also known as *Tribolium castaneum*, is a pest of food grains in storage. It also acts as an efficient model organism for investigation into the physiology, development, and application of entomology to coleopteran species. Numerous commercially significant crop pest species that cause significant agricultural losses belong to the order Coleoptera, making this group study even more crucial. The red flour beetle possesses traits that make it a good candidate for use as a genetic model organism, such as ease of rearing, rapid generation, high brood sizes, and ease of genetic manipulation. The red flour beetle has the potential to be used for genetic study, as shown by research using traditional mutations (Brown et al., 2000; 2 Lorenzen et al., 2003), whole-genome molecular mapping (Lorenzen et al., 2003), and RNA interference (Brown, 1999; Tomoyasu, 2004; Wang, 2007), and research on behavioural ecology, which has all shown the potential of *Tribolium* for genetic analysis. In certain respects, *T. castaneum* even surpasses *Drosophila* as an advanced model system for figuring out how development has changed throughout time when it comes to the functional genetic examination of fundamental biological issues. Although both the red-fleshed beetle and *Drosophila* are holometabolous insects, their morphologies as larvae and adults are fundamentally different. Even developmentally conserved traits like body segmentation are accomplished through quite distinct mechanisms. So, comparing the developmental processes of these two insects can answer several intriguing issues about the evolution of morphology and other traits. In addition to having a complete and annotated genome sequence, the red flour beetle also has systemic RNA interference (RNAi), transposon-based transformation systems and genetic maps for sensory and molecular markers, and chromosomal rearrangements that allow lethal mutations to be balanced in true-breeding populations. By inactivating gene functions at various developmental stages, it is feasible to study post-embryonic development as well as larval and adult physiology, including hormone control, host-parasite interactions, and pesticide resistance.

3.5. *Bombyx Mori* (Order: Lepidoptera; Family Bombycidae)

Bombyx mori (silkworms) is the sole producer of mulberry silk in the world, making it an economically important insect. Silkworm produces large offspring quickly; the larval stage lasts roughly 25 to 30 days. It possesses a large number of hereditary traits and 28 pairs of chromosomes. The genome of a silkworm is around 450 million base pairs (bp), or 1/6 of the human genome. Moreover, there are numerous mutant strains and phenotypic alterations in silkworms. There are no bioethical concerns associated with its death, whether it is hereditary or imposed (Chen et al., 2014). The midgut, fat body, silk gland, and haemolymph are just a few of the tissues and organs that can be easily removed from a silkworm due to its moderate body size. Because they are sensitive to chemicals including pharmaceuticals, heavy metals, and pesticides, as well as other qualities like their low cost, unique body composition, and whole genome sequencing, silkworms are a desirable model organism for evaluating environmental pollutants and public health.

3.6. *Scathophaga Stercoraria* (Order: Diptera; Family: Scathophagidae)

Due to their tendency to mate in dung, *Scathophaga stercoraria* is also known as the yellow dung fly. They typically consume nectar and tiny flies. The role of sex chromosomes in sexual assortment and conflict have been significant research topics in recent years. (Arnqvist and Rowe, 2005). These chromosomes are thought to be evolutionary hubs that are essential to the development of speciation because they alter so rapidly. (Presgraves, 2008). It has already been demonstrated that *S. stercoraria* has a large number of phenotypic features that are likely to be evaluated sexually based on factors including body size (Simmons & Ward, 1991) and male excitement behaviour (Mühlhäuser et al., 1996).

3.7. *Ormia Ochracea* (Order: Diptera; Family: Tachinidae)

Ormia ochracea belongs to a family of flies that all feed on the parasites of other arthropods. It stands out for both its very keen directional hearing and its parasitism of crickets. The female cricket is drawn to the male's singing and lays her eggs on or near it. Because of its size and distinctively linked ears. *O. ochracea* makes a perfect model for researching the neural circuits involved in sound localization. The phonotactic behaviour is rigorous and repeatable, and it offers a system for correlating the fly's response to the sound stimulus movement by movement.

4. Field of Studies for Model Insects

4.1. Neuroendocrinology

The neurological system and endocrine system collaborate to control function and maintain equilibrium through a process known as neuroendocrine integration. Numerous studies have highlighted the anatomical, functional, and developmental parallels between the neuroendocrine systems of vertebrates and arthropods (Hartenstein, 2006; De Loof et al., 2012). The bulk of hormones and neurohormones produced by animals are produced in the form of short polypeptides by proteolytic breakdown from larger protein precursors known as prohormones. Adipokinetic hormones, which resemble mammalian glucagon; insulin-like peptides (ILPs), which resemble mammalian tachykinins; numerous additional prominent instances of neuropeptides discovered in insects and beetles and their equivalents in vertebrates (Chowaski et al., 2017).

4.2. Food Intake

Feeding behaviour and energy homeostasis are subject to strict neuroendocrine and endocrine regulation (Monteleone et al., 2018). Studies on the beetles *T. castaneum* and *Zophobas atratus* have shown that sulfakinins (SK) inhibit foraging and regulate energy metabolism in these species and thus may be part of the insect brain-gut axis (Yu et al., 2013; Słocińska et al., 2015b). Impaired cholecystokinin (CCK) signalling in humans is associated with various diseases, such as obesity and depression (Coulter et al., 2017; Klockars et al., 2018). Therefore, SK may be an excellent source of potent drugs to treat these diseases in the future. Similarities between CCK and SK receptors have been shown to make this approach even more likely (Słocińska et al., 2015a; Yu et al., 2015).

4.3. Cardiovascular Disorders

Among all diseases, cardiovascular ailments (CVDs) are among the most prevalent. Therefore, it's critical to learn every detail possible about the processes that lead to specific diseases. According to Chowaski et al. (2016), there is a broad range of neurotransmitters, neuropeptides, and peptide hormones that influence contractility and control proper heart rhythm. Many of them share chemical and physical properties with mammalian peptides and hormones (Chowaski et al., 2017a).

4.4. Immunology

The primary components of an insect's immune system are innate immune mechanisms, which can be further separated into cellular and humoral responses. All cellular processes that involve the morphotic haemoglobin seen in insect haemolymph are included in the cellular response. Phagocytosis, nodulation, and encapsulation are listed as three primary cellular mechanisms (Strand, 2008; Urbaski et al., 2014). The similarity of fundamental immunological systems, including pathogen identification, immunodeficiency, and molecular pathways like Toll and JAK/STAT signalling, between beetles and model species in biomedical investigations is related to this (Altincicek et al., 2008; Vogel et al., 2014).

4.5. Biomonitoring

Most of the earth's ecosystems are impacted by heavy metals, chemical pesticides, and even radiation. To monitor and assess the condition of the environment and shifts in ecosystems, as well as to provide warning signals for impending ecological changes, many living organisms are used as models and bioindicators (Nouchi, 2012; Testi et al., 2012; Kalkan & Altu, 2015; Parmar et al., 2016; Siddig et al., 2016). Beetles are one of the species that have begun to draw attention as biological indicators of environmental toxins in the field of ecosystem monitoring (Zödl & Wittmann, 2003).

4.6. Toxicology and Pharmacology

Insects can be employed in preclinical drug studies as well as an early warning system to identify epigenetic risk factors linked to the effects of medications on future generations. At least two non-human animals must typically be studied as part of the preclinical study protocol. A growing number of research teams are seeking different research models by the 3 Rs: reduction, refinement, and replacement of vertebrates (Wilson-Sanders, 2011).

Table 1. Model insects and their application

Model Organism	Application	Reference
Tribolium castaneum	Molecular genetics, population genetics, epigenetics, RNAi target genes that are appropriate for agricultural insect infestations, CRISPR/Cas9-mediated genome editing, to assess olfaction and the influence of landscape structure on behaviour and ecology.	Farnworth et al., 2020, Feliciello et al., 2014, Gerken et al., 2018, Knorr et al., 2018, Kumar et al., 2018
Bombyx mori	Bacterial/Fungal/Virus infection model, Antibiotic drug screening model, Muscle contraction, Path, Gastrointestinal absorbability testogenicity test, Toxicity test.	Fujiyuki et al., 2012, Miyazaki et al., 2012, Matsumoto et al., 2012, Ueno et al., 2011, Matsumoto et al., 2014, Matsumoto et al., 2015, Kaito et al., 2002, Ishii et al., 2008, Fujiyuki et al., 2012, Hamamoto et al., 2004, Fujiyuki et al., 2012, Hamamoto et al., 2009, Hamamoto et al., 2004
Manduca sexta	Immune responses, Antiviral responses Antiviral responses, Pathogen recognition	Dunn & Drake, 1983, Beckage et al., 1994, Lavine & Beckage, 1996, Schleusener & Samuelson, 1996, Büyükgüzel et al., 2002, Yu et al., 2002
Galleria mellonella	Immune responses, Pathogen recognition, Signaling, AMPs	Warren & Huddleston 1962, Ratcliffe and Gagen 1976, Deverno et al., 1983, Dunphy & Downer, 1994, Yu et al., 2002

5. Insects as a Biological Model

Insects constitute a majority of the earth's biodiversity, about 58–65% (Hammond, 1992; Grimaldi & Engel, 2005). They exist everywhere, except the benthic zone (Grimaldi & Engel, 2005). They regulate soil fertility and structure, pollinate plants, disseminate seeds, cycle nutrients, affect the population dynamics of other species, and act as the primary food supply for other taxa (Majer, 1987). To maintain healthy and productive agricultural ecosystems, soil insects are essential (Cock et al., 2012). Despite the stark differences between the two environments' food web architectures, insects are almost always depicted as an important component in terrestrial and freshwater environments (Shurin & Seabloom, 2005). They play a variety of roles in the food web, such as being primary food sources for various predators, including vertebrates and invertebrates, pests of crops, predators of many significant agricultural pests, and pollinating the majority of plants. Insects are a great choice for additional research because they have evolved with distinctive characteristics, like their outer biochemical structure and ability to fly. Understanding biomechanics, climate change, developmental biology, ecology, evolution, genetics, paleolimnology, and physiology has significantly improved as a result of insect research. It is crucial to be able to use the aforementioned techniques outside of conventional genetic models since, in some circumstances, a conventional genetic model may not be the optimal model for a given subject. No animal has ever been known to endure the extremes that tardigrades can, including freezing to almost absolute zero and exposure to the vacuum of space. For example, the plesiomorphic trait of the arthropod stomatogastric neural system shared by all higher invertebrates, beginning with annelids, is of interest from an evolutionary perspective (Millott, 1943; Barna et al., 1997). As a result, it presents an excellent and appealing case for a comparative analysis of the neurological systems. Arthropods are used as model organisms for the development of motor patterns, which may be done by looking at arthropods that have a somatogastric ganglion present.

The comparative technique is based on the Darwinian concept that all life is ancestor-related. We can think of the relationships between organismic features that are inherited as homologous or homoplastic. In a homologous relationship, the two lineages' ancestors shared a characteristic, and their progeny carried it on. On the other hand, homoplastic traits have evolved separately; the common ancestral lineage lacks the property. Biologists can determine ancestral connections by contrasting various lineages, establishing frameworks and promoting adaptable justifications (Currie, 2012; Griffiths, 1996; Sansom, 2003), determining unidentified features, establishing genetic monitors (Ayala, 2009), and recognising general trends in the structure of life. Research on sexual conflict and selection uses *Callosobruchus maculatus* (bruchid beetle), *Chorthippus parallelus* (meadow grasshopper), *Drosophila* spp. (fruit flies), *Macrostomum lignano* (sand flatworm), *Gryllus bimaculatus* (a field cricket), and *Scathophaga stercoraria* (yellow dung fly). For tests and studies on hybrid zones, *Bombina bombina*, *Podisma* spp. (in the Alps), and *Caledia captiva* (in eastern Australia) are employed; for ecological genomics research, *Daphnia pulex* (an organism usually used as an environmental indicator) is used. Insect haemolymph contains several proteins and peptides with antibacterial activity. To find compounds that can replace antibiotics and antifungal drugs, haemolymph is a source. The fat body of the insect, together with haemocytes and, to some degree or extent, epithelial cells, is where antimicrobial peptides are mostly synthesised. 17 peptides were identified from *G. mellonella* caterpillars, according to the Antimicrobial Peptides Database. Many human pathogenic germs can attack insects' immune systems, which makes them an excellent choice for investigating microbial virulence determinants and their pathogenesis (Loh et al., 2013).

Rodents and lagomorphs, such as mice, rats, guinea pigs, and rabbits, are the most frequently used models in biomedical research because of how similar they are to humans both physiologically and immunologically. For logistical, financial, and ethical reasons, it is challenging to employ a significant number of these mammals in tests. As a result, invertebrates can be utilised in place of vertebrates in various types of testing, research, and educational settings (Niedźwiecka et al., 2015). Comparative genome studies reveal that many human genes have several homologues in invertebrates, including insects, which code for proteins that are part of pathogenic organism identification or signal transduction. These characteristics make it possible to use insects like *D. melanogaster M., Blattella germanica*, *Galleria mellonella*, *Culex quinquefasciatus*, and *Bombyx mori* in studies on host immunity, microbial pathogenicity, and the *in vivo* efficiency of antibiotics, fungicides, and other biologically active compounds.

Conclusion

There are currently numerous novel arthropod models available in addition to the classic D. melanogaster model for examining developmental genetic issues, as vertebrates and insects have similar innate immune responses and other factors; however, insects or their larvae do not entirely replace vertebrates in biomedical studies. This may be due to a lack of information on the full genome sequence of many insect species, which can be excellent model organisms, for example, G. mellonella, even though their research is substantially less expensive and time-consuming than research with vertebrates and can quickly assess a large number of infections or possible therapies. This may be due to a lack of a standardised rearing method for uniform results or a lack of complete genome sequences, and it can be a major limitation in the future. Therefore, it is vital to conduct more research in this field.

References

Adamski, Z., Bufo, S. A., Chowański, S., Falabella, P., Lubawy, J., Marciniak, P. & Rosiński, G. (2019). Beetles as model organisms in physiological, biomedical and environmental studies: a review. *Frontiers in Physiology*, *10*, 1-319.

Ankeny, R. A. & Leonelli, S. (2011). What's so special about model organisms? *Studies in History and Philosophy of Science Part A*, *42*, 313-323.

Arnqvist, G. & Rowe, L. (2005). *Sexual conflict*. Princeton University Press, *27*, 1-112.

Barna, J., Banyolgyi, T., Csoknya, M., Kiss, E. & Elekes, K. (1997). Neurochemical organization of the stomatogastric nervous system in oligochaeta. *Neurobiology*, *5*, 105-106.

Beckage, N. E., Tan, F. F., Schleifer, K. W., Lane, R. D. & Cherubin, L. L. (1994). Characterization and biological effects of *Cotesia congregata* polydnavirus on host larvae of the tobacco hornworm, *Manduca sexta*. *Archives of Insect Biochemistry and Physiology*, *26*, 165-195.

Beckage, N., Tan, F., Schleifer, K., Lane, R. & Cherubin, L. (1994). Characterization and biological effects of *Cotesia congregata* polydnavirus on host larvae of the tobacco hornworm, *Manduca sexta*. *Archives of Insect Biochemistry and Physiology*, *26*, 165–95.

Bombelli, P., Howe, C. J. & Bertocchini, F. (2017). Polyethylene bio-degradation by caterpillars of the wax moth *Galleria mellonella*. *Current Biology*, *27*, R292-R293.

Brown, S. J., DeCamillis, M., Gonzalez-Charneco, K., Denell, M., Beeman, R., Nie, W. & Denell, R. (2000) Implications of the Tribolium Deformed mutant phenotype for the evolution of Hox gene function. *Proceedings of the National Academy of Sciences*, *97*, 4510–4514.

Brown, S. J., Mahaffey, J. P., Lorenzen, M. L., Denell, R. E. & Mahaffey, J. W. (1999) Using RNAi to investigate orthologous homeotic gene function during development of distantly related insects. *Evolution and Development*, *1*, 11–15.

Büyükgüzel, K., Tunaz, H., Putnam, S. M. & Stanley, D. W. (2002). Prostaglandin biosynthesis by midgut tissue isolated from the tobacco hornworm, *Manduca sexta*. *Insect Biochemistry and Molecular Biology*, *32*, 435–443.

Cao, X., He, Y., Hu, Y., Wang, Y., Chen, Y. R., Bryant, B. & Jiang, H. (2015). The immune signaling pathways of *Manduca sexta*. *Insect biochemistry and molecular biology*, *62*, 64-74.

Chen, J., Xie, C., Tian, L., Hong, L., Wu, X. & Han, J. (2010). Participation of the p38 pathway in Drosophila host defense against pathogenic bacteria and fungi. *Proceedings of the National Academy of Sciences*, *107*, 20774-20779.

Chen, S., Rotaru, A. E., Liu, F., Philips, J., Woodard, T. L., Nevin, K. P. & Lovley, D. R. (2014). Carbon cloth stimulates direct interspecies electron transfer in syntrophic co-cultures. *Bioresource Technology*, *173*, 82-86.

Chowański, S., Lubawy, J., Urbański, A. & Rosiński, G. (2016). Cardioregulatory functions of neuropeptides and peptide hormones in insects. *Protein and Peptide Letters*. 23, 913–931.

Chowański, S., Spochacz, M., Szymczak, M. & Rosiński, G. (2017). Effect of biogenic amines on the contractile activity of visceral muscles in the beetle Tenebrio molitor. *Bulletin of Insectology* 70: 209–220.

Cock, M. J., Biesmeijer, J. C., Cannon, R. J., Gerard, P. J., Gillespie, D., Jimenez, J. J. and Raina, S. K. (2012). The positive contribution of invertebrates to sustainable agriculture and food security. *CABI Reviews*, 2012, 1-27.

Coulter, L., Ibrahimi, M., Patel, R. & Agius, M. (2017). Linking the psychosocial aetiology and neurobiology of unipolar depression. *Psychiatric Danubian*. 29, 441–446.

De Loof, A., Lindemans, M., Liu, F., De Groef, B. & Schoofs, L. (2012). Endocrine archaeology: do insects retain ancestrally inherited counterparts of the vertebrate releasing hormones GnRH, GHRH, TRH, and CRF? *General and Comparative Endocrinology*. 177, 18–27.

Deverno, P. J., Aston, W. P. & Chadwick, J. S. (1983). Transfer of immunity against Pseudomonas aeruginosa P11-1 in Galleria mellonella larvae. *Development and Comparative Immunology*, *7*, 423–434.

Diehl-Jones, W. L., Mandato, C. A., Whent, G. & Downer, R. G. (1996). Monoaminergic regulation of hemocyte activity. *Journal of Insect Physiology*, *42*, 13-19.

Dunn, P. E. & Drake, D. R. (1983). Fate of bacteria injected in naive and immunized larvae of the tobacco hornworm *Manduca sexta*. *Journal of Invertebrate Pathology*, *41*, 77–85.

Dunphy, G. B. & Downer, R. G. (1994). Octopamine, a modulator of the haemocytic nodulation response of non-immune *Galleria mellonella* larvae. *Journal of Insect Physiology*, *40*, 267-272.

Farnworth, M. S., Eckermann, K. N., Ahmed, H. M. M., Mühlen, D. S., He, B. & Bucher, G. (2020). The red flour beetle as model for comparative neural development: genome editing to mark neural cells in *Tribolium* brain development. *Methods in Molecular Biology*, *2047*, 191–217.

Feliciello, I., Akrap, I., Brajković, J., Zlatar, I. & Ugarković, Đ. (2014). Satellite DNA as a driver of population divergence in the red flour beetle *Tribolium castaneum*. *Genome Biology and Evolution*, *7*, 228–39.

Fujiyuki, T., Hamamoto, H., Ishii, K., Urai, M., Kataoka, K., Takeda, T., ... & Sekimizu, K. (2012). Evaluation of innate immune stimulating activity of polysaccharides using a silkworm (Bombyx mori) muscle contraction assay. *Drug Discoveries & Therapeutics*, *6*(2), 88-93.

Gerken, A. R., Scully, E. D. & Campbell. J. F. (2018). Red flour beetle (Coleoptera: Tenebrionidae) response to volatile cues varies with strain and behavioral assay. *Environmental Entomology*, *47*, 1252–65.

Grimaldi, D. & Engel, M. S. (2005). *Evolution of the Insects*. Cambridge University Press. *1*, 93-363.

Hamamoto, H., A. Tonoike, K. Narushima, R. Horie, & K. Sekimizu. (2009). Silkworm as a model animal to evaluate drug candidate toxicity and metabolism. *Comparative Biochemistry and. Physiology Part C: Toxicology and Pharmacology*. *149*, 334–339.

Hamamoto, H., K. Kurokawa, C. Kaito, K. Kamura, I. M. Razanajatovo, H. Kusuhara, T. Santa, and K. Sekimizu. (2004). Quantitative evaluation of the therapeutic effects of antibiotics using silkworms infected with human pathogenic microorganisms. *Antimicrobial Agents and Chemotherapy*, *48*, 774–779.

Hammond, P. (1992). Species inventory. *Global biodiversity: status of the earth's living resources*. Springer, Dordrecht, *1*, 17-39.

Hartenstein, V. (2006). The neuroendocrine system of invertebrates: a developmental and evolutionary perspective. *Journal of Endocrinology*. 190, 555–570.

Hunter, P. (2008). The paradox of model organisms: the use of model organisms in research will continue despite their shortcomings. *EMBO reports*, *9*, 717-720.

Ishii, K., Hamamoto, H., Kamimura, M. & Sekimizu, K. (2008). Activation of the silkworm cytokine by bacterial and fungal cell wall components via a reactive oxygen species-triggered mechanism. *Journal of Biological Chemistry*. *283*, 2185–2191.

Kaito, C., Akimitsu, N., Watanabe, H. & Sekimizu, K. (2002). Silkworm larvae as an animal model of bacterial infection pathogenic to humans. *Microbial Pathogenesis*. *32*, 183–190.

Kalkan, S. & Altuğ, G. (2015). Bio-indicator bacteria & environmental variables of the coastal zones: The example of the Güllük Bay, Aegean Sea, Turkey. *Marine Pollution Bulletin*. 95, 380–384.

Khyade, V. B. (2018). Review on biodegradation of plastic through waxworm (Order: Lepidoptera; Family: Pyralidae). *International Academic Journal of Economics*, *5*, 31-38.

Klockars, A., Levine, A. S. & Olszewski, P. K. (2018). Hypothalamic integration of the endocrine signaling related to food intake. *Current Topics in Behavioral Neurosciences, 1,* 239-269.

Knorr, E., Fishilevich, E., Tenbusch, L., Frey, M. L. F. & Rangasamy, M. (2018). Gene silencing in *Tribolium castaneum* as a tool for the targeted identification of candidate RNAi targets in crop pests. *Scientific Reports, 8,* 2061.

Kumar, H., Panigrahi, M., Chhotaray, S., Bhanuprakash, V. & Shandilya, R. (2018). Red flour beetle (*Tribolium castaneum*): from population genetics to functional genomics. *Veterinary World, 11,* 1043–46.

Kumar, P., Pandit, S. S., Steppuhn, A. & Baldwin, I. T. (2014). Natural history-driven, plant-mediated RNAi-based study reveals CYP6B46's role in a nicotine-mediated antipredator herbivore defense. *Proceedings of the National Academy of Sciences, 111,* 1245-1252.

Lavine, M. D. & Beckage, N. E. (1996). Temporal pattern of parasitism-induced immunosuppression in *Manduca sexta* larvae parasitized by *Cotesia congregata. Journal of Insect Physiology,* 42: 41-51.

Levy, A. & Currie, A. (2015). Model organisms are not (theoretical) models. *The British Journal for the Philosophy of Science. 66,* 327–348.

Loh, J. M., Adenwalla, N., Wiles, S. & Proft, T. (2013). *Galleria mellonella* larvae as an infection model for group A streptococcus. *Virulence, 4,* 419-428.

Lorenzen, M. D., Berghammer, A. J., Brown, S. J., Denell, R. E., Klingler, M. & Beeman, R. W. (2003) PiggyBac-mediated germline transformation in the beetle *Tribolium castaneum. Insect Molecular Biology, 12,* 433–440.

Lorenzen, M. D., Doyungan, Z., Savard, J., Snow, K., Crumly, L. R., Shippy, T. D., Stuart, J. J., Brown, S. J. & Beeman, R. W. (2005) Genetic linkage maps of the red flour beetle, *Tribolium castaneum,* based on bacterial artificial chromosomes and expressed sequence tags. *Genetics, 170,* 741–747.

Majer, J. (1987). The Role of Invertebrates in Conservation and Biological Survey. In *Proceedings of a workshop held during the 18th Scientific Conference of the Australian Entomological Society.* Department of Conservation and Land Management, 1: 5-99.

Matsumoto, Y., Ishii, M., Ishii, K., Miyaguchi, W., Horie, R., Inagaki, Y., Hamamoto, H., Tatematsu, K., Uchino, K. & Tamura, T. (2014). Transgenic silkworms expressing human insulin receptors for evaluation of therapeutically active insulin receptor agonists. *Biochemical and Biophysical Research Communications. 455,* 159–164.

Matsumoto, Y., Miyazaki, S., Fukunag, D. H., Shimizu, K., Kawamoto, S. & Sekimizu, K. (2012). Quantitative evaluation of cryptococcal pathogenesis and antifungal drugs using a silkworm infection model with *Cryptococcus neoformans. Journal of Applied Microbiology. 112,* 138–146.

Meng, X., Zhu, F. & Chen, K. (2017). Silkworm: a promising model organism in life science. *Journal of Insect Science, 17,* 1- 97.

Mikulak, E., Gliniewicz, A., Przygodzka, M. & Solecka, J. (2018). *Galleria mellonella* L. as model organism used in biomedical and other studies. *Epidemiological review. 72,* 57-73.

Millott, F. L. N. (1943). The visceral nervous system of the earthworm: I. Nerves controlling the tone of the alimentary canal. *Proceedings of the Royal Society of London. Series B-Biological Sciences, 131,* 271-295.

Miyazaki, S., Matsumoto, Y., Sekimizu, K. & Kaito, C. (2012). Evaluation of *Staphylococcus aureus* virulence factors using a silkworm model. FEMS *Microbiology Letters. 326,* 116–124.

Monteleone, A. M., Castellini, G., Volpe, U., Ricca, V., Lelli, L. & Monteleone, P. (2018). Neuroendocrinology and brain imaging of reward in eating disorders: a possible key to the treatment of anorexia nervosa and bulimia nervosa. *Progress in Neuro-Psychopharmacology and Biological Psychiatry, 80,* 132–142.

Mühlhäuser, C., Blanckenhorn, W. U. & Ward, P. I. (1996). The genetic component of copula duration in the yellow dung fly. *Animal Behaviour, 51,* 1401-1407.

Niedźwiecka, K. & Dyląg, M. (2015). Insects as model organisms to study the pathogenesis of fungal infections and evaluation of potential antimycotics. *Medycyna Doswiadczalna i Mikrobiologia, 67,* 133-139.

Nouchi, I. (2012). "Plants as bioindicators of air pollutants," in *Air Pollution and Plant Biotechnology: Prospects for Phytomonitoring and Phytoremediation,* eds. K. Omasa, H. Saji, S. Youssefian, and N. Kondo (Tokyo: Springer Science & Business Media). 41-100.

Pandey, U. B. & Nichols, C. D. (2011). Human disease models in Drosophila melanogaster and the role of the fly in therapeutic drug discovery. *Pharmacological Reviews, 63,* 411-436.

Parmar, T. K., Rawtani, D. & Agrawal, Y. (2016). Bioindicators: the natural indicator of environmental pollution. *Frontiers in Life Science. 9*, 110–118.

Presgraves, D. C. (2008). Sex chromosomes and speciation in Drosophila. *Trends in Genetics, 24*, 336-343.

Ratcliffe, N. A., Gagen, S. J. (1976). Cellular defense reactions of insect hemocytes in vivo, nodule formation and development in *Galleria mellonella* and *Pieris brassicae* larvae. *Journal of Invertebrate Pathology, 28*, 373–382.

Renwick, J. & Kavanagh, K. (2007). Insects as models for studying the virulence of fungal pathogens of humans. *New insights in medical mycology, 2*, 45-67.

Russell, J. J., Theriot, J. A., Sood, P., Marshall, W. F., Landweber, L. F., Fritz-Laylin, L. and Brunet, A. (2017). Non-model model organisms. *Biomedical Chromatography Biology*, 15: 1-31.

Schleusener, D. R. & Stanley-Samuelson, D. W. (1996). Phospholipase A2 in hemocytes of the tobacco hornworm, *Manduca sexta. Archives of Insect Biochemistry and Physiology, 33*, 63–74.

Shurin, J. B. & Seabloom, E. W. (2005). The strength of trophic cascades across ecosystems: predictions from allometry and energetics. *Journal of Animal Ecology, 74*, 1029-1038.

Siddig, A. A., Ellison, A. M., Ochs, A., Villar-Leeman, C. & Lau, M. K. (2016). How do ecologists select and use indicator species to monitor ecological change? Insights from 14 years of publication in Ecological Indicators. *Ecological indicators. 60*, 223–230.

Simmons, L. W. & Ward, P. I. (1991). The heritability of sexually dimorphic traits in the yellow dung fly *Scathophaga stercoraria* (L.). *Journal of Evolutionary Biology, 4*, 593-601.

Słocińska, M., Czubak, T., Marciniak, P., Jarmuszkiewicz, W. and Rosiński, G. (2015a). The activity of the nonsulfated sulfakinin Zopat-SK-1 in the neck-ligated larvae of the beetle *Zophobas atratus. Peptides. 69:* 127–132.

Słocińska, M., Marciniak, P., Jarmuszkiewicz, W. & Rosiński, G. (2015b). New metabolic activity of the nonsulfated sulfakinin Zopat-SK-1 in the insect fat body. *Peptides 68,* 157–163.

Soumya, M., Reddy, H., Nageswari, G. & Venkatappa, B. (2017). Silkworm (Bombyx mori) and its constituents: A fascinating insect in science and research. *Journal of Entomology and Zoology* Studies, *5*, 1701-1705.

Stankiewicz, M., Dąbrowski, M. & De Lima, M. E. (2012). Nervous system of Periplaneta americana cockroach as a model in toxinological studies: a short historical and actual view. *Journal of Toxicology, 2012*, 1-11.

Testi, A., Fanelli, G., Crosti, R., Castigliani, V. & D'Angeli, D. (2012). Characterizing river habitat quality using plant and animal bioindicators: a case study of Tirino River (Abruzzo Region, Central Italy). *Ecological Indicators, 20*, 24–33.

Theopold, U., Schmidt, O., Söderhäll, K. & Dushay, M. S. (2004). Coagulation in arthropods: defence, wound closure and healing. *Trends in immunology, 25*, 289-294.

Tomoyasu, Y. & Denell, R. E. (2004) Larval RNAi in Tribolium (Coleoptera) for analyzing adult development. *Development Genes and Evolution., 214*, 575–578.

Ueno, K. Y., Matsumoto, J. Uno, K., Sasamoto, K., Sekimizu, Y. and Chibana, H. (2011). Intestinal resident yeast, *Candida glabrata*, requires Cyb2p-mediated lactate assimilation to adapt in mouse intestine. *PLOS One, 6*, 65.

Wang, J. & Schweizer, H. C. (2008). A silkworm baculovirus model for assessing the therapeutic effects of antiviral compounds: characterization and application to the isolation of antivirals from traditional medicines. *Journal of General Virology, 89*, 188–194.

Wang, L., Wang, S., Li, Y., Paradesi, M. S. & Brown, S. J. (2007). BeetleBase: the model organism database for *Tribolium castaneum. Nucleic Acids Research, 35*, 476-479.

Wang, X., Fang, X., Yang, P., Jiang, X., Jiang, F., Zhao, D. & Kang, L. (2014). The locust genome provides insight into swarm formation and long-distance flight. *Nature Communications, 5*, 1-2957.

Warren, L. & Huddleston, P. (1962). Life history of the greater wax moth, *Galleria mellonella* L., in Arkansas. *Journal of the Kansas Entomological Society, 35*, 212–216.

Wilson-Sanders, S. E. (2011). Invertebrate models for biomedical research, testing, and education. *ILAR Journal. 52, 126*–152.

Yu, K. H., Kim, K. N., Lee, J. H., Lee, H. S., Kim, S. H., Cho, K. Y., Nam, M. H. & Lee, I. H. (2002). Comparative study on characteristics of lysozymes from the hemolymph of three lepidopteran larvae,

Galleria mellonella, Bombyx mori, Agrius convolvuli. Developmental and comparative immunology, 26, 707–713.

Yu, N., Nachman, R. J. & Smagghe, G. (2013). Characterization of sulfakinin and sulfakinin receptor and their roles in food intake in the red flour beetle Tribolium castaneum. *General and Comparative Endocrinology.* 188, 196–203.

Yu, N., Zotti, M. J., Scheys, F., Braz, A. S., Penna, P. H. & Nachman, R. J. (2015). Flexibility and extracellular opening determine the interaction between ligands and insect sulfakinin receptors. *Scientific Reports.* 5, 1-12627.

Zödl, B. & Wittmann, K. J. (2003). Effects of sampling, preparation and defecation on metal concentrations in selected invertebrates at urban sites. *Chemosphere* 52, 1095–1103.

Index

C

D

E

I

J

K

L

M

N

O

P

R

S